VISCOELASTIC
SOLIDS

CRC MECHANICAL ENGINEERING SERIES

Edited by Frank A. Kulacki, University of Minnesota

Published

Entropy Generation Minimization
Adrian Bejan, Duke University

The Finite Element Method Using MATLAB
Young W. Kwon, Naval Postgraduate School
Hyochoong Bang, Korea Aerospace Research Institute

Mechanics of Composite Materials
Autar K. Kaw, University of South Florida

Nonlinear Analysis of Structures
M. Sathyamoorthy, Clarkson University

Practical Inverse Analysis of Structures
David M. Trujillo, TRUCOMP
Henry R. Busby, Ohio State University

Viscoelastic Solids
Roderic Lakes, University of Wisconsin

To be Published

Fundamentals of Environmental Discharge Modeling
Lorin R. Davis, Oregon State University

Mechanics of Solids and Shells
Gerald Wempner, Georgia Institute of Technology

Mathematical and Physical Modeling of Materials: Processing Operations
Olusegun Johnson Ilegbusi, Northeastern University
Manabu Iguchi, Osaka University
Walter Wahnsiedler, ALCOA Technical Center

Mechanics of Fatigue
Vladimic V. Bolitin, Russian Academy of Sciences

VISCOELASTIC

SOLIDS

Roderic S. Lakes
University of Wisconsin
Wisconsin

CRC Press
Taylor & Francis Group
Boca Raton London New York

CRC Press is an imprint of the
Taylor & Francis Group, an **informa** business

In ancient rivalry with fellow spheres
the sun still sings its glorious song
and completes with tread of thunder
the journey it has been assigned.

—Goethe

Acknowledgments

I thank the University of Wisconsin—Madison and Cornell University, and the Vezzetti family for their hospitality during a portion of the time this book was in preparation. I am also grateful to my wife Diana for her patience and support. Finally I thank Professor J. B. Park and many students at the University of Iowa for their helpful suggestions.

Preface

This book is intended to be of use in a one semester graduate course on the theory of viscoelasticity. The material contained here has been used as a text in such a course. The objective is to make the subject accessible and useful to students in a variety of disciplines in engineering and experimental physical science. To that end, the coverage is intentionally broad. For those who pursue the subject via self-study, many references have been incorporated to provide some access to the literature. The subject may be profitably studied by undergraduate students, particularly those who are interested in vibration abatement, biomechanics, and the study of materials. Most of the book should be accessible to people who have completed an intermediate or an elementary course in the mechanics of deformable bodies. Exposure to elasticity theory, materials science, and vibration theory is helpful but not necessary.

A development of the theory is presented, including both transient and dynamic aspects, with emphasis on linear viscoelasticity. The structure of the theory is synthesized with the aim of developing physical insight. Methods for the solution of stress analysis problems in viscoelastic objects are developed and illustrated. Experimental methods for characterization of viscoelastic materials are explored in detail. Viscoelastic phenomena are described for a wide variety of materials including polymers, metals, high damping alloys, rock, piezoelectric materials, cellular solids, dense composite materials, and biological materials. To illustrate the sources of viscoelastic phenomena, we describe and analyze causal mechanisms, with cases of materials of extremely high damping and extremely low damping. The theory of viscoelastic composite materials is presented, with examples of various types of structure and the relationships between structure and mechanical properties. Many applications of viscoelasticity and viscoelastic materials are illustrated, with case studies and analysis of particular cases. Viscoelasticity is pertinent to applications as diverse as earplugs, gaskets, computer disks, satellite stability, medical diagnosis, injury prevention, vibration abatement, tire performance, sports, spacecraft explosions, and music. Examples are given of the use of viscoelastic materials in the prevention and alleviation of human suffering. Equations representing final results or those which are otherwise of particular importance are identified by boldface equation numbers.

Contents

About the Author

Roderic Lakes, Ph.D., is Professor in the Department of Engineering Physics, University of Wisconsin—Madison. He is primarily affiliated with the Engineering Mechanics Program and secondarily with the Biomedical Engineering Program and Materials Science Program as well as the Rheology Research Center at the University of Wisconsin. He was on the faculty at the University of Iowa, and served in association with the Department of Biomedical Engineering, the Department of Mechanical Engineering, and the Center for Laser Science and Engineering. This followed a year teaching at Tuskegee Institute in Alabama and two years as research associate at Yale University. He has held visiting faculty appointments at Queen Mary College, University of London, at the University of Wisconsin, Madison and at Cornell University.

He attended Columbia University for several summer sessions in Mathematics, earned the B.S. in Physics in 1969 at Rensselaer Polytechnic Institute, Troy, NY; he undertook graduate study at University of Maryland, College Park, Maryland and returned to Rensselaer to earn a Ph.D. in Physics. His dissertation topic was "Viscoelastic and Dielectric Relaxation in Bone".

Dr. Lakes' research interests include experimental mechanics, including viscoelastic spectroscopy, ultrasonics, and holographic interferometry; holographic and optical methods; characterization of materials such as fibrous composites, cellular solids, biomaterials, dissipative piezoelectrics, and human tissue; evaluation of structure–property relationships; development of materials with novel microstructures and novel properties; and holographic optical elements.

Dr. Lakes has published two books, 126 articles in rigorously reviewed archival journals, including three in *Science* and two in *Nature*. He has delivered 94 presentations, abstracts, extended abstracts, and invited lectures. He was the first to obtain substantive experimental evidence of Cosserat elasticity in structured materials such as bone and cellular solids. He found the first experimental evidence of cement line motion in bone and discovered slow compressional acoustic waves in bone. He has invented a new class of cellular materials with a negative Poisson's ratio and generated a U.S. patent on this invention. His recent work has been in the area of novel structured materials. As for research awards and other recognition, he has received the University Faculty Scholar Award, has seen materials with negative

Poisson's ratios cited worldwide, and has received the Burlington Northern Foundation Award for Faculty Achievement. He is listed in Who's Who in Frontiers of Science and Technology, Personalities of the West and Midwest, Men of Achievement, Who's Who in Technology Today, Who's Who in the Midwest, and American Men and Women of Science.

Dr. Lakes has served at Wisconsin on the Qualifying Examination Working Group and the Rheology Research Center Executive Committee, and at Iowa on the university faculty welfare committee and the ad hoc core curriculum subcommittee, and has chaired the teaching committee and the curriculum committee. National and international service includes review activity for many journals, including *Science*, *Nature*, and many other journals; funding agencies such as NSF, the Veterans Administration, and NIH; and membership in the ASME Joint Biomechanics Committee. Moreover, he serves on the Editorial Board of *Cellular Polymers (U.K.)*.

Dr. Lakes has been recognized by an Instructional Improvement Award and is a three-time recipient of the Outstanding Faculty Award of the Student Society of Biomedical Engineering at the University of Iowa.

chapter one

Introduction: phenomena

§1.1   Viscoelastic phenomena

Most engineering materials are described, for small strains, by Hooke's law of linear elasticity: stress σ is proportional to strain ε. In one dimension, Hooke's law is as follows:

$$\sigma = E\varepsilon, \tag{1.1.1}$$

with E as Young's modulus. Hooke's law for elastic materials can also be written in terms of a compliance J:

$$\varepsilon = J\sigma. \tag{1.1.2}$$

Consequently, the elastic compliance J is the inverse of the modulus E,

$$J = \frac{1}{E}. \tag{1.1.3}$$

By contrast to elastic materials, a viscous fluid under shear stress obeys $\sigma = \eta d\varepsilon/dt$, with η as the viscosity.

In reality all materials deviate from Hooke's law in various ways, for example, by exhibiting viscous-like as well as elastic characteristics. *Viscoelastic* materials are those for which the relationship between stress and strain depends on time. *Anelastic* solids represent a subset of viscoelastic materials: they have a unique equilibrium configuration and ultimately recover fully after removal of a transient load.

Some phenomena in viscoelastic materials are (i) if the stress is held constant, the strain increases with time (creep); (ii) if the strain is held constant, the stress decreases with time (relaxation); (iii) the effective stiffness depends on the rate of application of the load; (iv) if cyclic loading is applied, hysteresis (a phase lag) occurs, leading to a dissipation of mechanical energy; (v) acoustic waves experience attenuation; (vi) rebound of an object following an impact is less than 100%; and (vii) during rolling, frictional resistance occurs.

All materials exhibit some viscoelastic response. In common metals such as steel or aluminum as well as in quartz, at room temperature and at small strain, the behavior does not deviate much from linear elasticity. Synthetic polymers, wood, and human tissue as well as metals at high temperature display significant viscoelastic effects. In some applications, even a small viscoelastic response can be significant. To be complete, an analysis or design involving such materials must incorporate their viscoelastic behavior.

Knowledge of the viscoelastic response of a material is based on *measurement*. The mathematical formulation of viscoelasticity theory is presented in the following chapters with the aim of enabling prediction of the material response to arbitrary load histories. Even at present, it is not possible to calculate damping from models based on atomic and molecular models alone [1.1.1], although considerable understanding has evolved regarding causal mechanisms (Chapter 8) for viscoelastic behavior.

§1.2   *Motivations for studying viscoelasticity*

The study of viscoelastic behavior is of interest in several contexts. First, materials used for structural applications of practical interest may exhibit viscoelastic behavior which has a profound influence on the performance of that material. Materials used in engineering applications may exhibit viscoelastic behavior as an unintentional side effect. In applications, one may *deliberately* make use of the viscoelasticity of certain materials in the design process, to achieve a particular goal. Second, the mathematics underlying viscoelasticity theory is of interest within the applied mathematics community. Third, viscoelasticity is of interest in some branches of materials science, metallurgy, and solid-state physics since it is causally linked to a variety of microphysical processes and can be used as an experimental probe of those processes. Fourth, the causal links between viscoelasticity and microstructure are exploited in the use of viscoelastic tests as an inspection tool. Many applications of viscoelastic behavior are discussed in Chapter 10.

§1.3   *Transient properties: creep and relaxation*

§1.3.1   *Viscoelastic functions*

Creep is a slow, progressive deformation of a material under constant stress. In one dimension, suppose the history of stress σ as it depends on time t to be a step function beginning at time zero:

$$\sigma(t) = \sigma_0\, \mathcal{H}(t). \qquad\qquad (1.3.1)$$

$\mathcal{H}(t)$ is the unit Heaviside step function defined as 0 for t less than zero, 1 for t greater than zero, and ½ for t = 0 (see Appendix 1). The strain $\varepsilon(t)$ in a viscoelastic material will increase with time. The ratio

$$J(t) = \frac{\varepsilon(t)}{\sigma_0} \tag{1.3.2}$$

is called the creep compliance. In linearly viscoelastic materials, the creep compliance is independent of stress level. The intercept of the creep curve on the strain axis is ascribed by some authors to "instantaneous elasticity". However, no load can be physically applied instantaneously. If the loading curve is viewed as a mathematical step function, we remark that the region around zero time contains an infinite domain on a logarithmic scale, a topic we shall return to later. If the load is released at a later time, the strain will exhibit recovery, or progressive decrease of deformation. Strain in recovery may or may not approach zero, depending on the material. We remark that the recovery phase is not included in Eqs. 1.3.1 and 1.3.2 but will be treated in §2.2.

Elastic materials constitute a special case for which the creep compliance is

$$J(t) = J_0 \, \mathcal{H}(t),$$

with J_0 as a constant which is the elastic compliance. Elastic materials exhibit immediate "recovery" to zero strain following release of the load. Viscoelastic materials which exhibit complete recovery after sufficient time following creep or relaxation are called *anelastic*. *Viscous* materials constitute another special case in which the creep compliance is

$$J(t) = \eta \, t \, \mathcal{H}(t),$$

with η as the viscosity. Creep deformation in viscous materials is unbounded.

The creep response in Fig. 1.1 is shown beginning at the same time as the stress history, which is the cause. The corresponding functional form is $J(t) = j(t)\mathcal{H}(t)$, with $j(t)$ as a function defined over the entire time scale. This functional form for $J(t)$ follows from the physical concept of *causality*, meaning that the effect does not precede the cause.

Creep curves may exhibit three regions (Fig. 1.2): *primary* creep in which the curve is concave down, *secondary* creep in which deformation is proportional to time, and *tertiary* creep in which deformation accelerates until creep rupture occurs. Tertiary creep is always a manifestation of nonlinear viscoelasticity, and secondary creep is usually nonlinear as well. Although secondary creep is represented by a straight line in a plot of strain vs. time, that straight line has nothing whatever to do with linear viscoelasticity. Linear response involves a linear relationship between cause and effect:

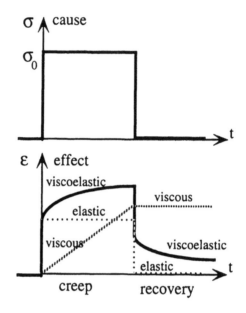

Figure 1.1   Creep and recovery. Stress σ and strain ε vs. time t.

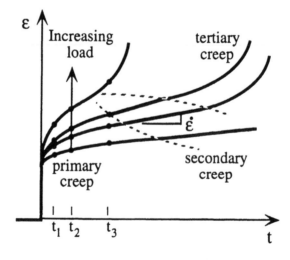

Figure 1.2   Regions of creep behavior. Strain ε vs. time t, for different load levels.

stress and strain at a given time in the case of creep. Specifically, data taken at different load levels may be compared by considering *isochronals* or data at the same time. Data points at times t_1, t_2, and t_3 are illustrated in Fig. 1.2. The nature of linear viscoelasticity and the distinction between linear and nonlinear behavior are presented in detail in §2.12 and §6.2.

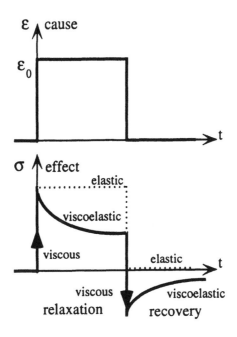

Figure 1.3   Relaxation and recovery.

Stress relaxation is the gradual decrease of stress when the material is held at constant strain. If we suppose the strain history to be a step function beginning at time zero:

$$\varepsilon(t) = \varepsilon_0 \mathcal{H}(t),$$

the stress $\sigma(t)$ in a viscoelastic material will decrease as shown in Fig. 1.3. The ratio

$$E(t) = \frac{\sigma(t)}{\varepsilon_0} \qquad (1.3.3)$$

is called the relaxation modulus. In linear materials, it is independent of strain level, so $E(t)$ is a function of time alone.

The symbol E for stiffness in uniaxial tension and compression is used in subsequent sections since the introductory presentations are restricted to one dimension. Creep and relaxation can occur in shear or in volumetric deformation as well. The relaxation function for shear is called $G(t)$. For volumetric deformation, the elastic bulk modulus is called B (also called K). A corresponding relaxation function $B(t)$ may be defined as above, but with the stress as a hydrostatic stress. A distinction is made in the creep compliances, $J_G(t)$ for creep in shear, $J_E(t)$ for creep in extension, and $J_B(t)$ for creep in volumetric deformation.

The relaxation curve is drawn as decreasing with time and the creep curve is drawn as increasing with time; it is natural to ask whether this must be so. To address this question, let us consider only *passive* materials and perform a thought experiment. A passive material is one without any external sources of energy; for a passive mechanical system, the only energy stored in the material is strain energy, and in a dynamic system, kinetic energy. An analytical definition of a passive material is given in §2.3. Let an initially unstrained specimen be deformed in creep under deadweight loading. A material which *raises* the weight can do so only by performing positive work on it; this is impossible in a passive material. So for passive materials $J(t)$ is an increasing function.

At times it is convenient to present normalized, dimensionless forms of the relaxation or creep functions:

$$E(t) = E(\infty) + \{E(0) - E(\infty)\}e(t), \tag{1.3.4}$$

$$J(t) = J(0) + \{J(\infty) - J(0)\}j(t), \tag{1.3.5}$$

in which $e(t)$ is the normalized relaxation function and $j(t)$ is the normalized creep function. These are defined so that $j(t)$ increases from zero to one over the full time scale and $e(t)$ decreases from one to zero.

A distinction may be made between aging and nonaging materials: in aging materials, properties change with time, typically time as measured following the formation or transformation of the material. Concrete, for example, is an aging material. The discussion here is for the most part restricted to nonaging materials.

§1.3.2 Solids and liquids

In the modulus formulation, a viscoelastic solid is a material for which $E(t)$ tends to a finite, nonzero limit as time t increases to infinity; in a viscoelastic liquid, $E(t)$ tends to zero.

In the compliance formulation, a viscoelastic solid is a material for which $J(t)$ tends to a finite limit as time t increases to infinity; in a viscoelastic liquid, $J(t)$ increases without bound as t increases.

The time scale extends from zero to infinity. In practice, creep or relaxation procedures in certain regions of the time scale are difficult to accomplish. For example, the region 10^{-10} to 10^{-2} sec is effectively inaccessible to most kinds of transient experiment since the load can be applied only so suddenly. Observation of the behavior of materials at long times is limited by the patience and ultimately by the lifetime of the experimenter. In this vein one may define [1.3.1] the dimensionless Deborah number D:

$$D \equiv \frac{\text{time of creep or relaxation}}{\text{time of observation}}. \tag{1.3.6}$$

If D is large, we perceive a material to be a solid even if it ultimately relaxes to zero stress. The difficulty in discriminating solids from liquids is a result of the finite lifetime of the human experimenter. Longer term observations are also possible, as described by the Biblical prophetess Deborah: "The mountains flowed before the Lord" [1.3.2]. The original language may be translated as "flowed" [1.3.1] but some translations give "quaked" [RSV] or "melted" [KJV]. The flow of mountains is extremely slow, so that they appear solid to human observers, but are observed to flow before God. There is also an intermediate time scale of interest to engineers: we may wish materials used in structures to behave as solids over the lifespan of human civilizations, which is longer than that of individual people. We shall return to that problem in discussions of experimental methods and of applications.

§1.4   Dynamic response to sinusoidal load

If a stress $\sigma(t)$ varying sinusoidally (Fig. 1.4) in time t,

$$\sigma(t) = \sigma_0 \sin(2\pi\nu t) \tag{1.4.1}$$

of frequency ν (in cycles per second or Hertz [Hz]) is applied to a linearly viscoelastic material, the strain

$$\varepsilon(t) = \varepsilon_0 \sin(2\pi\nu t - \delta) \tag{1.4.2}$$

will also be sinusoidal in time but will lag the stress by a phase angle δ. The *period* T of the waveform is the time required for one cycle: $T = 1/\nu$.

The phase angle is related to the time lag Δt between the sinusoids by $\delta = 2\pi(\Delta t)/T$. To see this, the argument in Eq. 1.4.2 may be written

$$2\pi\nu t - \delta = 2\pi\nu t - \frac{2\pi\nu\delta}{2\pi\nu} = 2\pi\nu\left(t - \frac{\delta}{2\pi\nu}\right) = 2\pi\nu(t - \Delta t). \tag{1.4.3}$$

So

$$\Delta t = \frac{\delta}{2\pi\nu}, \tag{1.4.4}$$

with

$$\nu = \frac{1}{T},$$

$$\delta = \frac{2\pi\Delta t}{T}. \tag{1.4.5}$$

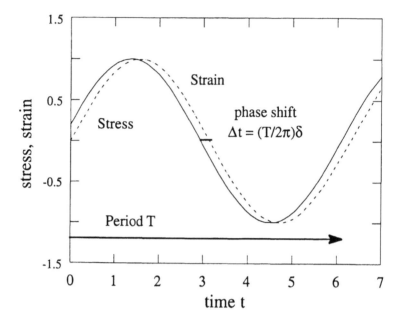

Figure 1.4 Stress and strain vs. time t (in arbitrary units) in dynamic loading of a viscoelastic material.

As a result of the phase lag between stress and strain, the dynamic stiffness can be treated as a complex number:

$$\frac{\sigma}{\varepsilon_0} = E' + iE''. \qquad (1.4.6)$$

The single and double primes designate the real and imaginary parts; they do not represent derivatives. The loss angle δ is a dimensionless measure of the viscoelastic damping of the material. The dynamic functions E', E'', and δ depend on frequency. The tangent of the loss angle is called the *loss tangent*: $\tan \delta$. In an elastic solid, $\tan \delta = 0$. The relationship between the transient properties $E(t)$ and $J(t)$ and the dynamic properties E', E'', and $\tan \delta$ is developed in §3.2. Dynamic viscoelastic behavior, in particular $\tan \delta$ and its consequences, is at times referred to as *internal friction* or as *mechanical damping*.

§1.5   *Demonstration of viscoelastic behavior*

Several commonly available materials may be used to demonstrate viscoelastic behavior. For example, Silly Putty®, sold as a toy, may be formed into a long rod and hung from a support so that it is loaded under its own weight. It will creep without limit, behaving as a liquid. It will also bounce like a rubber ball, behaving nearly elastically at sufficiently high strain rate. Another example is a foam used for earplugs [1.5.1]. This foam can be compressed substantially, and will recover most of the deformation in a period of a minute or so; simple creep experiments and demonstrations can also be performed with this material. As for the rate of decay of vibration, an aluminum tuning fork can be used to demonstrate free decay of vibration. Following an impact, the fork is audible for many seconds, hence for thousands of cycles, demonstrating the low loss tangent of aluminum. A similarly shaped fork made of a material with a higher tan δ, such as a stiff plastic or wood (or a plastic ruler mounted as a cantilever and set into vibration), will damp out its vibrations much more quickly.

§1.6   Other works on viscoelasticity

The purpose of the present work is to provide an introduction to the theory of viscoelasticity with emphasis on the linear theory of both transient and dynamic behavior, solution of stress analysis problems, experimental methods, the properties of real materials, causal mechanisms, analysis of composites, and applications. The interested reader is referred to other works dealing with viscoelastic materials [1.6.1–1.6.28].

Works emphasizing the mathematical aspects of the subject include the following. Christensen [1.6.1] presents the linear theory and includes the solution of advanced problems of research interest. Renardy et al. [1.6.2] as well as Gurtin and Sternberg [1.6.3] present a postulational approach to the linear theory, emphasizing the proof of theorems. The works of Bland [1.6.4] and Flügge [1.6.5] are early introductions of the linear theory, the former emphasizing mechanical models with springs and dashpots. Golden and Graham [1.6.6] emphasize viscoelastic stress analysis, specifically methods for the solution of difficult boundary value problems not amenable to simple methods. Haddad [1.6.7] emphasizes the linear theory and provides examples of analysis of composites. Pipkin [1.6.8] presents a variety of pedagogic examples as well as treatment of viscoelastic fluids. Gross [1.6.9] presents interrelations among the viscoelastic functions and a systematic conceptual organization of these functions. Findley et al. [1.6.10] and Lockett [1.6.11] deal with nonlinear viscoelasticity and transient behavior. Shames and Cozzarelli [1.6.12] deal with viscoelasticity in the general context of inelastic phenomena; this work includes a discussion of nonlinear viscoelasticity and of plasticity. Tschoegl [1.6.13] presents an exhaustive treatment of the linear theory, with emphasis on constitutive equations. Gittus [1.6.14], Rabotnov [1.6.15], and Kraus [1.6.16] deal with creep, principally of metals to be used for structural applications. An article by Bert [1.6.17] contains a review of work on dynamic properties and their determination.

Works dealing with the viscoelastic properties of particular classes of materials include books on polymers by Ferry [1.6.18]; McCrum et al. [1.6.19]; Aklonis and MacKnight [1.6.20]; and Ward and Hadley [1.6.21]. McCrum et al. [1.6.19] introduce structural aspects of polymers, molecular theories, and many experimental results for viscoelastic and dielectric relaxation in polymers. Ward and Hadley [1.6.21] present many aspects of polymer mechanics including linear and nonlinear viscoelasticity, as well as fracture. The classic works of Zener [1.6.22] and Nowick and Berry [1.6.23] deal with metals as crystalline materials; Nowick and Berry [1.6.23] deal with other crystalline materials as well. Creus [1.6.24] discusses viscoelasticity and material aging with application to concrete structures and includes examples of viscoelastic finite element analysis and other numerical techniques. Zinoviev and Ermakov [1.6.25] present analyses of viscoelasticity of fibrous and laminated composites. Lodge [1.6.26] presents viscoelastic liquids.

Works dealing with causal mechanisms include those of Ferry [1.6.18] and McCrum et al. [1.6.19] on molecular theories for polymers; Zener [1.6.22] and Nowick and Berry [1.6.23] on crystalline materials, principally metals; and Nabarro and de Villiers [1.6.27] on the physics of creep including plasticity of crystalline materials.

Works dealing largely with applications of viscoelastic materials include volumes on creep by Pomeroy [1.6.28] and Kraus [1.6.16]; the latter deals primarily with metals.

§1.7   *Historical aspects*

An awareness of viscoelastic behavior dates back at least to the late 18th century. Coulomb (1736–1806) reported studies of the torsional stiffness of wires by a torsional vibration method [1.7.1, 1.7.2]. He also discussed damping of vibration and demonstrated experimentally that the principal cause of it was not air resistance but was a characteristic of the wire itself. As for creep and relaxation [1.7.3], Vicat in 1834 surveyed the sagging of wires and of suspension bridges. Weber [1.7.5] and Kohlrausch [1.7.6] found deviations from perfect elasticity in galvanometer suspensions. Creep behavior was observed in silk threads under load [1.7.7]. On release of torque on the galvanometer suspension, the instrument did not return to zero immediately; instead it converged gradually. This creep recovery was referred to as "elastic after effect"; see also Zener [1.6.22]. Viscoelastic behavior has been studied by eminent figures such as Boltzmann, Coriolis, Gauss, and Maxwell [1.7.3]. Early mathematical modeling of relaxation processes included a "stretched exponential" formalism (discussed in Chapter 2) which was used to model creep in silk, glass fibers, and rubber [1.7.3]. Maxwell [1.7.8] developed a relaxation analysis of gas viscosity which is also applicable to viscoelasticity. The integral representation of Boltzmann [1.7.9] for the stress-strain relationship forms the basis of the linear theory of viscoelasticity as it is currently understood [1.7.10]. The theory of integral equations and of functional analysis as developed by Volterra [1.7.11] provides much of the mathematical underpinning. Intensive study of viscoelasticity began in the 1930s as polymeric materials assumed technological importance. Leaderman first suggested that an increase in temperature has the effect of contracting the time scale of creep in polymers. Data at different temperatures were used by

Tobolsky and Andrews in 1945 to prepare "master curves" to infer behavior over extended time scales. Ferry [1.7.14] soon after presented a theoretical interpretation of the temperature effect. Plazek [1.7.15] reviewed many experimental efforts in connection with temperature dependence of polymer viscoelasticity, particularly a polyisobutylene provided by the U.S. National Bureau of Standards. As for terminology, reference to a "history" of strain is due to Maxwell [1.7.16], while reference to "hereditary" phenomena is due to Volterra [1.7.17].

§1.8 Summary

Viscoelastic behavior manifests itself in creep, or continued deformation of a material under constant load; and in stress relaxation, or progressive reduction in stress while a material is under constant deformation. These observed phenomena are the basis for the constitutive equations developed in the next section, and for experimental methods for characterizing materials, and are of concern in applications of materials.

Examples

Example 1.1

A creep curve is fitted well by a straight line. Does that mean the material is linearly viscoelastic?

Answer

To have linear viscoelasticity, it is necessary that the strain response *at a given time* is proportional to the load. To make a plot of stress vs. strain at constant time, a set of creep curves, each one obtained under a different stress, is required.

Example 1.2

The creep compliance is defined as the time-dependent ratio of strain to stress in response to a stress which is a step function in time. Why not define the viscoelastic response in terms of a stress–strain curve? After all, the stress–strain curve is a standard way of expressing the elastic, plastic, and fracture characteristics of a material.

Answer

In a viscoelastic material, stress, strain, and time are coupled while in an elastic or an elastic–plastic material; there is no time dependence. One can interpret the initial part of a stress–strain curve in terms of linear viscoelasticity as presented in §2.5. However, experiments which provide a

stress–strain curve are ordinarily conducted at a constant strain rate. Therefore, as time increases, so does strain. At sufficiently large strains, material
nonlinearity, damage, and fracture occur. The creep and relaxation procedures decouple the dependence on time from any dependence on load level,
simplifying interpretation.

Problems

1.1 How will the sound from tuning forks made of brass, aluminum, and
    plastic differ?
1.2 Discuss five materials which you have observed to be viscoelastic.
1.3 Under what conditions may an aging material be regarded as non-
    aging?
1.4 Can you think of a material which is not passive?
1.5 How does a linearly viscoelastic material differ from an elastoplastic
    material?

References

1.1.1 Plunkett, R., "Damping analysis: an historical perspective", in *M3D: Mechanics
      and Mechanisms of Material Damping*, ed. V. K. Kinra and A. Wolfenden, ASTM,
      Philadelphia, PA, 1992, ASTM STP 1169, pp. 562–569.
1.3.1 Reiner, M., "The Deborah number", *Phys. Today* 17, 62, 1969.
1.3.2 Judges 5:5.
1.5.1 EAR, Division of Cabot Corp., 7911 Zionsville Rd., Indianapolis, IN.
1.6.1 Christensen, R. M., *Theory of Viscoelasticity*, Academic, NY, 1982.
1.6.2 Renardy, M., Hrusa, W., and Nohel, W. J., *Mathematical Problems in Viscoelas-
      ticity*, Longman, Essex, UK; J. Wiley, NY, 1987.
1.6.3 Gurtin, M. E. and Sternberg, E., "On the linear theory of viscoelasticity", *Arch.
      Rational Mech. Anal.* 11, 291–356, 1962.
1.6.4 Bland, D. R., *The Theory of Linear Viscoelasticity*, Pergamon, Oxford, 1960.
1.6.5 Flügge, W., *Viscoelasticity* Blaisdell, Waltham, MA, 1967.
1.6.6 Golden, J. M. and Graham, G. A. C., *Boundary Value Problems in Linear Vis-
      coelasticity*, Springer Verlag, Berlin, 1988.
1.6.7 Haddad, Y. M., *Viscoelasticity of Engineering Materials*, Chapman and Hall,
      London, 1995.
1.6.8 Pipkin, A. C., *Lectures on Viscoelasticity Theory*, Springer Verlag, Heidelberg;
      George Allen and Unwin, London, 1972.
1.6.9 Gross, B., *Mathematical Structure of the Theories of Viscoelasticity*, Hermann,
      Paris, 1968.
1.6.10 Findley, W. N., Lai, J. S., and Onaran, K., *Creep and Relaxation of Nonlinear
       Viscoelastic Materials*, North-Holland, Amsterdam, 1976.
1.6.11 Lockett, F. J., *Nonlinear Viscoelastic Solids*, Academic, NY, 1972.
1.6.12 Shames, I. H. and Cozzarelli, F. A., *Elastic and Inelastic Stress Analysis*, Prentice
       Hall, Englewood Cliffs, NJ, 1992.
1.6.13 Tschoegl, N. W., *The Phenomenological Theory of Linear Viscoelastic Behavior*,
       Springer Verlag, Berlin, 1989.

1.6.14 Gittus, J., *Creep, Viscoelasticity and Creep Fracture in Solids*, J. Wiley, Halstead, NY; Applied Science, London, 1975.

1.6.15 Rabotnov, Yu. N., *Creep Problems in Structural Members*, North-Holland, Amsterdam, 1969.

1.6.16 Kraus, H., *Creep Analysis*, J. Wiley, NY, 1980.

1.6.17 Bert, C. W., "Material damping: an introductory review of mathematical models, measures, and experimental techniques", *J. Sound Vib.*, 29, 129–153, 1973.

1.6.18 Ferry, J. D., *Viscoelastic Properties of Polymers*, 2nd ed., J. Wiley, NY, 1970.

1.6.19 McCrum, N. G., Read, B. E., and Williams, G., *Anelastic and Dielectric Effects in Polymeric Solids*, Dover, NY, 1991.

1.6.20 Aklonis, J. J. and MacKnight, W. J., *Introduction to Polymer Viscoelasticity*, J. Wiley, NY, 1983.

1.6.21 Ward, I. M. and Hadley, D. W., *Mechanical Properties of Solid Polymers*, J. Wiley, NY, 1993.

1.6.22 Zener, C., *Elasticity and Anelasticity of Metals*, University of Chicago Press, IL, 1948.

1.6.23 Nowick, A. S. and Berry, B. S., *Anelastic Relaxation in Crystalline Solids*, Academic, NY, 1972.

1.6.24 Creus, G. J., *Viscoelasticity-Basic Theory and Applications to Concrete Structures*, Springer Verlag, Berlin, 1986.

1.6.25 Zinoviev, P. A. and Ermakov, Y. N., *Energy Dissipation in Composite Materials*, Technomic, Lancaster, PA, 1994.

1.6.26 Lodge, A. S., *Elastic Liquids*, Academic, NY, 1964.

1.6.27 Nabarro, F. R. N. and de Villiers, H. L., *The Physics of Creep*, Taylor and Francis, London, 1995.

1.6.28 Pomeroy, C. D. ed., *Creep of Engineering Materials*, Mechanical Engineering Publishers, Ltd., London, 1978.

1.7.1 Coulomb, C. A., "Recherches théoretiques et experimentales sur la force de torsion et sur l'élasticité des fils de métal", *Mém. Acad. Sci.*, 1784.

1.7.2 Timoshenko, S. P. , *History of Strength of Materials*, Dover, NY, 1983.

1.7.3 Scher, H., Shlesinger, M. F., and Bendler, J. T., "Time-scale invariance in transport and relaxation", *Phys. Today*, 44, 26–34, 1991.

1.7.4 Vicat, L. T., "Note sur l'allongement progressif du fil de fer soumis à diverses tensions", *Ann., Ponts Chausées, Mém. Docum* 7, 1834.

1.7.5 Weber, W., "Über die Elastizität der Seidenfäden", *Ann. Phys. Chem. (Poggendorf's)* 34, 247–257, 1835.

1.7.6 Kohlrausch, R., "Ueber das Dellmann'sche Elektrometer", *Ann. Phys. (Leipzig)* 72, 353–405, 1847.

1.7.7 Weber, W., "Über die Elastizität fester Körper", *Ann. Phys. Chem. (Poggendorf's)* 54, 247–257, 1841.

1.7.8 Maxwell, J. C., "On the dynamical theory of gases", *Phil. Trans. Royal Soc. London* 157, 49–88, 1867.

1.7.9 Boltzmann, L., Zur Theorie der Elastischen Nachwirkungen, *Sitzungsber. Kaiserlich Akad. Wissen Math. Naturwissen.* 70, 275–306, 1874.

1.7.10 Markovitz, H., "Boltzmann and the beginnings of linear viscoelasticity", *Trans. Soc. Rheol.* 21, 381–398, 1977.

1.7.11 Volterra, V., *Theory of Functionals and of Integral and Integro-differential Equations*, Dover, NY, 1959.

1.7.12 Leaderman, H., *Elastic and Creep Properties of Filamentous Materials and Other High Polymers*, Textile Foundation, Washington, DC, 1943.

1.7.13 Tobolsky, A. V. and Andrews, R. D., "Systems manifesting superposed elastic and viscous behavior", *J. Chem. Phys.* 13, 3–27, 1945.

1.7.14 Ferry, J. D., "Mechanical properties of substances of high molecular weight. VI. Dispersion in concentrated polymer solutions and its dependence on temperature and concentration", *J. Am. Chem. Soc.* 72, 3746–3752, 1950.

1.7.15 Plazek, D. J., "Oh, thermorheological simplicity, wherefore art thou?", *J. Rheol.* 40, 987–1014, 1996.

1.7.16 Maxwell, J. C., *Constitution of Bodies*, Scientific Papers, 1877, pp. 616–624.

1.7.17 Volterra, V., "Sur la Théorie Mathématique des Phénomènes Héréditaires", *J. Math. Pures Appl.* 7, 249–298, 1928.

chapter two

Constitutive relations

§2.1   Introduction

For some applications of viscoelastic materials, it is sufficient to understand creep and relaxation properties. For example, at times structural elements are maintained under steady load or constant extension. In other applications the response to an arbitrary load or strain history is required. To predict this response, *constitutive equations* which incorporate all possible responses are of use. Various mathematical tools are used in the development of these equations. The viscoelastic functions in the equations are obtained from *experiment*.

§2.2   Prediction of the response of linearly viscoelastic materials

The creep and relaxation properties described above in §1.3 permit one to predict the response of the material to a step function stress or strain. To predict the response of the material to *any* history of stress or strain (that is, stress or strain as a function of time), constitutive equations are developed.

The following is restricted to isothermal deformation in one dimension. The symbol E is used to represent an elastic modulus; however, the analysis applies equally to shear deformation corresponding to a shear modulus G or volumetric deformation corresponding to a bulk modulus B or K.

To develop the constitutive equation for linear materials, we use the *Boltzmann superposition principle*, which states that the effect of a compound cause is the sum of the effects of the individual causes. First we consider the strain associated with a relaxation and recovery experiment, with the intention to use the idea of linearity as embodied in the Boltzmann superposition principle to predict the resulting stress history. Recall that stress relaxation refers to the time variation of stress $\sigma_0 = \varepsilon_0 E(t)$ in response to a strain history which is a step function $\mathcal{H}(t)$ in time (Appendix 1): $\varepsilon_0 \mathcal{H}(t)$; E(t) is the relaxation modulus. The assumed strain may be written as a superposition of a step up followed by a step down (Fig. 2.1a):

$$\varepsilon(t) = \varepsilon_0[\mathcal{H}(t) - \mathcal{H}(t - t_1)] \tag{2.2.1}$$

so the stress is, from the Boltzmann superposition principle (Fig. 2.1b),

$$\sigma(t) = \varepsilon_0[E(t) - E(t - t_1)]. \tag{2.2.2}$$

Here the stress due to a delayed step in strain $\varepsilon_0\mathcal{H}(t - t_1)$ follows the same form of time history $\varepsilon_0E(t - t_1)$ as the stress $\varepsilon_0E(t)$ due to an earlier step in strain, $\varepsilon_0\mathcal{H}(t)$ only delayed in time. This is only true if the material in question is *nonaging*; that is, its mechanical properties do not change with time. Aging materials are considered later. The stress may or may not recover to zero as time t becomes large, depending on the material. Recovery is also called elastic aftereffect [1.6.23].

Now consider an arbitrary history of strain $\varepsilon(t)$ as a function of time t, assumed to be zero prior to time zero, as shown in Fig. 2.2.

Consider a segment of this history from time $t - \tau$ to time $t - \tau + \Delta\tau$ (with τ as a time variable; it plays the same role as the constant t_1 in the above recovery example), as shown in Fig. 2.2. The segment of strain history may be written

$$\varepsilon(t) = \varepsilon(\tau)[\mathcal{H}(t - \tau) - \mathcal{H}(t - \tau + \Delta\tau)]. \tag{2.2.3}$$

The strain at time t "now" from this pulse is zero.

As in Eq. 2.2.2 the increment of stress at time t due to the strain pulse in the past is, from the Boltzmann superposition principle,

$$d\sigma(t) = \varepsilon(\tau)[E(t - \tau) - E(t - \tau + \Delta\tau)]. \tag{2.2.4}$$

Since

$$\frac{dE(t - \tau)}{d\tau} = \lim_{\Delta t \to 0} \frac{E(t - \tau + \Delta\tau) - E(t - \tau)}{\Delta\tau},$$

this stress increment may be written

$$d\sigma(t) = -\varepsilon(\tau)\frac{dE(t - \tau)}{d\tau}. \tag{2.2.5}$$

The entire strain history may be decomposed into such pulses. The stress at time t is the summation of the stress effects of each of these pulses. We invoke the principle of causality and include only pulses prior to or up to

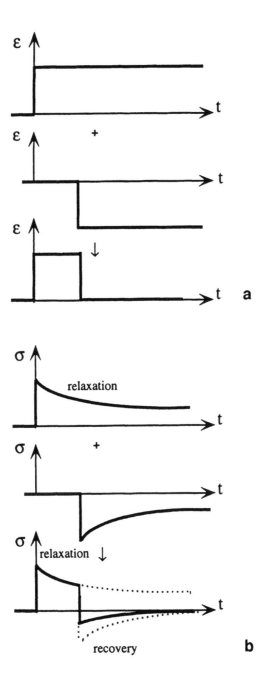

Figure 2.1 (a) Summation of step functions of strain ε vs. time t, to synthesize a box function strain history for relaxation and recovery. (b) The recovery curve of stress σ is generated from a superposition of individual relaxation curves if the material is linear.

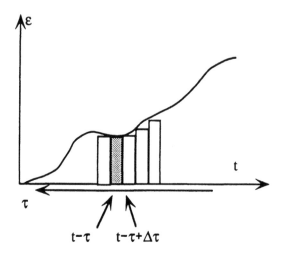

Figure 2.2 Analysis of arbitrary strain history: decomposition into pulse functions. Strain $\varepsilon(t)$ as a function of time t.

the present time. Pulses in the future are considered to have no effect in the present. In the limit as the pulse width $\Delta\tau$ becomes small the summation converges to an integral:

$$\sigma(t) = -\int_0^t \varepsilon(\tau)\frac{dE(t-\tau)}{d\tau}\,d\tau + E(0)\varepsilon(t). \tag{2.2.6}$$

The second term arises as follows. After one strain pulse or an arbitrary number of strain pulses, the strain is zero. However, in an arbitrary strain history, the final strain $\varepsilon(t)$ at time t (now) is not in general zero. The final increment of stress from zero to $\varepsilon(t)$ is $E(0)\varepsilon(t)$ since the stress has no time to relax.

Integrate by parts to obtain the final form of the *Boltzmann superposition integral*. Here t is time and τ is a time variable of integration:

$$\sigma(t) = \int_0^t E(t-\tau)\frac{d\varepsilon(\tau)}{d\tau}\,d\tau. \tag{2.2.7}$$

If the role of stress and strain are interchanged and the above arguments repeated, a complementary relation may be obtained.

$$\varepsilon(t) = \int_0^t J(t-\tau)\frac{d\sigma(\tau)}{d\tau}\,d\tau. \tag{2.2.8}$$

Consequently, if the response of a material to step stress or strain has been determined experimentally, the response to *any* load history can be found for the purpose of analysis or design. Observe, however, that if we input the same step strain in Eq. 2.2.7 as was used in the definition of E(t), the derivative does not exist at the step discontinuity, so that the stress cannot be obtained. The difficulty can be overcome by writing the strain history and its derivatives as *distributions* such as the *delta function* (Example 2.1), which are more general than functions, or by expressing the Boltzmann integral as a Stieltjes integral (see below) and avoiding the use of derivatives.

An alternative way of obtaining the Boltzmann superposition integral makes use of a Stieltjes integral rather than a Riemann integral. The strain history is decomposed into step functions (Fig. 2.3) rather than pulse functions. Each step component in strain gives rise to a relaxing component in stress, in view of the definition of the relaxation function.

$$\sigma(t) = \int_0^t E(t-\tau)\,d\varepsilon.$$

Here the strain history need not be a differentiable function of time. If it is a differentiable function, the differential $d\varepsilon$ can be written $d\varepsilon = (d\varepsilon/d\tau)\,d\tau$ to obtain the Boltzmann superposition integral in Eq. 2.2.7. The Boltzmann superposition integral can also be derived using the concepts of functional analysis [1.6.1] or general principles of continuum mechanics [2.2.1].

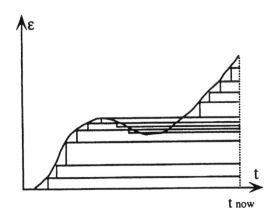

Figure 2.3 Representation of the strain history $\varepsilon(t)$ by a sum of step functions, leading in the limit to a Stieltjes integral.

§2.3 Restrictions on the viscoelastic functions; fading memory

The relaxation and creep functions are not arbitrary functions. Restrictions on the form of these functions have been derived based on physical principles. For example, the stored energy density W for elastic materials of modulus E at strain ε in one dimension has the form

$$W(t) = \frac{1}{2} E \varepsilon^2,$$
(2.3.1)

and for linearly viscoelastic materials it has the form [2.3.1]

$$W(t) = \frac{1}{2} \int_{-\infty}^{t} \int_{-\infty}^{t} E(2t - \tau - \eta) \frac{d\varepsilon}{d\tau} \frac{d\varepsilon}{d\eta} \, d\tau d\eta$$
(2.3.2)

in which τ and η are time variables. In three dimensions, for dilatation, ε is interpreted as ε_{kk} and E as the bulk relaxation modulus; for shear, ε is interpreted as the deviatoric strain e_{ij} and E as the shear relaxation modulus $2\mu(t)$.
    The corresponding dissipation inequality is [2.3.1]

$$-\frac{1}{2} \int_{-\infty}^{t} \int_{-\infty}^{t} \frac{\partial}{\partial t} E(2t - \tau - \eta) \frac{d\varepsilon}{d\tau} \frac{d\varepsilon}{d\eta} \, d\tau d\eta \geq 0.$$
(2.3.3)

This embodies the thermodynamic principle that the dissipation rate of energy is nonnegative.
    If one requires the energy density to be nonnegative during relaxation (step-strain history), then

$$E(t) \geq 0.$$
(2.3.4)

Similarly, to satisfy the dissipation inequality during relaxation,

$$\frac{dE(t)}{dt} \leq 0.$$
(2.3.5)

The relaxation function therefore decreases monotonically with time as was informally demonstrated in §1.3. Alternatively one may define [2.3.2, 2.3.3] an energy function w such that

$$w(t) = \int_{-\infty}^{t} \sigma(\tau) \frac{d\varepsilon}{d\tau} \, d\tau.$$
(2.3.6)

A *passive* system is one for which

$$w(t) \geq 0$$
(2.3.7)

for all t. The physical interpretation is that of a physical system which can absorb energy but which is unable to generate energy within itself. A *dissipative* material is defined [2.3.4, 2.3.5] as one for which the energy function w(t) ≥ 0, for any path starting from the virgin state (of no prior deformation). Dissipativity has been demonstrated to be a consequence of the second law of thermodynamics [2.3.4]. Moreover, [2.3.3] for a dissipative material, w(t) ≥ 0 for every closed strain path. Closed strain paths are of particular interest in understanding the dynamic response, as discussed further in §3.4.

In a material which exhibits *fading memory* the magnitude of the effect of a more recent cause (such as a strain) is greater than the magnitude of the effect of a cause of identical magnitude in the distant past. For fading memory it is sufficient [2.3.1] that

$$\left|\frac{dE(t)}{dt}\right|_{t=t_1} \le \left|\frac{dE(t)}{dt}\right|_{t=t_2} \quad \text{for } t_1 > t_2 > 0. \tag{2.3.8}$$

To demonstrate this, observe in the alternate form of the Boltzmann integral, Eq. 2.2.6, that the strain history in the integral is weighted by the time derivative of the relaxation modulus. Consequently, if that derivative decreases with time, the material exhibits fading memory. Therefore

$$\frac{d^2E(t)}{dt^2} \ge 0; \tag{2.3.9}$$

that is, the relaxation curve is concave up.

To have a *unique* solution of boundary value problems for viscoelastic materials, it is a necessary condition that the initial value of the relaxation functions be positive [1.6.3, 2.3.4].

§2.4   Relation between creep and relaxation

A relationship between the creep function J(t) and the relaxation function E(t) is desired. These functions appear in the Boltzmann superposition integrals above. Such integral equations may be manipulated with the aid of integral transforms (Appendix 2) such as the Laplace transform (Appendix 3). Specifically, Laplace transformation of a linear integral equation or a linear differential equation converts it into an algebraic equation.

We use the derivative theorem and convolution theorem for the Laplace transform to convert constitutive Eqs. 2.2.7 and 2.2.8 to $\sigma(s) = s\,E(s)\varepsilon(s)$ and $\varepsilon(s) = s\,J(s)\sigma(s)$, respectively. Here s is the transform variable.

So

$$\frac{\sigma(s)}{\varepsilon(s)} = sE(s) \text{ and } \frac{\sigma(s)}{\varepsilon(s)} = \frac{1}{sJ(s)}. \tag{2.4.1}$$

By setting these stress–strain ratios equal,

$$E(s)J(s) = \frac{1}{s^2}. \tag{2.4.2}$$

By taking the inverse transform, using the convolution theorem and the relation $\mathscr{L}[t] = 1/s^2$,

$$\int_0^t J(t-\tau)E(\tau)d\tau = \int_0^t E(t-\tau)J(\tau)d\tau = t. \tag{2.4.3}$$

The relationship is implicit. Explicit relationships can be developed via Laplace transformation provided a specific analytical form is given for $E(t)$ or $J(t)$; examples are given in subsequent sections. If a viscoelastic function is known in numerical form, the integral can be approximated as a sum, and the calculation performed numerically (Appendix 4). Approximate interrelations can also be developed as is done in §4.3.

§2.5   Stress vs. strain for constant strain rate

Suppose the material is subjected to a constant strain rate history beginning at time zero: $\varepsilon(t) = \{0$ for $t < 0$, Rt for $t \geq 0\}$, with R as the strain rate. By substituting in the Boltzmann integral, and changing variables, with T as a variable of integration,

$$\sigma(t)\int_0^t E(t-\tau)\frac{d\varepsilon}{d\tau}d\tau = R\int_0^t E(t-\tau)d\tau = -R\int_0^t E(T)dT. \tag{2.5.1}$$

So by Liebnitz' rule (see Appendix 1)

$$\frac{d\sigma(t)}{dt} = R\,E(t), \tag{2.5.2}$$

so the *slope* of the stress–strain curve of a linearly viscoelastic material decreases with time, and hence with strain (Fig. 2.4). The stress–strain curve is nonlinear even though the material is assumed to be linearly viscoelastic. The reason is that although the stress and strain are linearly related at a particular time t, the constant strain rate test involves strain and the time changing simultaneously. Viscoelastic materials depend on time in their stiffness.

One cannot distinguish from a single stress–strain curve (assuming it is concave down) whether the material is linearly viscoelastic, nonlinearly viscoelastic, nonlinearly elastic, or nonlinearly inelastic. These possibilities could be distinguished by performing a series of creep experiments at

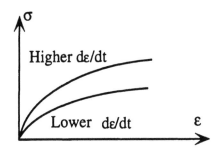

Figure 2.4   Stress σ vs. strain ε of a *linearly* viscoelastic material in response to constant strain rate.

different stress levels or a series of relaxation experiments at different strain levels as is further discussed in §6.2.

§2.6   Particular creep and relaxation functions

§2.6.1   Exponentials and mechanical models

Among the simplest transient response functions are those which involve exponentials. In relaxation we may consider

$$E(t) = E_0 e^{-t/\tau_r}, \tag{2.6.1}$$

with τ_r called the relaxation time. In creep, the corresponding relation is

$$J(t) = J_0(1 - e^{-t/\tau_c}), \tag{2.6.2}$$

with τ_c called the creep or retardation time.

Exponential response functions arise in simple discrete mechanical models composed of springs, which are perfectly elastic ($\sigma_s = E\varepsilon_s$); and dashpots, which are perfectly viscous ($\sigma_d = \eta d\varepsilon_d/dt$, with η as viscosity). The dashpot may be envisaged as a piston–cylinder assembly in which motion of the piston causes a viscous fluid to move through an aperture. Some people find spring–dashpot models to be useful in visualizing how viscoelastic behavior can arise; however, they are not necessary in understanding or using viscoelasticity theory.

Warning. Spring–dashpot models have a pedagogic role; however, real materials in general are not describable by models containing a small number of springs and dashpots.

The *Maxwell model* consists of a spring and dashpot in series. Since deformation is assumed quasistatic, inertia is neglected and the force or stress is the same in both elements; the total deformation or strain is the sum of the strains. So for the Maxwell model,

$$\frac{d\varepsilon}{dt} = \frac{d\varepsilon_s}{dt} + \frac{d\varepsilon_d}{dt} = \frac{1}{E}\frac{d\sigma}{dt} + \frac{\sigma}{\eta}. \tag{2.6.3}$$

This differential equation can be rewritten in terms of the relaxation time $\tau \equiv \eta/E$ as

$$E\frac{d\varepsilon}{dt} = \frac{d\sigma}{dt} + \frac{\sigma}{\tau}. \tag{2.6.4}$$

If we input step strain, the relaxation response is found to be

$$E(t) = E_0 e^{-t/\tau_r}; \tag{2.6.5}$$

however, if we input step stress, the creep response is

$$J(t) = \frac{1}{E} + \frac{t}{\eta}. \tag{2.6.6}$$

This form is unrealistic for primary creep since the predicted creep response is a linear function of time, in contrast to curves which are observed experimentally.

Now the *Voigt model*, also called the Kelvin model, consists of a spring and dashpot in parallel so that they both experience the same deformation or strain and the total stress is the sum of the stresses in each element. So

$$\sigma = E\varepsilon + \eta\frac{d\varepsilon}{dt}, \tag{2.6.7}$$

or

$$\frac{\sigma}{E} = \varepsilon + \tau_c\frac{d\varepsilon}{dt},$$

in which $\tau_c = \eta/E$ is referred to as the retardation time. The creep response of the Voigt model is

$$J(t) = \frac{1}{E}\left(1 - e^{-t/\tau_c}\right), \tag{2.6.8}$$

but the relaxation response is a constant plus a delta function.

More realistic behavior involving a single exponential in both creep and relaxation can be modeled by the *standard linear solid* [1.6.22] which contains

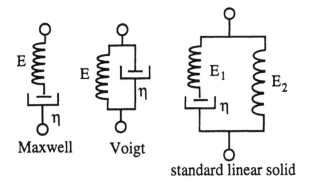

Figure 2.5 Spring–dashpot models. E represents spring stiffness and η represents dashpot viscosity.

three elements (e.g., Fig. 2.5). The left side of the model is a Maxwell model for which

$$E_1 \frac{d\varepsilon}{dt} = \frac{d\sigma_1}{dt} + \frac{\sigma_1}{\tau}, \qquad (2.6.9)$$

as above with $\tau \equiv \eta/E_1$.
    For the right side of the model, $\sigma_2 = E_2\varepsilon$, so

$$E_2 \frac{d\varepsilon}{dt} = \frac{d\sigma_2}{dt}. \qquad (2.6.10)$$

The total load (written as a stress) is the sum of the loads on the left and right side of the diagram:

$$\sigma = \sigma_1 + \sigma_2. \qquad (2.6.11)$$

Construct from the above

$$\frac{d\varepsilon}{dt}(E_1 + E_2) = \frac{d\sigma_1}{dt} + \frac{\sigma_1}{\tau} + \frac{d\sigma_2}{dt}, \qquad (2.6.12)$$

and add $\tau^{-1}E_2\varepsilon = \tau^{-1}\sigma_2$ to both sides to eliminate the individual stresses and to obtain the differential equation for the standard linear solid.

$$\frac{d\varepsilon}{dt}(E_1 + E_2) + \frac{\varepsilon E_2}{\tau} = \frac{\sigma}{\tau} + \frac{d\sigma}{dt}. \qquad (2.6.13)$$

To obtain the *relaxation function* of the standard linear solid, take the Laplace transform of the differential equation to obtain the following:

$$\left(E_1 + E_2\right)\left(s\varepsilon(s) - \varepsilon(0)\right) + \frac{\varepsilon s E_2}{\tau} = \frac{\sigma(s)}{\tau} + s\sigma(s) - \sigma(0). \qquad (2.6.14)$$

The strain and stress histories are assumed to begin after time zero, which may be considered to occur in the remote past. Then the surface terms vanish, simplifying the analysis. So

$$\sigma(s) = \frac{\left[s\left(E_1 + E_2\right) - \dfrac{E_2}{\tau}\right]\varepsilon(s)}{\left(s + \dfrac{1}{\tau}\right)}. \qquad (2.6.15)$$

But, from above (§2.2),

$$\sigma(s) = sE(s)\varepsilon(s)$$

is the Boltzmann integral in the Laplace plane, so

$$E(s) = \frac{E_2}{s} + \frac{E_1}{\left(s + \dfrac{1}{\tau}\right)}. \qquad (2.6.16)$$

By transforming back to obtain the relaxation modulus, we identify $\tau = \tau_r$, the relaxation time.

$$E(t) = E_2 + E_1 e^{-t/\tau_r}. \qquad \textbf{(2.6.17)}$$

So the relaxation function for the standard linear solid is a decreasing exponential.

To obtain the *creep function* corresponding to $E(t) = E_2 + E_1 e^{-t/\tau_r}$, consider the Laplace transform

$$E(s) = \frac{E_2}{s} + \frac{E_1}{\left(s + \dfrac{1}{\tau_r}\right)}, \qquad (2.6.18)$$

and recall that, from Eq. 2.4.2,

$$E(s)J(s) = \frac{1}{s^2} \text{ so } J(s) = \frac{1}{s^2 E(s)}, \qquad (2.6.19)$$

so

$$J(s) = \frac{1}{sE_2 + E_1 \dfrac{s^2}{s + \dfrac{1}{\tau_r}}} = \frac{s + \dfrac{1}{\tau_r}}{sE_2\left(s + \dfrac{1}{\tau_r}\right) + E_1 s^2}$$

(2.6.20)

$$= \frac{s + \dfrac{1}{\tau_r}}{s^2(E_2 + E_1) + E_2 \dfrac{s}{\tau_r}} = \frac{1}{E_2 + E_1} \frac{s + \dfrac{1}{\tau_r}}{s^2 + \dfrac{E_2}{E_2 + E_1} \dfrac{s}{\tau_r}}$$

Write this in the form

$$J(s) = \frac{P}{s} + \frac{Q}{\left(s + \dfrac{1}{R\tau_r}\right)} = \frac{P\left(s + \dfrac{1}{R\tau_r}\right) + Qs}{s\left(s + \dfrac{1}{R\tau_r}\right)},$$

(2.6.21)

and compare terms to find P, Q, and R.

But $P = 1/E_2$ by consideration of the asymptotic behavior for large time, so

$$J(s) = \frac{\dfrac{1}{E_2 R\tau_r} + \left(\dfrac{1}{E_2} + Q\right)s}{s\left(s + \dfrac{1}{R\tau_r}\right)}.$$

(2.6.22)

Compare with the above,

$$R = \frac{E_1 + E_2}{E_2},$$

and

$$Q = \frac{1}{E_1 + E_2} - \frac{1}{E_2} = -\frac{E_1}{E_2(E_1 + E_2)}.$$

(2.6.23)

By transforming back to obtain the creep function corresponding to the exponential relaxation function in Eq. 2.6.17,

$$J(t) = \frac{1}{E_2} - \frac{E_1}{E_2(E_1 + E_2)} e^{-t/\tau_c}, \tag{2.6.24}$$

with

$$\tau_c = \tau_r \frac{(E_1 + E_2)}{E_2}, \tag{2.6.25}$$

called the creep or retardation time.

Observe that the retardation time is *not* equal to the relaxation time; it is larger. Physical insight is developed in Chapter 4 in connection with the distinction between creep and relaxation. A comparative graph is shown in Fig. 4.4. The ratio of retardation to relaxation times depends on the *relaxation strength* defined as the change in stiffness during relaxation divided by the stiffness at long time:

$$\Delta = \frac{E_1}{E_2}. \tag{2.6.26}$$

So,

$$\tau_c = \tau_r(1 + \Delta). \tag{2.6.27}$$

If the relaxation strength is small ($E_1 \ll E_2$), then the retardation and relaxation times are approximately equal.

We may also write the relaxation strength as

$$\Delta = \frac{E(0) - E(\infty)}{E(\infty)}. \tag{2.6.28}$$

By analysis of Eq. 2.6.24 for $J(t)$, the relaxation strength in the compliance formulation is obtained as

$$\Delta = \frac{J(\infty) - J(0)}{J(0)}. \tag{2.6.29}$$

The single exponential response functions do not well approximate the behavior of most real materials. Specifically, a single exponential relaxation function undergoes most of its relaxation over about one decade (a factor of ten) in time scale. Real materials relax or creep over many decades of time scale (see Chapter 7). It is of course possible to generalize the differential equations to include derivatives of arbitrarily high order n. For example, differential operators may be defined as follows [1.6.3] in terms of the time derivative operator:

$$D = \frac{d}{dt}.$$

The constitutive equation can then be written as a polynomial of the differential operator:

$$P(D) = \sum_{k=0}^{N} p_k \frac{d^k}{dt^k}, \quad Q(D) = \sum_{k=0}^{N} q_k \frac{d^k}{dt^k}. \tag{2.6.30}$$

Suppose [1.6.3]

$$q_r = \sum_{n=r}^{N} p_n \frac{d^{n-r}E}{dt^{n-r}}\Big|_0, \tag{2.6.31}$$

with these derivatives to be evaluated for an argument of zero and that the stress and strain meet initial conditions

$$\sum_{r=k}^{N} p_r \frac{d^{r-k}\sigma}{dt^{r-k}}\Big|_0 = \sum_{r=k}^{N} q_r \frac{d^{r-k}\varepsilon}{dt^{r-k}}\Big|_0, \tag{2.6.32}$$

in which these derivatives are to be evaluated for an argument of zero. Suppose further that the material obeys the linear differential equation

$$P(D)\,\sigma(t) = Q(D)\,\varepsilon(t). \tag{2.6.33}$$

Apply a Laplace transform to the differential equation:

$$P(s)\,\sigma(s) = Q(s)\,\varepsilon(s). \tag{2.6.34}$$

The stress–strain ratio in the Laplace plane,

$$\frac{\sigma(s)}{\varepsilon(s)} = \frac{Q(s)}{P(s)},$$

when compared with the Laplace transformed Boltzmann integral, Eq. 2.4.1, gives

$$E(s) = \frac{Q(s)}{sP(s)}. \tag{2.6.35}$$

The relaxation function $E(t)$ is obtained by inversion of the transform by partial fractions. The explicit form, based on the Laplace transform of a quotient of polynomials, is

$$E(t) = \sum_{n=0}^{N} \frac{Q\left(c_n e^{tc_n}\right)}{\lim_{s \to c_n} \frac{sP(s)}{s - c_n}}.$$

The transient response function associated with a linear differential equation of order n is a sum of n exponentials. It may be written more simply as the following [1.6.1].

$$E(t) = \sum_{n=0}^{N} \lambda_n e^{-t/\tau_n}. \tag{2.6.36}$$

Differential equations such as Eq. 2.6.33 can be considered to arise from complicated assemblages of springs and dashpots; however, such a visualization is hardly useful since the springs and dashpots are not identified with physical features in the material.

§2.6.2 *Exponentials and internal causal variables*

The purpose of this segment is to show how single exponentials in the transient response can arise via the coupling between strain and a physical quantity, known as an internal variable, which can be shown to relax by a flow process when perturbed. Examples of internal variables are temperature, electric field, and pore pressure of fluid in a porous material (see Chapter 8).

Suppose that the strain ε depends linearly on the stress σ and on one internal variable β, then [1.6.23]

$$\varepsilon(\sigma,\beta) = J_0\sigma + c\beta, \tag{2.6.37}$$

with c as a coupling coefficient which depends on the material. Assume further that the internal variable, when perturbed, approaches its equilibrium value β_e (assumed to be zero for zero stress; $\beta_e = \sigma b$ with b characteristic of the material) according to a first-order equation

$$\frac{d\beta}{dt} = -\frac{(\beta - \beta_e)}{\tau}, \tag{2.6.38}$$

with τ as the relaxation time. Combining these relations to eliminate β yields the following:

$$\varepsilon + \tau\frac{d\varepsilon}{dt} = \tau J_0\frac{d\sigma}{dt} + \sigma(J_0 + Kb). \tag{2.6.39}$$

This is equivalent, with different symbols for the constants, to the differential equation, Eq. 2.6.13, for the standard linear solid described above, for which the relaxation and creep functions contain single exponentials. Analyses for specific kinds of internal variables are given in the discussion of relaxation mechanisms in Chapter 8.

If there are multiple internal variables, then a relaxation function as a discrete sum of exponentials is expected.

$$E(t) = \sum_{n=0}^{N} \lambda_n e^{-t/\tau_n}. \tag{2.6.40}$$

Here the λ's are weighting coefficients. One may also envisage a distribution or *spectrum* $H(\tau)$ of relaxation times τ and associated exponential terms, as follows (see §4.2):

$$E(t) - E_e = \int_{-\infty}^{\infty} H(\tau) e^{-t/\tau} \, d\ln\tau = \int_{0}^{\infty} \frac{H(\tau)}{\tau} e^{-t/\tau} \, d\tau. \tag{2.6.41}$$

§2.6.3 Fractional derivatives

Simple constitutive equations covering many decades of time scale have been constructed using fractional derivatives [2.6.1], which are operators that generalize the order of differentiation to fractional orders α with $0 < \alpha < 1$. The concept of fractional derivative can be visualized via the Laplace transform relation (with vanishing initial values and initial slopes of $f(t)$):

$$\mathcal{L}\left\{\frac{d^\alpha f(t)}{dt^\alpha}\right\} = s^\alpha F(s). \tag{2.6.42}$$

This relation can be demonstrated for integer order via the definition of the Laplace transform. For fractional order it may be regarded as a definition of the fractional derivative.

Fractional derivatives can also be defined by the following integral representation [2.6.1]:

$$\frac{d^\alpha x(t)}{dt^\alpha} \equiv \frac{1}{\Gamma(1-\alpha)} \frac{d}{dt} \int_{0}^{t} \frac{x(t)}{(t-\tau)^\alpha} \, d\tau. \tag{2.6.43}$$

A constitutive equation of the form

$$\sigma(t) = E\tau^{-\alpha} \frac{d^\alpha \varepsilon(t)}{dt^\alpha}, \tag{2.6.44}$$

with E as a stiffness constant, τ as a time constant, and α as a material parameter, can be used to obtain the relaxation function following Koeller [2.6.2] in notation due to Rabotnov [2.6.3]. The strain is set equal to a Heaviside step function $\mathcal{H}(t)$. The relaxation function is obtained via an alternate form of the fractional derivative:

$$\frac{d^{-\alpha}x(t)}{dt^{-\alpha}} \equiv \int_0^t \frac{(t-\tau)^{\alpha-1}}{\Gamma(\alpha)} x(\tau)\, d\tau. \tag{2.6.45}$$

$$E(t) = E\tau^{-\alpha}\frac{d^{\alpha}\mathcal{H}(t)}{dt^{\alpha}} = E\tau^{-\alpha}\int_0^t \frac{(t-\tau)^{-\alpha}}{\Gamma(1-\alpha)} \mathcal{H}(t-\tau)\,\delta(\tau)\, d\tau. \tag{2.6.46}$$

$$E(t) = E\tau^{-\alpha}\frac{t^{-\alpha}}{\Gamma(1-\alpha)} \mathcal{H}(t). \tag{2.6.47}$$

So the relaxation function is a power law in time. Terms of this type can be superposed to obtain sufficient generality to model much material behavior. It is found [2.6.1] that only one or two fractional derivative terms in the constitutive relation suffice to describe behavior of actual materials over many decades of time scale.

§2.6.4 Power law behavior

For some materials [1.6.8, 2.6.1] the relaxation function is approximated by $E(t) = At^{-n}$. The plot of this relaxation function on a log–log scale is a straight line. The relaxation can be expressed in terms of percent per decade. Let us use the above Laplace transform relations to determine $J(t)$. By taking the Laplace transform of $E(t)$,

$$E(s) = \frac{A\Gamma(-n+1)}{s^{-n+1}}, \tag{2.6.48}$$

but $E(s)J(s) = 1/s^2$ (Eq. 2.4.2) is the relation between creep and relaxation properties, from above, so

$$J(s) = \frac{1}{s^2 E(s)} = \frac{1}{A\Gamma(-n+1)}s^{-n+1-2} = \frac{1}{A\Gamma(1-n)}\frac{1}{s^{n+1}}. \tag{2.6.49}$$

But

$$\frac{1}{s^{n+1}} = \frac{\mathcal{L}[t^n]}{\Gamma(n+1)}, \tag{2.6.50}$$

$$J(t) = \frac{1}{A\Gamma(1-n)\,\Gamma(n+1)}\,t^n. \qquad (2.6.51)$$

So the creep function is also a power law.

Andrade creep is a particular form of power law for which $J(t) = J_0 + At^{1/3}$. It was originally developed for metals at large strain but is also used to model creep in glassy polymers [1.6.21].

§2.6.5 Stretched exponential

Relaxation in complex materials with strongly interacting constituents often follows [2.6.4–2.6.8] the stretched exponential or KWW (after Kohlrausch, Williams, and Watts) form

$$E(t) = \left(E_0 - E_\infty\right) e^{-(t/\tau_r)^\beta} + E_\infty, \qquad (2.6.52)$$

with $0 < \beta \le 1$, E_0 and E_∞ as constants, and τ_r as a characteristic relaxation time. Behavior of this type is commonplace [2.6.7]. The relaxation of the stretched exponential occurs over a wider range of time on a logarithmic scale than does the Debye exponential form ($\beta = 1$) which occurs mostly within one decade, a factor of ten in the time scale, as shown in Fig. 2.6. Dynamic behavior corresponding to this relaxation function can be obtained analytically for $\beta = 0.5$ [2.6.9] but for general values of β numerical methods are appropriate. In viscoelastic solids, $E_\infty > 0$ and $\tan \delta$ forms a broad peak with $\tan \delta \approx \nu^{-n}$ for frequencies ν well above the peak, which can be demonstrated numerically using the transformation relations developed in §3.2.

§2.6.6 Distinguishing among viscoelastic functions

To make meaningful comparisons, it is considered helpful to use logarithmic time scales in the plotting of relaxation results. A comparison of single exponential (Debye) and power law relaxations is shown on a linear scale over one decade in Fig. 2.7a and on a logarithmic scale over eight decades in Fig. 2.7b. The curves appear to be similar when plotted on a linear scale, since only about one decade (a factor of ten) of time is readily viewed. They differ considerably when viewed on a logarithmic scale.

§2.7 Effect of temperature

The development thus far has been under the assumption of isothermal conditions. The viscoelastic functions of materials can be expected to depend on temperature T, as well as time t, e.g., the relaxation function may be written $E = E(t, T)$. Envisage a material in which the viscoelasticity arises

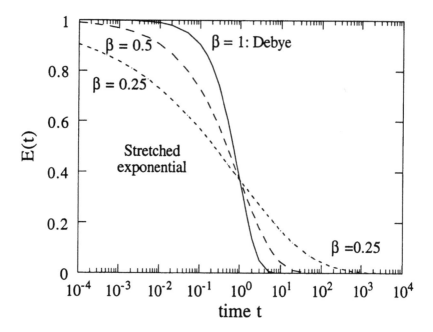

Figure 2.6 Stretched exponential relaxation functions. Arbitrary units of stiffness and time.

from a molecular rearrangement process which occurs under stress, or from a diffusion process under stress. The speed of such processes depends on the speed of molecular motion of which temperature is a measure. If all the processes contributing to the viscoelasticity of the material are accelerated to the same extent by a temperature rise, then we have a relaxation function of the following form:

$$E(t, T) = E\left(\zeta, T_0\right), \text{with } \zeta = \frac{t}{a_T(T)}, \tag{2.7.1}$$

in which ζ is called the reduced time, T_0 is the reference temperature, and $a_T(T)$ is called the shift factor. For such materials a change in temperature stretches or shrinks the effective time scale. Since viscoelastic properties are usually plotted vs. log time or log frequency, a temperature change for such materials corresponds to a horizontal *shift* of the material property curves on the log time or of frequency axis. Materials which behave according to Eq. 2.7.1 obey *time–temperature superposition* and are called *thermorheologically simple* [2.7.1].

The shift factor $a_T(T)$ depends on temperature and on the material as discussed in Chapter 7. Many materials exhibit temperature-dependent creep behavior which follows the *Arrhenius* relation [1.6.23],

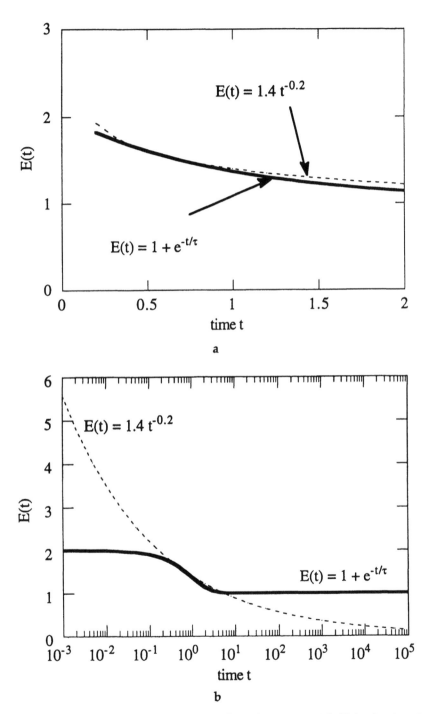

Figure 2.7 Comparison of power law and single exponential (Debye) relaxation functions on (a) a linear time scale over one decade and (b) over a logarithmic time scale over eight decades. Arbitrary units of stiffness and time.

$$\tau^{-1} = \nu_0 \exp\left\{-\frac{U}{kT}\right\}, \tag{2.7.2}$$

with τ as a time constant, ν_0 as a characteristic frequency, T as the absolute temperature, U as the *activation energy*, and k as the Boltzmann constant. This form gives rise to thermorheologically simple behavior with the following shift factor a_T as is shown in Example 6.9:

$$\ln a_T = \frac{U}{k}\left\{\frac{1}{T_2} - \frac{1}{T_1}\right\}. \tag{2.7.3}$$

The time–temperature shift for polymers tends to follow the empirical WLF equation [1.6.18] (after Williams, Landel, and Ferry [2.7.2]),

$$\log a_T = -\frac{C_1\left(T - T_{ref}\right)}{C_2 + \left(T - T_{ref}\right)}, \tag{2.7.4}$$

in which T_{ref} is the reference temperature and the logarithm is of base ten. In polymers near the glass transition temperature, $a_T(T)$ varies sufficiently rapidly with temperature that an isothermal approximation may not be warrantable in the presence of small temperature variations [1.6.6]. The constants C_1 and C_2 depend on the particular polymer. Even so, the behavior of different amorphous polymers is sufficiently similar in normalized coordinates that Ferry has proposed the following "universal" constants, for T_{ref} taken as T_g the glass transition temperature: $C_1 = 17.44$ and $C_2 = 51.6$.

Figure 2.8 shows representative relaxation data over three decades of time, and for several temperatures. The curve constructed by time–temperature shifts is called a *master curve*. The following procedure is used. A temperature called the reference temperature, is chosen, for which there is no shift. The master curve is generated by horizontally shifting the experimental curve for temperatures just above or below the reference temperature until they coincide. The directions of the shifts along the log time axis in Fig. 2.8 are shown by arrows. Then the curve for the next higher or lower temperature is shifted, and so on until all the data are analyzed. If overlap cannot be achieved by this procedure, then the material is not thermorheologically simple. The master curve obtained from shifting the three decade curves in Fig. 2.8 is shown in Fig. 2.9. The ten decade range of this master curve comes about as a result of the shift of the individual curves along the abscissa. For some materials, particularly polymers, a modest change in temperature can give rise to a very large change in the relaxation times, as shown in Fig. 2.10. The numbers used to generate this WLF plot are representative of a Hevea rubber [1.6.18], with the simplification of assuming a single relaxation time process.

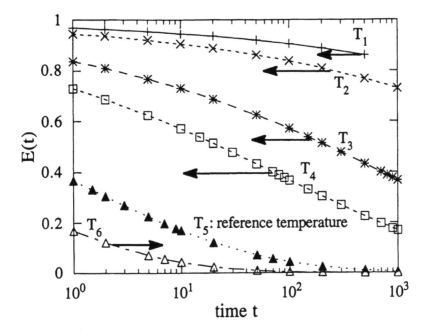

Figure 2.8   Relaxation curves over three decades of time, and for several tempera-tures, with $T_3 > T_2 > T_1$. Arbitrary units of stiffness and time.

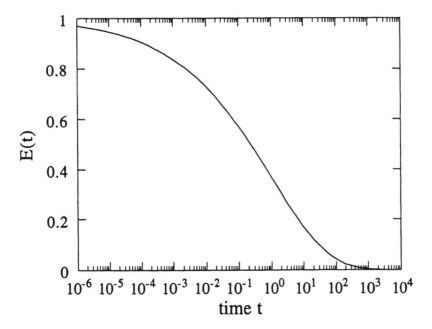

Figure 2.9   Master curve generated by shifting the above relaxation curves. Arbi-trary units of stiffness and time.

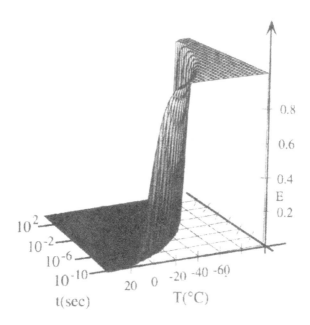

Figure 2.10   Plot of the WLF equation for shift of relaxation times, assuming $T_{ref} = -73°C$, $C_1 = 16.8$, and $C_2 = 53.6$; and assuming a single exponential relaxation process.

Many amorphous polymers are thermorheologically simple; however, crystalline polymers and most composite materials are not thermorheologically simple. The different phases each have different temperature dependencies of their viscoelastic behavior. In any material which has a multiplicity of relaxation mechanisms, thermorheological simplicity is not to be expected. Moreover, if a phase transformation (such as melting or freezing) occurs in any part of the temperature range of interest and in any portion of the material, the material will not be thermorheologically simple. If the material burns or decomposes in part of a temperature range, the material will clearly not be thermorheologically simple.

To experimentally determine if a material is thermorheologically simple, one may perform a set of creep or relaxation tests at different temperatures, and plot the results. If the various curves can be made to overlap by horizontal shifts on the log time axis, the material is considered thermorheologically simple. If the curves do not overlap, the material is not thermorheologically simple. The concept of time–temperature superposition has been of greatest use in polymers. Plazek [2.7.3] observes that given data within a fairly narrow experimental window (three decades or less) the test for thermorheological simplicity can only be definitive in its failure. The experiment is capable of demonstrating thermorheological complexity but it cannot demonstrate simplicity. The reason is that deviations can and do appear if the material is examined over a greater number of decades of time or frequency.

§2.8   Three-dimensional linear constitutive equation

In three dimensions, Hooke's law of linear elasticity is given by

$$\sigma_{ij} = C_{ijkl}\varepsilon_{kl}, \tag{2.8.1}$$

with C_{ijkl} as the elastic modulus tensor, and the usual Einstein summation convention assumed in which repeated indices are summed over [2.8.1–2.8.3]. In other words,

$$\sigma_{ij} = \sum_{k=1}^{3}\sum_{l=1}^{3} C_{ijkl}\varepsilon_{kl}.$$

There are 81 components of C_{ijkl}, but by taking into account the symmetry of the stress and strain tensors, only 36 of them are independent. If the elastic solid is describable by a strain energy function, the number of independent elastic constants is reduced to 21. An elastic modulus tensor with 21 independent constants describes an anisotropic material with the most general type of anisotropy, triclinic symmetry. Materials with orthotropic symmetry are invariant to reflections in two orthogonal planes and are describable by nine elastic constants. Materials with axisymmetry, also called transverse isotropy or hexagonal symmetry, are invariant to 60° rotations about an axis and are describable by five independent elastic constants. Materials with cubic symmetry are describable by three elastic constants. Isotropic materials, with properties independent of direction are describable by two independent elastic constants.

Examples of materials with these symmetry classes are as follows. A unidirectional fibrous material may have orthotropic symmetry if the fibers are arranged in a rectangular packing, or hexagonal symmetry if the fibers are packed hexagonally. Wood has orthotropic symmetry while human bone has hexagonal symmetry. A composite with an equal number of fibers in three orthogonal directions is cubic. A fabric with fibers in a square array is two dimensionally cubic. Table salt crystals are cubic. Glass, cast glassy polymers, and cast polycrystalline metals exhibit approximately isotropic behavior.

For linearly viscoelastic materials the following is obtained by conducting for each component arguments identical to those given for one dimension:

$$\sigma_{ij}(t) = \int_{0}^{t} C_{ijkl}(t-\tau)\frac{d\varepsilon_{kl}}{d\tau}\,d\tau. \tag{2.8.2}$$

This constitutive equation is sufficiently general to accommodate any degree of anisotropy. Observe that each independent component of the modulus tensor can have a different time dependence.

Many practical materials are approximately isotropic (their properties are independent of direction). For isotropic elastic materials, the constitutive equation is

$$\sigma_{ij} = \lambda \varepsilon_{kk} \delta_{ij} + 2\mu \varepsilon_{ij},$$
(2.8.3)

in which λ and μ are the two independent Lamé elastic constants, δ_{ij} is the Kronecker delta (1 if $i = j$, 0 if $i \neq j$), and $\varepsilon_{kk} = \varepsilon_{11} + \varepsilon_{22} + \varepsilon_{33}$. Engineering constants such as Young's modulus E, shear modulus G, and Poisson's ratio ν can be extracted from these tensorial constants. Specifically (see also Appendix 7) [2.8.3],

$$G = \mu,$$
(2.8.4)

$$E = \frac{G(3\lambda + 2G)}{(\lambda + G)},$$
(2.8.5)

$$\nu = \frac{\lambda}{2(\lambda + G)}.$$
(2.8.6)

As for the interpretation of C_{1111} for isotropic elastic materials we note that [2.8.1, 2.8.2]

$$C_{1111} = \lambda + 2G = 2G \left[\frac{\nu}{1-2\nu} + 1 \right],$$
(2.8.7)

with G (also called μ) as the shear modulus, λ as the other Lamé modulus, and ν as Poisson's ratio (see Example 5.9). Observe that $C_{1111} \neq E$, unless Poisson's ratio is zero. For comparison, the bulk modulus is

$$B = 2G \frac{1+\nu}{3(1-2\nu)}.$$
(2.8.8)

In isotropic viscoelastic solids,

$$\sigma_{ij}(t) = \int_0^t \lambda(t-\tau) \, \delta_{ij} \, \frac{d\varepsilon_{kk}}{d\tau} \, d\tau + \int_0^t 2\mu(t-\tau) \frac{d\varepsilon_{ij}}{d\tau} \, d\tau.$$
(2.8.9)

There are two independent viscoelastic functions which can have different time dependence. As in the case of elasticity, we may consider engineering functions rather than tensorial ones, for example, $E(t)$, $G(t) = \mu(t)$, $\nu(t)$, and $B(t)$.

§2.9 Aging materials

Aging materials are those in which the material properties themselves change with time. The origin of the time scale for aging is the time of creation of the material. This time may be the time of mixing in the case of concrete, the time of initiation of polymerization or the time of last solidification in the case of polymers, or the birth of an organism in the case of biological materials. In a linearly viscoelastic material which ages, the constitutive equation for one spatial dimension is

$$\varepsilon(t) = \int_0^t J(t, \tau) \frac{d\sigma}{d\tau} \, d\tau. \tag{2.9.1}$$

The creep function in this case is a function of two variables, and it can be represented as a surface or as a family of curves. In some classes of material [2.9.1, 2.9.2], aging time t_a slows the characteristic retardation times τ_c by a factor:

$$\tau_c = \tau_0 t_a^{\mu}, \tag{2.9.2}$$

in which μ is the Struik shift factor and τ_0 is a retardation time. In such cases, the effect of aging is represented by a shift of the creep curves along the log time axis.

We may consider a general creep function of the form $J(t, \tau)$, which allows aging, and impose the restriction that over a time scale of interest the material properties do not depend on time. So the creep function is unchanged by a shift on the time axis. This corresponds to [1.6.5], $J(t + t_0, \tau + t_0) = J(t, \tau)$, or $J(t, \tau) = J(t - \tau)$. This is the form for nonaging materials developed in §2.2.

§2.10 Dielectric and other forms of relaxation

Materials have other physical properties, not only mechanical properties. For example [2.10.1], the strain ε_{ij} can depend on stress σ_{kl} via the elastic compliance S_{ijkl}, on electric field E_k (in piezoelectric materials with modulus tensor d_{kij} at constant temperature), and on temperature ΔT (in most materials, via the thermal expansion α_{ij}). Moreover, the electric displacement vector D_i depends on electric field via K_{ij} which is the dielectric tensor at constant stress and temperature, and in some materials it can depend on temperature ΔT via the pyroelectric effect; and p is the pyroelectric coefficient at constant stress. These phenomena are incorporated in the following linear constitutive equations (without relaxation):

$$\varepsilon_{ij} = S_{ijkl} \, \sigma_{kl} + d_{kij} \, E_k + \alpha_{ij} \, \Delta T, \tag{2.10.1}$$

$$D_i = d_{ijk}\, \sigma_{jk} + K_{ij}\, E_j + p_i\, \Delta T. \tag{2.10.2}$$

Here the usual Einstein summation convention over repeated subscripts is used. The above equations are linear and they incorporate no relaxation.

Relaxation can occur in these physical properties in real materials. The linear constitutive equations with relaxation can be constructed as follows by using the same arguments given above, for each tensor element in the above relations:

$$\varepsilon_{ij}(t) = \int_0^t S_{ijkl}(t-\tau)\frac{d\sigma_{kl}}{d\tau}\, d\tau + \int_0^t d_{kij}(t-\tau)\frac{dE_k}{d\tau}\, d\tau + \int_0^t \alpha_{ij}(t-\tau)\frac{d\Delta T}{d\tau}\, d\tau,$$

$$\tag{2.10.3}$$

$$D_i(t) = \int_0^t d_{ijk}(t-\tau)\frac{d\sigma_{jk}}{d\tau}\, d\tau + \int_0^t K_{ij}(t-\tau)\frac{dE_j}{d\tau}\, d\tau + \int_0^t p_i(t-\tau)\frac{d\Delta T}{d\tau}\, d\tau.$$

$$\tag{2.10.4}$$

Time dependence of K_{ij} is referred to as dielectric relaxation. Relaxation can also occur in the magnetic properties of materials, in optical properties, and in strain–optical retardation [2.10.2]. Electrical and magnetic properties, as well as viscoelastic properties (see Chapter 3), can also be described in the frequency domain.

§2.11 Adaptive and "smart" materials

Materials can be time dependent in ways other than viscoelastic response. For example, a structural member can gradually become stiffer and stronger if additional substance is added in response to heavy loading. It can become less dense and weaker if material is removed in response to minimal loading. Biological materials such as bone [2.11.1, 2.11.2, 2.11.3] and tendon behave in this way. Such behavior is called adaptive elasticity. Constitutive equations have been developed for the adaptive elasticity of bone [2.11.4, 2.11.5]. The stress–strain relation is a modification of Hooke's law in that the proportionality between stress σ and strain ε is dependent on the volume fraction of material present. The change in stiffness of the bone is assumed to depend on its porosity:

$$\sigma_{ij} = (\xi_0 + e)C_{ijkl}(e)\varepsilon_{kl}. \tag{2.11.1}$$

Specifically, a change e in the solid volume fraction of a porous material with respect to a reference volume fraction ξ_0 drives the adaptation process:

$$\frac{de}{dt} = a(e) + A_{ij}(\varepsilon_{ij}). \tag{2.11.2}$$

Here a and A are coefficients which depend on the type of bone. Conventional engineering materials such as steel, aluminum, concrete, and polymers are *not* adaptive.

"Smart" materials are those which can respond in an active way to mechanical loads. A smart material may contain embedded sensors which provide information concerning the strain, temperature, or degree of damage in the material [2.11.6]. Sensor information can inform the user that an overload condition is present or that repair is necessary. To achieve a true smart response, the sensor data are input to a control system, which may contain a microprocessor that controls actuators within the material to reduce the deformation or repair the damage. Sensors and the actuators are discussed in the context of experimental methods in §6.3. For example, piezoelectric materials generate an electrical signal when strained and also deform in response to electrical excitation. Piezoelectric materials have been used both as sensors and as actuators. Another class of actuator material is electrorheological fluids, which have a viscosity that can be controlled by an electric field [2.11.7]. They are fluid–solid composites. A representative electrorheological fluid contains particles of corn starch dispersed in silicone oil.

Ideally, a smart material would emulate such biological characteristics as self repair, adaptability to conditions, self-assembly, homeostasis, and capacity for regeneration. Thus far, no synthetic material has such characteristics. Currently there are few examples of smart materials; the subject is largely a research topic. Some examples are given in §10.6.

§2.12 *Effect of nonlinearity*

In this section, constitutive equations are considered other than that of linear viscoelasticity. The aim is to gain the ability to distinguish between several different types of mechanical response; experimental aspects are discussed in §6.2. We remark that a linear material obeys linear integral equations or linear differential equations. For a material to be linear, it is necessary but not sufficient for the measured creep or relaxation function to be independent of stress or strain.

Consider first elastic behavior. Linear elasticity is expressed by Hooke's law, Eq. 1.1.1 in one dimension and Eq. 2.8.1 in three dimensions. In one dimension, nonlinear elasticity may be expressed,

$$\sigma = f(\varepsilon)\varepsilon, \tag{2.12.1}$$

with $f(\varepsilon)$ allowed to be a nonlinear function of strain. Time is not involved in either linear or nonlinear elasticity, so there is no creep or relaxation, the material recovers fully and instantaneously, and the path for unloading is identical to the path for loading. Isochronals, or data at constant time, taken from creep curves, all coincide in an elastic material, whether it be linear or nonlinear.

Consider an elastoplastic material. At sufficiently small strain the material behaves elastically, so there is no time-dependent creep or relaxation behavior and the material recovers fully and instantaneously. If the yield point is exceeded, there is still no time-dependent creep or relaxation behavior since time is not included in elastoplasticity. However, recovery is incomplete: there is some residual strain and the path for unloading differs from the path for loading. The residual strain is constant in time.

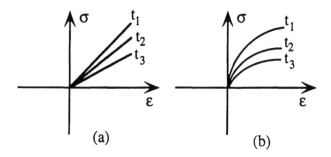

Figure 2.11   Stress σ vs. strain ε for nonlinear materials. In linear materials (a) the isochronals are straight lines, and in nonlinear materials (b) the isochronals are curved.

Nonlinear viscoelasticity gives rise to curved isochronals in the stress–strain diagram in Fig. 2.11b. Several constitutive equations are available for the modeling of nonlinear viscoelasticity. The simplest of these are restricted to describing creep. A simple equation commonly used is the Bailey-Norton relation intended to model primary and secondary creep:

$$\varepsilon(t) = A \, \sigma^m \, t^n. \tag{2.12.2}$$

Creep formulations of this type do not account for recovery or history effects.

The following simple nonlinear relation allows for prediction of history dependence. This single-integral form is called nonlinear superposition, which allows the relaxation function to depend on strain level:

$$\sigma(t) = \int_0^t E(t - \tau, \varepsilon(\tau)) \frac{d\varepsilon}{d\tau} \, d\tau. \tag{2.12.3}$$

A similar equation may be written in the compliance formulation:

$$\varepsilon(t) = \int_0^t J(t - \tau, \sigma(\tau)) \frac{d\sigma}{d\tau} \, d\tau. \tag{2.12.4}$$

If a series of relaxation tests is done at different strain levels, relaxation will be observed but the functional form of the relaxation curves will depend on the strain level. This type of nonlinear equation has been used to model the behavior of soft tissues [2.12.1]. Eq. 2.12.3 describes a specific kind of non-linearity. It predicts, for example, that recovery occurs at the same rate as relaxation. This may not be true in all materials.

A single integral nonlinear relation which has been used for rubbery materials is the BKZ relation of Bernstein, Kearsley, and Zapas [2.12.2]:

$$\sigma(t) = \int_0^t A(t-\tau)f(\varepsilon_g(\tau))\, d\tau, \tag{2.12.5}$$

in which $A(t)$ is a function and ε_g is a generalized strain measure appropriate for large deformation. For elastomers the relaxation behavior takes the form

$$\frac{\sigma(t)}{\lambda^2 - \lambda^{-1}} = \left(\lambda^2 - 1\right)\left\{\frac{1}{2}A_1(t) + A_2(t)\right\} + \frac{1}{\lambda}\left\{A_1(t) + A_2(t)\right\} + A_3 - A_1(t),$$

$$\tag{2.12.6}$$

in which λ is the extension ratio defined as the ratio of deformed length to initial length, $A_1(t)$ and $A_2(t)$ are functions of time, and A_3 is a constant. A strain measure appropriate for finite (large) deformations, in terms of the extension ratio, is [1.6.21]

$$\varepsilon_g = \frac{1}{2}\left(\lambda^2 - 1\right). \tag{2.12.7}$$

Again, Eq. 2.12.6 describes a specific kind of nonlinearity.

A creep formulation used by Schapery [2.12.3] was derived using principles of irreversible thermodynamics. There is a stress-induced shift in the time scale, as follows:

$$\varepsilon(t) = J_0 g_0 \sigma(t) + g_1 \int_0^t J(\zeta(t) - \zeta_\tau(\tau)) \frac{dg_2\sigma(\tau)}{d\tau}\, d\tau.$$

Here J_0 is a time-independent compliance; J is a creep compliance which depends on stress indirectly through the reduced time variables $\zeta(t)$ and $\zeta_\tau(\tau)$ defined below; and g_0, g_1, and g_2, are material properties which are functions of stress.

$$\zeta(t) = \int_0^t \frac{d\xi}{a_\sigma[\sigma(\xi)]},$$

$$\zeta_\tau(\tau) = \int_0^\tau \frac{d\xi}{a_\sigma[\sigma(\xi)]}.$$

The quantity $a_\sigma[\sigma(\xi)]$ is a shift factor which depends on stress; this, in turn, depends on time. This formulation, since it contains five functions and a constant, is quite adaptable [1.6.10].

More general equations for nonlinear viscoelasticity have been proposed [2.12.4–2.12.7]. In one dimension for small strain, in the modulus formulation,

$$\sigma(t) = \int_0^t E_1(t - \tau) \frac{d\varepsilon}{d\tau} \, d\tau + \int_0^t \int_0^t E_2(t - \tau_1, t - \tau_2) \left(\frac{d\varepsilon}{d\tau_1} \right) \left(\frac{d\varepsilon}{d\tau_2} \right) d\tau_1 d\tau_2 + \dots$$

$$(2.12.8)$$

In the compliance formulation one may write

$$\varepsilon(t) = \int_0^t J_1(t - \tau) \frac{d\sigma}{d\tau} \, d\tau + \int_0^t \int_0^t J_2(t - \tau_1, t - \tau_2) \left(\frac{d\sigma}{d\tau_1} \right) \left(\frac{d\sigma}{d\tau_2} \right) d\tau_1 d\tau_2 + \dots$$

$$(2.12.9)$$

Here τ_1, τ_2 are time variables of integration. To experimentally distinguish materials obeying the above two equations, simple creep or relaxation tests are inadequate, since both equations predict relaxation behavior which can depend on strain level. They may, however, be distinguished by the use of strain histories containing more than one step. For example, a relaxation and recovery test contains two steps, and a creep and recovery test is similar. In linear materials and in materials obeying nonlinear superposition, initial recovery follows the same functional form as the preceding relaxation or creep. In materials obeying multiple integral formulations such as Eq. 2.12.8, recovery can differ significantly from relaxation. The kernels E_1, E_2, and E_3 in the higher order terms in Eq. 2.12.8 can be extracted by performing a series of experiments with multiple steps at different strain levels and with different time lapses between the steps. Such experiments (§6.2) are laborious.

The multiple integral formulation [2.12.6, 2.12.7] of nonlinear viscoelasticity is applicable to a broad range of nonlinear systems in addition to viscoelastic materials.

§2.13 Summary

Constitutive equations for viscoelastic materials have been developed. The concept of linearity as embodied in the Boltzmann superposition principle was used to obtain an integral equation, the Boltzmann superposition integral, as the constitutive equation for linear viscoelasticity. Manipulations of such equations are facilitated by the Laplace transform. We find that in linearly viscoelastic materials, the stress–strain curve for constant strain rate is not a straight line; other types of experiment are usually more informative. A brief consideration of simple causes for relaxation led to single exponential creep and relaxation functions and to the concept of spectra.

Examples

Example 2.1

Suppose the stress follows a step function in time, $\sigma_0 \mathcal{H}(t)$. Obtain the strain history using the Boltzmann integral and delta functions.

Solution

The step occurs at zero; therefore, to include the response to the full stress history the time scale for integration must begin prior to zero.

$$\varepsilon(t) = \int_{-\infty}^{t} J(t-\tau)\frac{d\sigma(\tau)}{d\tau}\, d\tau = \sigma_0 \int_{-\infty}^{t} J(t-\tau)\,\delta(\tau)\, d\tau = \sigma_0 J(t). \qquad \text{(E2.1.1)}$$

The last expression follows via the sifting property of the delta function. Use of the Dirac delta function permits one to handle the singularity within the integral to obtain the creep response.

Example 2.2

Suppose the stress follows an impulse function in time, $A\delta(t)$. Obtain the strain history using the Boltzmann integral.

Solution

$$\varepsilon(t) = \int_{-\infty}^{t} J(t-\tau)\frac{d\sigma(\tau)}{d\tau}\, d\tau = A \int_{-\infty}^{t} J(t-\tau)\frac{d\delta(\tau)}{d\tau}\, d\tau = A\frac{dJ(t)}{dt}. \qquad \text{(E2.2.1)}$$

The derivative of the delta is called the doublet (Appendix 1),

$$\frac{d\delta(\tau)}{d\tau} = \psi.$$

The doublet sifts the negative of the derivative, as follows:

$$\int_{-\infty}^{\infty} f(x) \frac{d\delta(x-a)}{dx}\, dx = -\frac{df(x)}{dx}\Big|_{x=a} \qquad (E2.2.2)$$

To demonstrate the sign, perform the substitution $T = t - \tau$, $dT = -d\tau$, so

$$\varepsilon(t) = A \int_{-\infty}^{t} J(t-\tau) \frac{d\delta(\tau)}{d\tau}\, d\tau = A \int_{t-T=-\infty}^{t-T=t} J(T) \frac{d\delta(t-T)}{(-dT)}(-1)\, dT. \qquad (E2.2.3)$$

Moreover, the doublet is an odd function so $\psi(-x) = -\psi(x)$, so

$$\varepsilon(t) = A \frac{dJ(t)}{dt}. \qquad (E2.2.4)$$

Therefore, the doublet sifts the derivative of the creep function from the integral.

Example 2.3

Suppose a material is subjected to zero stress and strain for time t less than zero, followed by a constant strain ε_0 for $0 < t < t_1$, and then by zero stress for $t > t_1$. Obtain an expression for the stress and strain for all times and consider the particular case of the strain at infinite time.

Solution

The stress and strain for $t < 0$ are given as zero. For $0 < t < t_1$, the strain is constant, so that stress relaxation occurs and the stress is $\sigma(t) = \varepsilon_0 E(t)$. To evaluate strain for $t > t_1$, take a Laplace transform of the Boltzmann integral,

$$\sigma(t) = \int_0^t E(t-\tau) \frac{d\varepsilon(\tau)}{d\tau}\, d\tau. \qquad (E2.3.1)$$

$$\frac{\sigma(s)}{s} = E(s)\, \varepsilon(s), \qquad (E2.3.2)$$

so

$$\int_0^t \sigma(\tau)\, d\tau = \int_0^t \varepsilon(\tau)\, E(t-\tau)\, d\tau. \qquad (E2.3.3)$$

Substitute $\sigma(t) = \varepsilon_0 E(t)$ to obtain for $t > t_1$,

$$\varepsilon_0 \int_0^{t_1} E(\tau) \, d\tau = \int_0^t \varepsilon(\tau) \, E(t - \tau) \, d\tau. \qquad \text{(E2.3.4)}$$

This is an implicit form for the strain. It may be deconvoluted numerically if needed.

To obtain the limit for long time, recognize that

$$\int_0^t \varepsilon(\tau) \, E(t - \tau) \, d\tau = \int_0^t \varepsilon(t - \tau) \, E(\tau) \, d\tau. \qquad \text{(E2.3.5)}$$

$$\frac{\varepsilon(\infty)}{\varepsilon_0} = \frac{\displaystyle\int_0^{t_1} E(\tau) \, d\tau}{\displaystyle\int_0^{\infty} E(\tau) \, d\tau}. \qquad \text{(E2.3.6)}$$

The denominator on the right may be thought of as an asymptotic viscosity η_{asymp} for the following reason. This viscosity is infinite for a viscoelastic solid. Since viscosity is defined as the ratio of stress to strain rate, consider a constant strain rate history with a strain rate R:

$$\sigma(t) = \int_0^t E(t - \tau) \frac{d\varepsilon(\tau)}{d\tau} \, d\tau = R \int_0^t E(t - \tau) \, d\tau. \qquad \text{(E2.3.7)}$$

For long time, the ratio of stress to strain rate, and hence the asymptotic viscosity is

$$\eta_{asymp} = \frac{\sigma(t)}{R} = \int_0^{\infty} E(t - \tau) \, d\tau. \qquad \text{(E2.3.8)}$$

We remark that Hopkins and Kurkjian [E2.3.1] discussed this problem in the context of asymptotic viscosity in glass fibers; experiments were adduced in which glass fibers were rolled about a cylinder for several months, and then unwound. Their curvature persisted for years.

Example 2.4

Suppose a material is subjected to zero stress and strain for time t less than zero, followed by a constant strain ε_0 for $0 < t < t_1$, followed by zero stress for $t > t_1$. Find an explicit expression for the strain for $t \gg t_1$.

Solution

The stress history for $t \gg t_1$ may be regarded as a pulse

$$\sigma(t) = k\,\delta(t),\text{ with }k = \sigma_{avg}t_1 \qquad (E2.4.1)$$

in which

$$\sigma_{avg} = \frac{1}{t_1}\int_0^{t_1}\sigma(\tau)\,d\tau = \frac{1}{t_1}\varepsilon_0\int_0^{t_1}E(\tau)\,d\tau, \qquad (E2.4.2)$$

since during $0 < t < t_1$, we have stress relaxation. The response to a pulse stress is $\varepsilon(t) = k\dfrac{dJ(t)}{dt}$ from Example 2.2, so

$$\varepsilon(t) = \varepsilon_0\int_0^{t_1}E(\tau)\,d\tau\,\frac{dJ(t)}{dt},\text{ for }t \gg t_1. \qquad (E2.4.3)$$

To evaluate asymptotic behavior as time becomes large we can decompose the creep function into the sum of a transient, a constant, and a linear function of time:

$$J(t) = J_{trans}(t) + J(\infty) + t/\eta_{asymp}, \qquad (E2.4.4)$$

so for a sufficiently long time after the transient has effectively vanished, $dJ(t)/dt$ approaches $1/\eta_{asymp}$, so

$$\varepsilon\,(t \to \infty) \approx \frac{k}{\eta_{asymp}} = \frac{\varepsilon_0}{\eta_{asymp}}\int_0^{t_1}E(\tau)\,d\tau. \qquad (E2.4.5)$$

This result is identical with the long time limit in Example 2.3. For a viscoelastic solid, $1/\eta_{asymp} = 0$, so only the transient term contributes to the time dependence.

Example 2.5

Show that

$$J(0)\,E(t) + \int_0^t E(t-\tau)\frac{dJ(\tau)}{d\tau}\,d\tau = 1.$$

Solution

Take Laplace transformations of the Boltzmann integrals for creep and relaxation Eqs. 2.2.7 and 2.2.8 to obtain:

$$\varepsilon(s) = sJ(s)\sigma(s) \text{ and } \sigma(s) = sE(s)\varepsilon(s). \tag{E2.5.1}$$

By substituting,

$$s^2J(s)E(s) = 1, \text{ or } sJ(s)E(s) = (1/s).$$

By taking the inverse Laplace transform,

$$J(0) E(t) + \int_0^t E(t-\tau)\frac{dJ(\tau)}{d\tau} d\tau = 1, \tag{E2.5.2}$$

as desired.

Example 2.6

Show that $J(t)E(t) \le 1$.

Solution

Begin with the exact result obtained above,

$$1 = J(0) E(t) + \int_0^t E(t-\tau)\frac{dJ(\tau)}{d\tau} d\tau, \tag{E2.6.1}$$

so

$$1 \ge J(0) E(t) + \int_0^t E(t)\frac{dJ(\tau)}{d\tau} d\tau, \tag{E2.6.2}$$

since for $\tau > 0$,

$$E(t - \tau) > E(t), \tag{E2.6.3}$$

since E is a decreasing function.

$$1 \ge J(0)E(t) + E(t) [J(t) - J(0)] = E(t)J(t), \tag{E2.6.4}$$

since E is a decreasing function. So

$$J(t)E(t) \leq 1, \tag{E2.6.5}$$

as desired.

Example 2.7

Consider the following constitutive equations with A, B, m, and n as constants. Discuss the physical meaning of the following equations:

$$\text{(a)} \quad \sigma(t) = \int_0^\infty E(t-\tau) \frac{d\varepsilon(\tau)}{d\tau} \, d\tau.$$

$$\tag{E2.7.1}$$

$$\text{(b)} \quad \sigma(t) = \int_0^t \left[(t-\tau)^{-n} A + (t-\tau)^{-m} B\varepsilon(t-\tau) \right] \frac{d\varepsilon(\tau)}{d\tau} \, d\tau.$$

Solution

For (a) the integral includes times greater than t, so that the material is responding to strains at times in the future. Such behavior is called *acausal*, and has not been physically observed. By contrast the Boltzmann integral represents causal behavior.

For (b) the modulus function in the [ ] brackets depends on $(t - \tau)$ only, not an arbitrary function of t and τ, so the material is nonaging. The second term confers strain dependence so that the material is a particular type of nonlinearly viscoelastic solid.

Example 2.8

Determine the strain response to a stress history which is triangular in time:

$\sigma(t) = 0$ for $t < 0$,

$\sigma(t) = (\sigma_0/t_1)t$ for $0 < t < t_1$,

$\sigma(t) = 2\sigma_0 - (\sigma_0/t_1)t$ for $t_1 < t < 2t_1$, and

$\sigma(t) = 0$ for $2t_1 < t < \infty$.   (E2.8.1)

Plot $\varepsilon(t)$ and $\sigma(\varepsilon)$ for the particular creep function $J(t) = J_0(1 - e^{-t/\tau_c})$.

Solution

Substitute in the Boltzmann integral. The slopes are piecewise constant.

$$\varepsilon(t) = 0 \qquad\qquad\qquad\qquad \text{for } t < 0,$$

$$\varepsilon(t) = \frac{\sigma_0}{t_1} \int_0^t J(t-\tau)\, d\tau \qquad\qquad \text{for } 0 < t < t_1,$$

$$\varepsilon(t) = \frac{\sigma_0}{t_1} \left\{ \int_0^{t_1} J(t-\tau)\, d\tau - \int_{t_1}^t J(t-\tau)\, d\tau \right\} \qquad \text{for } t_1 < t < 2t_1, \text{ and}$$

$$\varepsilon(t) = \frac{\sigma_0}{t_1} \left\{ \int_0^{t_1} J(t-\tau)\, d\tau - \int_{t_1}^{2t_1} J(t-\tau)\, d\tau \right\} \qquad \text{for } 2t_1 < t < \infty.$$

$$(E2.8.2)$$

To generate a plot for the given creep function, substitute the given creep function and decompose the exponential as follows:

$$\int_a^b \left[1 - \exp\left\{ -\frac{t-\tau}{\tau_c} \right\} \right] d\tau = b - a - \tau_c \left[\exp\left\{ -\frac{t-b}{\tau_c} \right\} - \exp\left\{ -\frac{t-a}{\tau_c} \right\} \right].$$

The strain history for $\tau_c = 10$ time units is shown in Fig. E2.8a and the stress–strain relation, in Fig. E2.8b.

Example 2.9

Obtain the creep compliance for the Voigt model from the differential equation via Laplace transforms.

Solution

For the Voigt model, the differential equation is

$$\sigma = E\varepsilon + \eta \frac{d\varepsilon}{dt}. \qquad\qquad (E2.9.1)$$

Take a Laplace transform:

$$\sigma(s) = E\varepsilon(s) + \eta s\varepsilon(s). \qquad\qquad (E2.9.2)$$

But the time constant is $\tau = \eta/E$, so

$$\sigma(s) = E\varepsilon(s)(1 + \tau s). \qquad\qquad (E2.9.3)$$

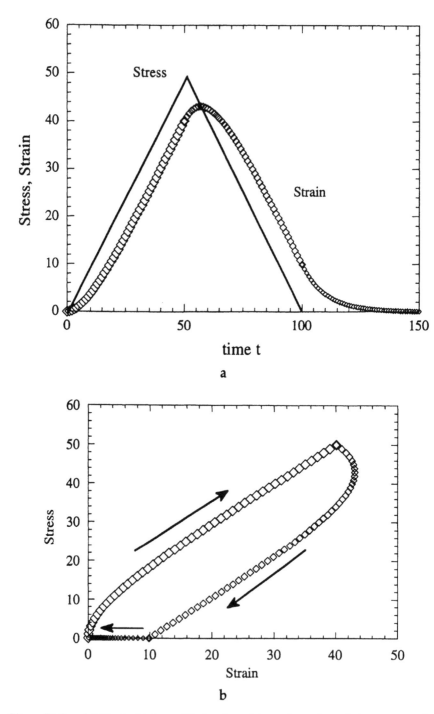

Figure E2.8   (a) Strain history $\varepsilon(t)$ due to a triangular pulse of stress, assuming $J(t) = J_0(1 - e^{-t/10})$. (b) Stress vs. strain due to a triangular pulse of stress.

But

$$J(s)E(s) = 1/s^2 \qquad \text{(E2.9.4)}$$

from Eq. 2.4.2, so

$$J(s) = \frac{1}{sE(1+\tau s)}. \qquad \text{(E2.9.5)}$$

$$J(s) = \frac{1}{E}\frac{\dfrac{1}{\tau}}{s\left(s+\dfrac{1}{\tau}\right)} = \frac{1}{E}\frac{s-s+\dfrac{1}{\tau}}{s\left(s+\dfrac{1}{\tau}\right)} = \frac{1}{E}\left\{\frac{s+\dfrac{1}{\tau}}{s\left(s+\dfrac{1}{\tau}\right)} - \frac{s}{s\left(s+\dfrac{1}{\tau}\right)}\right\}$$

$$\qquad \text{(E2.9.6)}$$

$$= \frac{1}{E}\left\{\frac{1}{s} - \frac{1}{\left(s+\dfrac{1}{\tau}\right)}\right\}.$$

Take the inverse transform:

$$J(t) = \frac{1}{E}\left(1-e^{-t/\tau}\right). \qquad \text{(E2.9.7)}$$

An alternate solution may be developed by constructing sJ(s) from Eq. E2.9.5 for J(s).

$$sJ(s) = \frac{1}{E\tau}\frac{1}{\left(s+\dfrac{1}{\tau}\right)}, \qquad \text{(E2.9.8)}$$

so following an inverse transform,

$$\frac{dJ(t)}{dt} = \frac{1}{E\tau}e^{-t/\tau}. \qquad \text{(E2.9.9)}$$

By integrating, and recognizing from the Voigt model that the limiting compliance at infinite time is 1/E,

$$J(t) = \frac{1}{E}\left(1-e^{-t/\tau}\right),$$

as above.

Example 2.10

Why bother with integral equations when the mathematics of differential equations is more familiar to most people?

Answer

It is indeed possible to describe viscoelastic behavior with differential equations. The drawback is that a differential description of material behavior over a wide range of time or frequency usually requires differential equations having high order and containing many terms. Integral equations are simpler for such a situation.

Example 2.11

Find the stress rate corresponding to a constant strain rate history beginning at time zero. Use Laplace transforms.

Solution

$$\varepsilon(t) = qt\mathcal{H}(t). \qquad\qquad (E2.11.1)$$

Take a Laplace transform,

$$\varepsilon(s) = q/s^2, \qquad\qquad (E2.11.2)$$

but the Laplace transform of the Boltzmann integral is

$$\sigma(s) = sE(s)\varepsilon(s), \text{ so } \sigma(s) = qE(s)/s, \text{ so } s\sigma(s) = qE(s). \qquad (E2.11.3)$$

By transforming back,

$$\frac{d\sigma}{dt} = q\,E(t). \qquad\qquad (E2.11.4)$$

Example 2.12

Suppose a linearly viscoelastic material is subjected to the following stress history:

$\sigma(t) = 0$ for $t < t_1$,

$\sigma(t) = \sigma_0$ for $t_1 < t < 2t_1$,

$\sigma(t) = 0$ for $2t_1 < t < 3t_1$,

$\sigma(t) = -\sigma_0$ for $3t_1 < t < 4t_1$, and

$\sigma(t) = 0$ for $4t_1 < t < \infty$.

Determine the strain response in general. For $t \gg 4t_1$, develop an approximate simple solution for the strain response.

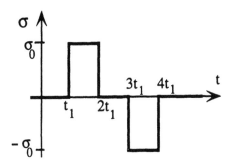

Figure E2.12   Given stress history.

Solution

Write the stress history as a sum of Heaviside step functions:

$$\sigma(t) = \sigma_0[\mathcal{H}(t - t_1) - \mathcal{H}(t - 2t_1) - \mathcal{H}(t - 3t_1) + \mathcal{H}(t - 4t_1)].$$

Apply the Boltzmann superposition principle. If the stress follows a step function in time, the strain is proportional to the creep compliance. Since the stress is written as a sum of step functions, the strain follows a sum of creep curves,

$$\varepsilon(t) = \sigma_0[J(t - t_1) - J(t - 2t_1) - J(t - 3t_1) + J(t - 4t_1)],$$

with the creep function understood to have a value of zero for a negative argument.

To obtain an approximate solution for $t > 4t_1$, rewrite this, with shifted time variables $t' = t - 2t_1$ and $t'' = t - 4t_1$, using the definition of the derivative, as

$$\varepsilon(t) = \sigma_0 t_1 \left[\frac{J(t' + t_1) - J(t')}{t_1} - \frac{J(t'' + t_1) - J(t'')}{t_1} \right].$$

The primes represent shifted variables, not derivatives.

In the limit as $t_1 \to 0$ (compared with t),

$$\varepsilon(t) \approx \sigma_0 t_1 \left[\frac{dJ(t'+2t_1)}{dt'} - \frac{dJ(t'')}{dt''} \right] = \sigma_0 t_1 \left[\frac{dJ(t''+2t_1)}{dt'} - \frac{dJ(t'')}{dt''} \right]$$

$$= \sigma_0 t_1^2 \left\{ \frac{1}{2t_1} \left[\frac{dJ(t''+2t_1)}{dt''} - \frac{dJ(t'')}{dt''} \right] \right\}.$$

$$\varepsilon(t) \approx \sigma_0 2t_1^2 \frac{d^2 J(t-4t_1)}{dt^2}.$$

As $t_1 \to 0$ in comparison with t, $J(t - 4t_1) \to J(t)$. Observe that the sifting property of the doublet functional can also be used to obtain a second derivative approximation.

Problems

2.1 Predict the strain response of a linearly viscoelastic material to a stress which is approximated as a doublet (derivative of a delta function) in time, assuming any needed material properties are known.

2.2 A linearly viscoelastic material is subjected to the following history: for $t < 0$, $\sigma = 0$, $\varepsilon = 0$; for $0 < t < t_1$, $\varepsilon = \varepsilon_0$; and for $t > t_1$, $\sigma = 0$. Find the stress and strain at all times. Discuss the physical significance of this stress and strain history. How could it be applied?

2.3 Determine analytically the stress–strain curve for a viscoelastic solid of known properties under constant stress rate.

2.4 Perform an informal experiment to determine the creep properties of a foam earplug (or other available viscoelastic material). A sophisticated electronic apparatus is not required to do a meaningful creep test. Use available materials to build a simple apparatus. Although the experiment is informal, careful diagrams should be made of your apparatus, and the procedure should be described systematically. Plot creep compliance $J(t)$ vs. log time. Is a simple spring-dashpot model appropriate? Estimate error bars for your creep compliance.

2.5 Predict the strain response of a linearly viscoelastic material of known but arbitrary properties to a stress history proportional to $A\delta(t) - B\delta(t + t_1)$ in which t is time and A, B, and t_1 are positive constants.

2.6 A durometer is a device used to measure the stiffness of rubbery materials. The durometer has a flat surface which is pressed to the rubber specimen. A protruding probe causes an indentation in the rubber. The observer reads the value of the indentation from a dial gage, which reads from 0 to 100, linked to the probe. A spring within the durometer provides the indenting force exerted by the probe. If the rubber is viscoelastic, the dial reading changes with time after the durometer is pressed to the rubber specimen. In Example 6.1 it is shown that if the specimen material is so stiff that most of the deformation occurs in the spring, the spring displacement as a function of time follows the creep function of the material.

Suppose that the specimen material is very compliant, so that most of the deformation occurs in the specimen and little occurs in the spring, and the durometer reading is near zero. For that case, show that the normalized spring displacement $x_1(t)$ follows the time dependence of the relaxation function.

2.7 Must recovery at zero stress following some arbitrary stress history always be monotonic?

2.8 Show that

$$J(t) \geq \frac{t}{\int_0^t E(t)\, d\tau}.$$

Hint: use $\int_0^t E(t-\tau)\, J(\tau)\, d\tau = t.$

2.9 Does any physical law require a material to exhibit fading memory? If not, give an example of a material which does not exhibit fading memory.

2.10 Plot the strain response for time $t > 0$ of a linearly viscoelastic material subjected to the stress history given in Problem 2.5 for the cases of a standard linear solid, a power–law material, and a material which exhibits stretched-exponential transient properties. Discuss.

References

2.2.1 Coleman, B. D. and Noll, W., "Foundations of linear viscoelasticity", *Rev. Mod. Phys.* 33, 239–249, 1961.

2.3.1 Christensen, R. M., "Restrictions upon viscoelastic relaxation functions and complex moduli", *Trans. Soc. Rheol.* 16, 603–614, 1972.

2.3.2 Zemanian, A. H., *Distribution Theory and Transform Analysis*, Dover, NY, 1965.

2.3.3 Day, W. A., "Restrictions on relaxation functions in linear viscoelasticity", *Q. J. Mech. Appl. Math.* 24, 487–497, 1971.

2.3.4 Fabrizio, M. and Morro, A., *Mathematical Problems in Linear Viscoelasticity*, SIAM, Philadelphia, PA, 1992.

2.3.5 Gurtin, M. E. and Herrera, I., "On dissipation inequalities and linear viscoelasticity", *Q. Appl. Math.* 23, 235–245, 1965.

2.6.1 Torvik, P. J. and Bagley, R. L., "On the appearance of fractional derivatives in the behavior of real materials", *J. Appl. Mech.* 51, 294–298, 1984.

2.6.2 Koeller, R. C., "Applications of fractional calculus to the theory of viscoelasticity", *J. Appl. Mech.* 51, 299–307, 1984.

2.6.3 Rabotnov, Yu. N., *Elements of Hereditary Solid Mechanics*, Mir Publishers, Moscow, 1980.

2.6.4 Nutting, P. G., "A new general law of deformation", *J. Franklin Inst.* 191, 679–685, 1921.

2.6.5 Kohlrausch, R., "Ueber das Dellmann'sche Elektrometer", *Ann. Phys. (Leipzig)* 72, 353–405, (1847).

2.6.6 Jonscher, A. K., "The 'universal' dielectric response", *Nature* 267, 693–679, 1977.

2.6.7 Palmer, R. G., Stein, D. L., Abrahams, E., and Anderson, P. W., "Models of hierarchically constrained dynamics for glassy relaxation", *Phys. Rev. Lett.* 53, 958–961, 1984.

2.6.8 Ngai, K. L., "Universality of low-frequency fluctuation, dissipation, and relaxation properties of condensed matter." I, *Comments Solid State Phys.* 9, 127–140, 1979.

2.6.9 Williams, G. and Watts, D. C., "Non-symmetrical dielectric relaxation behaviour arising from a simple empirical decay function", *Trans. Faraday Soc.* 66, 80–85, 1970.

2.7.1 Schwarzl, F. and Staverman, A. J., "Time-temperature dependence of linear viscoelastic behavior", *J. Appl. Phys.* 23, 838–843, 1952.

2.7.2 William, M. L., Landel, R. F., and Ferry, J. D., "The temperature dependence of relaxation mechanisms in amorphous polymers and other glass forming liquids", *J. Am. Chem. Soc.* 77, 3701–3707, 1955.

2.7.3 Plazek, D. J., "Oh, thermorheological simplicity, wherefore art thou?", *J. Rheol.* 40, 987–1014, 1996.

2.8.1 Sokolnikoff, I. S., *Mathematical Theory of Elasticity*, Krieger, Malabar, FL, 1983.

2.8.2 Timoshenko, S. P. and Goodier, J. N., *Theory of Elasticity*, McGraw-Hill, NY, 1982.

2.8.3 Fung, Y. C., *Foundations of Solid Mechanics*, Prentice Hall, Englewood Cliffs, NJ, 1968.

2.9.1 Hodge, I. M., "Physical aging in polymeric glasses", *Science* 267, 1945–1947, 1995.

2.9.2 Struik, L. C. E., *Physical Aging in Amorphous Polymers and Other Materials*, Elsevier Scientific, Amsterdam and New York, 1978.

2.10.1 Nye, J. F., *Physical Properties of Crystals*, Clarendon, Oxford, 1976.

2.10.2 Stein, R. S., "Rheo-optical studies of rubbers", *Rubber Chem. Technol.* 49, 458–535, 1976.

2.11.1 Wolff, J., *Das Gesetz der Transformation der Knochen*, Hirschwald, Berlin, 1892.

2.11.2 Frost, H. M., *Bone Remodelling and Its Relation to Metabolic Bone Diseases*, C. Thomas, Springfield, IL, 1973.

2.11.3 Currey, J., *The Mechanical Adaptations of Bones*, Princeton University Press, NJ, 1984.

2.11.4 Cowin, S. C. and Hegedus, D. H., "Bone remodeling I: theory of adaptive elasticity", *J. Elast.* 6, 313–326, 1976.

2.11.5 Hegedus, D. H. and Cowin, S. C., "Bone Remodeling II: Small strain adaptive elasticity", *J. Elast.* 6, 337–352, 1976.

2.11.6 Culshaw, B., *Smart Structures and Materials*, Artech, Boston, 1996.

2.11.7 Gandhi, M. V. and Thompson, B. S., *Smart Materials and Structures*, Chapman and Hall, London, 1992.

2.12.1 Fung, Y. C., "Stress strain history relations of soft tissues in simple elongation", in *Biomechanics: Its Foundations and Objectives*, ed. Y. C. Fung, N. Perrone, and M. Anliker, Prentice Hall, Englewood Cliffs, NJ, 1972.

2.12.2 Bernstein, B., Kearsley, E. A., and Zapas, L. J, "A study of stress relaxation with finite strain", *Trans. Soc. Rheol.* 7, 391–410, 1963.

2.12.3 Schapery, R. A., "On the characterization of nonlinear viscoelastic materials", *Polym. Eng. Sci.* 9, 295, 1969.

2.12.4 Green, E. and Rivlin, R. S., "The mechanics of non-linear materials with memory", *Arch. Rational Mech. Anal.* 1, 1–21, 1957.

2.12.5 Ward, I. M. and Onat, E. T., "Non-linear mechanical behaviour of oriented polypropylene", *J. Mech. Phys. Solids* 11, 217–229, 1963.

2.12.6 Schetzen, M., *The Volterra and Wiener Theories of Nonlinear Systems*, J. Wiley, NY, 1980.

2.12.7 Lockett, F. J., *Nonlinear Viscoelastic Solids*, Academic, NY, 1972.

E2.3.1 Hopkins, I. L. and Kurkjian, C. R., "Relaxation spectra and relaxation processes in solid polymers and glasses", in *Physical Acoustics II*, ed. W. P. Mason, Academic, NY, 1965.

chapter three

Dynamic behavior

§3.1   Introduction and rationale

The response of viscoelastic materials to sinusoidal load is developed in this chapter, and this response is referred to as dynamic. The dynamic behavior is of interest since viscoelastic materials are used in situations in which the damping of vibration or the absorption of sound is important. The results developed in this section have bearing on applications of viscoelastic materials, on experimental methods for their characterization, and on development of physical insight regarding viscoelasticity.

The frequency of the sinusoidal load on an object or structure may be so slow that inertial terms do not appear (the subresonant regime), or high enough that resonance of structures made of the material occurs. At sufficiently high frequency, dynamic behavior is manifested as wave motion. This distinction between ranges of frequency does not appear in the classical continuum description of a homogeneous material since the continuum view deals with differential elements of material.

§3.2   The linear dynamic response functions

In this section we obtain the dynamic material behavior (for a strain which is sinusoidal in time) from the transient behavior, assumed to be known. Let the history of strain ε to be sinusoidal (Figs. 1.4 and 3.1). In complex exponential form (discussed in Appendix 5), $\varepsilon(t) = \varepsilon_0\, e^{i\omega t}$. ω is the angular frequency in radians per second and is related to the frequency ν in cycles per second (Hz) by $\omega = 2\pi\nu$.

We make use of the Boltzmann superposition integral, with the lower limit taken as $-\infty$ since a true sinusoid has no starting point:

$$\sigma(t) \int_{-\infty}^{t} E(t-\tau)\frac{d\varepsilon}{d\tau}\, d\tau. \tag{3.2.1}$$

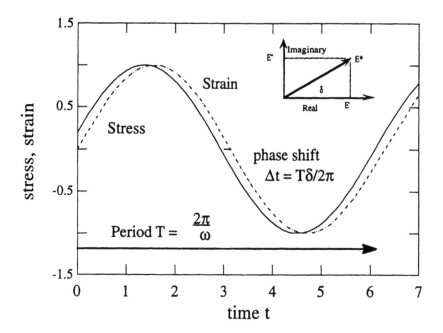

Figure 3.1   Stress and strain vs. time, with a phase shift. Arbitrary units of stress and time.

To achieve explicit convergence of the integral, decompose the relaxation function into the sum

$$E(t) \equiv \acute{E}(t) + E_e,$$

with $E_e = \lim_{t \to \infty} E(t)$ called the "equilibrium" modulus (in the context of polymers), and substitute the strain history in Eq. 3.2.1. Recall (§1.3) that for solids, $E_e > 0$ and for liquids, $E_e = 0$.

$$\sigma(t) = E_e \varepsilon_0 e^{i\omega t} + i\omega \varepsilon_0 \int_{-\infty}^{t} \acute{E}(t-\tau) \, e^{i\omega t} \, d\tau. \qquad (3.2.2)$$

Make the substitution $t' = t - \tau$. The prime represents a shifted variable, not a derivative.

$$\sigma(t) = \varepsilon_0 e^{i\omega t} \left[E_e + \omega \int_0^\infty \acute{E}(t') \sin \omega t' dt' + i\omega \int_0^\infty \acute{E}(t') \cos \omega t' dt' \right]. \qquad (3.2.3)$$

If the strain is sinusoidal in time, so is the stress. The stress and strain are no longer in phase as anticipated in Fig. 1.4. The stress–strain relation becomes

$$\sigma(t) = E^*(\omega)\varepsilon(t) = (E' + i\, E'')\varepsilon(t), \tag{3.2.4}$$

in which, with

$$\acute{E}(t) = E(t) - E(\infty),$$

$$E'(\omega) \equiv E_e + \omega\int_0^\infty \acute{E}(t')\sin\omega t'\, dt' \tag{3.2.5}$$

is called the storage modulus,

$$E''(\omega) \equiv \omega\int_0^\infty \acute{E}(t')\cos\omega t'\, dt' \tag{3.2.6}$$

is called the loss modulus, and the loss tangent (dimensionless) is

$$\tan\delta(\omega) \equiv \frac{E''(\omega)}{E'(\omega)}. \tag{3.2.7}$$

Equations 3.2.5 and 3.2.6 give the dynamic, frequency-dependent mechanical properties in terms of the relaxation modulus. E' is the component of the stress–strain ratio in phase with the applied strain; E'' is the component 90° out of phase. The single and double primes do not represent derivatives; they are a conventional notation for the real and imaginary parts, respectively. Equations 3.2.5 and 3.2.6 are one-sided Fourier sine and cosine transforms. The relations may be inverted [1.6.18] to obtain

$$E(t) = E_e + \frac{2}{\pi}\int_0^\infty \frac{(E' - E_e)}{\omega}\sin\omega t\, d\omega, \tag{3.2.8}$$

$$E(t) = E_e + \frac{2}{\pi}\int_0^\infty \frac{E''}{\omega}\cos\omega t\, d\omega. \tag{3.2.9}$$

Physically the quantity δ represents the phase angle between the stress and strain sinusoids. The dynamic stress–strain relation can be expressed as

$$\sigma(t) = \left| E * (\omega) \right| \varepsilon_0 e^{i(\omega t + \delta)} \tag{3.2.10}$$

with $E^* \equiv E' + i\, E''$.

One can also consider the dynamic behavior in the compliance formulation. Since in the modulus formulation

$$\sigma(t) = E^*(\omega)\varepsilon(t),$$

determined above, is an algebraic equation for sinusoidal loading, the strain is

$$\varepsilon(t) = \frac{1}{E * (\omega)} \sigma(t).$$

But the complex compliance J^* is defined by the equation

$$\varepsilon(t) = J^*(\omega)\sigma(t), \tag{3.2.11}$$

with

$$J^* = J' - i\, J''. \tag{3.2.12}$$

So, the relationship between the dynamic compliance and the dynamic modulus is

$$J * (\omega) = \frac{1}{E * (\omega)}. \tag{3.2.13}$$

This is considerably simpler than the corresponding relation for the transient creep and relaxation properties. Note that $\tan \delta(\omega) = J''(\omega)/J'(\omega)$.

In the compliance formulation, as with the modulus formulation, exact interrelations are available [1.6.18] between the creep and dynamic functions, as follows. Here, $J_e = \lim_{t \to \infty} J(t)$ is the "equilibrium" compliance at infinite time. Recall (§1.3) that for solids, $J_e < \infty$ and for liquids, J_e diverges:

$$J'(\omega) = J_e - \omega \int_0^\infty \left[J_e - J(t') \right] \sin \omega t'\, dt', \tag{3.2.14}$$

$$J''(\omega) = \omega \int_0^\infty \left[J_e - J(t') \right] \cos \omega t'\, dt'. \tag{3.2.15}$$

These are for solids; for liquids, terms involving the asymptotic viscosity appear.

The frequency range extends from zero to infinity. Frequencies below 0.1 Hz are associated with seismic waves and with Fourier spectra of quasistatic events. Vibration of structures and solid objects occur from about 0.1 Hz to 10 kHz depending on the size of the structure. Stress waves from 20 Hz to 20 kHz are perceived as sound. Waves above 20 kHz are referred to as ultrasonic; ultrasonic frequencies between 1 and 10 MHz are commonly used in the nondestructive evaluation of engineering materials and for diagnostic ultrasound in medicine. Frequencies above about 10^{12} Hz correspond to molecular vibration and represent an upper limit for stress waves in real solids. Due to the wide range of frequency often considered to be of interest, plots of properties vs. log frequency are commonly used. A factor of ten on such a scale is referred to as a decade.

Plots of dynamic viscoelastic functions may assume a variety of forms. First, one may plot the dynamic properties vs. frequency, with the frequency scale ordinarily shown logarithmically. Second, one may plot the imaginary part vs. the real part: E'' vs. E' or J'' vs. J'. Such a plot is often used in dielectric relaxation studies and is referred to as a "Cole–Cole plot". A single relaxation time process gives a semicircle in a Cole–Cole plot. Third, one may plot stiffness vs. loss: $|E^*|$ vs. tan δ, a stiffness–loss map.

§3.2.1 Dynamic stress–strain relation

§3.2.1.1 The loss tangent, tan δ

Consider now the relation between stress and strain in dynamic loading of a linearly viscoelastic material. First, the shape of the stress–strain curve is determined, and a relation for the loss tangent is developed. Above, we have considered stress and strain as they depend on time; see also §1.4. Suppose we have

$$\varepsilon = B \sin \omega t, \qquad (3.2.16)$$

and

$$\sigma = D \sin (\omega t + \delta). \qquad (3.2.17)$$

These are parametric equations for an elliptical Lissajous figure (Fig. 3.2). The elliptical shape is a consequence of linearly viscoelastic behavior. For some materials, the curve of stress vs. strain in dynamic loading is not elliptical, but has pointed ends; this behavior is a manifestation of material nonlinearity.

The loop is called a *hysteresis loop. Hysteresis* in general refers to a lag between cause and effect. In some contexts, such as the study of metals, a specific view of hysteresis is taken, that it represents a frequency-independent damping, as discussed in Chapter 7.

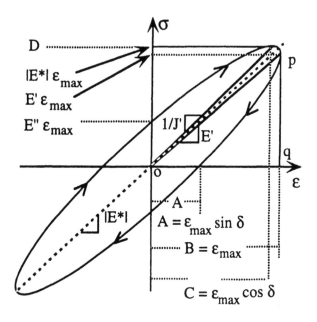

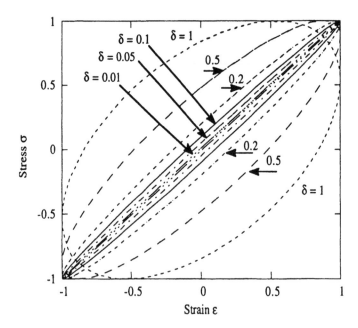

Figure 3.2   Stress σ vs. strain ε for a linearly viscoelastic material under oscillatory loading. (top) Illustration of slopes and intercepts. A material with a rather large value of tan δ ≈ 0.4 is shown for illustration. The material could be a viscoelastic rubber (see Chapter 7). (bottom) Elliptical Lissajous figures for various δ. For most structural metals, δ < 0.01 (see Chapter 7).

To interpret the dimensions and intercepts of the figure, first let the intercept with the ε axis be called A, assumed to occur at time t_1. Then A = B sin ωt_1 and

$$0 = D \sin(\omega t_1 + \delta) = D \sin \omega t_1 \cos \delta + D \cos \omega t_1 \sin \delta. \qquad (3.2.18)$$

By substituting

$$\sin \omega t_1 = \frac{A}{B} \text{ and } \cos \omega t_1 = \frac{1}{B}\sqrt{B^2 - A^2}$$

into the last expression,

$$0 = D\frac{A}{B}\cos\delta + \frac{D}{B}\left(\sqrt{B^2 - A^2}\right)\sin\delta. \qquad (3.2.19)$$

By squaring the last expression,

$$A^2\cos^2\delta = B^2\sin^2\delta - A^2\sin^2\delta, \qquad (3.2.20)$$

so

$$A^2(\sin^2\delta + \cos^2\delta) = B^2\sin^2\delta,$$

so that

$$\sin\delta = \frac{A}{B}. \qquad (3.2.21)$$

In Eqs. 3.2.16 and 3.2.17 one may also let $\omega t = -\delta$, so $\sigma = 0$. Then ε_A = B sin $(-\delta)$ at the intercept. However, the ellipse is symmetrical, so A = B sin δ, giving Eq. 3.2.21.

Now, with the aim of obtaining a different relation between δ and the parameters of the ellipse, consider the point of maximum stress, which occurs when

$$\sin(\omega t + \delta) = 1. \qquad (3.2.22)$$

Then

$$\omega t + \delta = \frac{\pi}{2}, \qquad (3.2.23)$$

so, referring to Fig. 3.2, C = $\varepsilon(\sigma_{max})$,

$$C = \varepsilon(\sigma_{max}) = B\left(\frac{\pi}{2} - \delta\right) = B\left[\sin\frac{\pi}{2}\cos(-\delta) + \cos\frac{\pi}{2}\sin(-\delta)\right], \quad (3.2.24)$$

so

$$\varepsilon(\sigma_{max}) = B\cos\delta, \quad\quad\quad (3.2.25)$$

Since

$$A = B\sin\delta,$$

$$\tan\delta = \frac{A}{C}. \quad\quad\quad \mathbf{(3.2.26)}$$

The width of the elliptical Lissajous figure, then, is a measure of the loss angle δ of a linearly viscoelastic material.

§3.2.1.2   The storage modulus E′

To obtain a relation for the storage modulus E' in connection with the elliptical stress–strain curve, suppose $\varepsilon = \varepsilon_{max}\sin\omega t$. Then the stress σ at the maximum strain ε_{max} is

$$\sigma = E'\varepsilon_{max}\sin\omega t + E''\varepsilon_{max}\cos\omega t.$$

For ε_{max}, $\omega t = \pi/2$, and then

$$\sigma = E'\varepsilon_{max},$$

so the slope of the line from the origin to the point of maximum strain is E', as shown in Fig. 3.2.

The stress value proportional to the storage modulus is also marked on the ordinate in Fig. 3.2.

§3.2.1.3   The storage compliance J′

To obtain a relation for the storage compliance J' in connection with the elliptical stress–strain curve, suppose

$$\sigma = \sigma_{max}\sin\omega t.$$

Then the strain at the maximum stress is

$$\varepsilon = (J' - i J'')\sigma_{max}.$$

$$\varepsilon = J'\sigma_{max}\sin\omega t - J''\sigma_{max}\cos\omega t.$$

For σ_{max},

$$\omega t = \pi/2,$$

then

$$\varepsilon = J'\sigma_{max} \text{ or } \sigma_{max} = \varepsilon/J'.$$

So the slope of the line from the origin to the point of maximum stress is $1/J'$, as shown in Fig. 3.2. Observe that $1/J' \geq E'$; equality occurs if the material is elastic ($\delta = 0$).

§3.2.1.4 The loss modulus E″
Consider time $t = 0$, so $\varepsilon = 0$, so

$$\sigma(t) = D \sin \delta,$$

but

$$\sigma(t) = \varepsilon_{max}[E' \sin \omega t + E'' \cos \omega t],$$

so for $t = 0$,

$$\sigma(0) = \varepsilon_{max} E'',$$

which is marked as an intercept on the ordinate in Fig. 3.2.

§3.2.1.5 The dynamic modulus |E*|
Also observe that $\sigma_{max} = |E^*|\varepsilon_{max}$, so again referring to Fig. 3.2, this modulus corresponds to a line of intermediate slope between that for E′ and that for $1/J'$.

§3.2.2 Standard linear solid
The *dynamic functions* of the standard linear solid, which exhibits a relaxation that is exponential in time,

$$E(t) = E_2 + E_1 e^{-t/\tau}, \tag{3.2.27}$$

with τ, called the relaxation time, are obtained from Eqs. 3.2.5 and 3.2.6 developed above and from the following standard integrals:

$$\int_0^\infty e^{-ax} \sin bx\, dx = \frac{b}{a^2+b^2} \quad \int_0^\infty e^{-ax} \cos bx\, dx = \frac{a}{a^2+b^2}.$$

So, for the standard linear solid,

$$E'(\omega) = E_2 + E_1 \frac{\omega^2 \tau_r^2}{1 + \omega^2 \tau_r^2},$$ (3.2.28)

$$E''(\omega) = E_1 \frac{\omega \tau_r}{1 + \omega^2 \tau_r^2}.$$ (3.2.29)

These forms may also be obtained by substitution of a sinusoidal stress history in the differential equations shown in §2.6. The shape of $E''(\omega)$ is referred to as a *Debye peak*.

In terms of the relaxation strength $\Delta = E_1/E_2$, we have

$$\tan \delta = \frac{\Delta}{\sqrt{1+\Delta}} \frac{\omega \tau_m}{1 + \omega^2 \tau_m^2},$$ (3.2.30)

with

$$\tau_m = \tau_r \sqrt{1+\Delta},$$ (3.2.31)

as is demonstrated in Example 3.10. A plot of several viscoelastic functions corresponding to the standard linear solid is shown in Fig. 4.2. We remark that if the loss is small, the peak value of $\tan \delta$ is $(1/2)\Delta$.

§3.3 Kramers–Kronig relations

Since the dynamic properties E' and E'' (or J' and J'') of a linear material are derived from the same transient function, they cannot be independent. The relationships between them are known as the Kramers–Kronig relations, in honor of the scientists who developed them in connection with electromagnetic phenomena [3.3.1, 3.3.2]; see also [3.3.3].

To derive the Kramers–Kronig relations, we recognize that the dynamic response is a Fourier transform of a causal response in the time domain [3.3.1, 3.3.4]. Causality entails an effect which cannot precede the cause; causality was implicitly included in the definitions of the creep and relaxation functions. Under these conditions the dynamic compliance function is analytic in the complex ω plane. We apply Cauchy's integral theorem:

$$\oint_c \frac{f(z)}{z-a}\, dz = 2\pi i\, f(a) \cdot 1/2 \quad \text{if } a \text{ is on contour } c \text{ (principal part)};$$

$$1 \text{ if } a \text{ inside } c;$$ (3.3.1)

$$0 \text{ if } a \text{ outside } c.$$

Here f(z) is a complex analytic function of a complex variable z. Now define

$$\zeta = \xi + i\eta;\ f = u + iv.$$

Cauchy's integral theorem is used to evaluate the integral using a contour along the real line and avoiding the singularity at z = a via a semicircular arc. The integral is considered as a principal part, denoted by $\wp$. To calculate the principal part, one evaluates the integral by integrating almost up to the singularity; then one continues integrating from just after the singularity. Finally, one takes the limit as the distance q from the singularity tends to zero:

$$\wp \int_0^\infty f(x)\,dx = \lim_{q\to 0} \int_0^{x-q} f(x)\,dx + \lim_{q\to 0} \int_{x+q}^\infty f(x)\,dx. \tag{3.3.2}$$

Then

$$f(\zeta) = \frac{1}{i\pi}\wp \int_c \frac{f(z)}{z-\zeta}\,dz. \tag{3.3.3}$$

Decompose this into its real and imaginary parts:

$$u(\xi, 0) = \frac{1}{\pi}\wp \int_{-\infty}^\infty \frac{v(x, 0)}{x-\xi}\,dx, \quad v(\xi, 0) = \frac{-1}{\pi}\wp \int_{-\infty}^\infty \frac{u(x, 0)}{x-\xi}\,dx. \tag{3.3.4}$$

$$u(\xi, 0) = \frac{1}{\pi}\left[\wp \int_0^\infty \frac{v(x, 0)}{x-\xi}\,dx + \wp \int_{-\infty}^0 \frac{v(x, 0)}{x-\xi}\,dx\right]$$
$$= \frac{1}{\pi}\left[\wp \int_0^\infty \frac{v(x, 0)}{x-\xi}\,dx - \wp \int_0^{-\infty} \frac{v(x, 0)}{x-\xi}\,dx\right]. \tag{3.3.5}$$

But f is the Fourier transform of a real function, so

$$v(x) = -v(-x),$$

so

$$u(\xi, 0) = \frac{1}{\pi}\left[\wp \int_0^\infty \frac{v(x, 0)}{x-\xi}\,dx + \wp \int_0^\infty \frac{v(x, 0)}{x+\xi}\,dx\right]$$
$$= \frac{1}{\pi}\left[\wp \int_0^\infty \frac{v(x)(x+\xi)+v(x)(x-\xi)}{x^2-\xi^2}\,dx\right] = \frac{2}{\pi}\left[\wp \int_0^\infty \frac{xv(x)}{x^2-\xi^2}\,dx\right]. \tag{3.3.6}$$

By similar arguments and with u(−x) = u(x)

$$v(\xi) = \frac{2\xi}{\pi}\left[\wp \int_0^\infty \frac{u(x)}{x^2 - \xi^2}\,dx \right].$$

(3.3.7)

Let the analytic function f(z) be the complex dynamic compliance J*(ω), with proper consideration of the limiting behavior at infinity for an analytic function.

Then the Kramers–Kronig relations for the dynamic compliance are [1.6.18, 1.6.23]

$$J'(\omega) - J'(\infty) = \frac{2}{\pi}\left[\wp \int_0^\infty \frac{\varpi J''(\varpi)}{\varpi^2 - \omega^2}\,d\varpi \right],$$

(3.3.8)

and

$$J''(\omega) = \frac{2\omega}{\pi}\left[\wp \int_0^\infty \frac{J'(\varpi) - J'(\infty)}{\omega^2 - \varpi^2}\,d\varpi \right].$$

(3.3.9)

These expressions indicate that the storage and loss components of the dynamic compliance are intimately related: the presence of loss entails *dispersion* which is a variation of compliance with frequency. Moreover, the relationship is nonlocal in that the presence of loss at one frequency is linked to dispersion over the entire frequency range. Illustrations of the relation between loss and dispersion may be found in Examples 3.4, 3.5, 3.10, in Figs. 4.1 and 4.2, and in Chapter 7.

Alternate approaches to deriving the Kramers–Kronig relations are available [1.6.23, 3.3.5]. For example, one may use the fact that the Fourier transform of the step function $\mathcal{H}(t)$ is

$$\mathcal{F}\{\mathcal{H}(t)\} = \lim_{\varepsilon \to 0^+} \frac{i}{(\omega + i\varepsilon)} = \wp\frac{i}{\omega} + \pi\delta(\omega).$$

Again assume causality; the transient response function must have the form J(t) = j(t)$\mathcal{H}$(t), in which j(t) is defined for the entire time scale. Take a Fourier transform of the above product, using the convolution theorem. Since the value of j(t) for t < 0 is immaterial, one may choose j(−|t|) = j(|t|) so that J*(ω) is pure real, to get Eq. 3.3.9. Similarly, one may choose j(−|t|) = −j(|t|) so that J*(ω) is pure imaginary, to get Eq. 3.3.8. In a related vein, Golden and Graham [1.6.6] remark that causality means the relaxation function E(t) is defined for t > 0, so Re E* is even in frequency, Im E* is odd in frequency.

§3.4 Energy storage and dissipation

The elliptical plot of stress vs. strain in dynamic loading of a linear material encloses some area. The area within the ellipse represents an energy per volume dissipated in the material, per cycle. Let us examine the mechanical damping as it relates to the storage and dissipation of energy. To that end, consider [1.6.18, 1.6.19]:

$$\sigma = \sigma_0 \sin \omega t, \text{ and } \varepsilon = \varepsilon_0 \sin (\omega t - \delta). \tag{3.4.1}$$

The period $T = 1/\nu = (2\pi)/\omega$ corresponds to one full cycle. The energy stored over one full cycle is zero since the material returns to its starting configuration. To find the stored energy, integrate over 1/4 cycle, or $t = \pi/(2\omega)$.

$$\int_0^{\varepsilon_0} \sigma d\varepsilon = \int_0^{\pi/2\omega} \sigma \frac{d\varepsilon}{dt} dt$$

$$= \omega \varepsilon_0 \sigma_0 \int_0^{\pi/2\omega} \left[\cos \omega t \sin \omega t \cos \delta + \sin^2 \omega t \sin \delta \right] dt \tag{3.4.2}$$

$$= \varepsilon_0 \sigma_0 \left[\frac{\cos \delta}{2} + \frac{\pi \sin \delta}{4} \right].$$

The first term represents stored energy; the second term (which vanishes for $\delta = 0$) represents dissipated energy. Since $E' = |E^*| \cos \delta$, the energy stored elastically in the material is

$$W_s = \int_0^{\varepsilon_0} E' \varepsilon \, d\varepsilon = \frac{1}{2} E' \varepsilon_0^2. \tag{3.4.3}$$

This stored energy corresponds to the area in triangle opq in Fig. 3.2. The stored energy is calculated for a quarter cycle rather than a full cycle since the material returns to its original condition after a full cycle.

The dissipated energy for a full cycle is proportional to the area within the ellipse in Fig. 3.2. The dissipated energy over a quarter cycle is determined by taking one quarter of the integral over a full cycle. Recall $E'' = |E^*| \sin \delta$.

$$W_d = \frac{1}{4} \int_0^{2\pi/\omega} \sigma \frac{d\varepsilon}{dt} dt = \frac{\pi}{4} E'' \varepsilon_0^2. \tag{3.4.4}$$

So

$$\frac{W_d}{W_s} = \frac{\pi}{2} \tan \delta.$$ (3.4.5)

The physical meaning of the loss tangent is associated with the ratio of energy dissipated to the energy stored in dynamic loading. This is why it is considered a measure of "internal friction". Some authors use a measure Ψ known as the *specific damping capacity,*

$$\Psi = 2\pi \tan \delta.$$ (3.4.6)

Ψ refers to the energy ratio for a full cycle. The quantity Ψ is meaningful for nonlinear materials as well as linear ones since the energies can be calculated from the stress–strain loop even if its shape is not elliptical.

Since there are several possible ways of defining the stored energy, there are also several expressions for Ψ [3.4.1]. In the above, the stored elastic component of energy was used, in harmony with the results of Ferry [1.6.18] and Graesser and Wong [3.4.2]. One can also consider as stored energy the total strain energy at maximum deformation [3.4.3] or the total kinetic plus potential energies in vibrating systems with inertia [3.4.4].

Restrictions on the dynamic viscoelastic functions can be derived [3.4.5] from various energy principles. For example, the requirement of a nonnegative rate of dissipation of energy gives the conclusion $E'' \geq 0$. Demonstration of $E' \geq 0$ requires the assumption of both nonnegative stored energy and nonnegative rate of dissipation of energy. Although experiments show $dE'(\omega)/d\omega \geq 0$, analytical demonstration of such a requirement has proven elusive [3.4.5]. Energy relations may also be considered in connection with the elliptical stress–strain curve. Since the dissipated energy per cycle is $\pi E'' \varepsilon_0^2$, a negative value of E'' would correspond to a gain of mechanical energy per cycle, and a traversal of the elliptical curve in the opposite direction. In passive materials (see §2.3), there is no external or internal source of energy which could supply this gain. Under these assumptions, we have $E'' \geq 0$.

Energy concepts can be introduced as follows [3.4.6]. One may assume that work must be done to deform a viscoelastic solid from its virgin state (prior to the application of any stress). Constitutive laws which have the following property are called *dissipative:*

$$\int_0^t \sigma_{ij}(\tau)\,\varepsilon_{ij}(\tau)\,d\tau \geq 0,$$

for all sufficiently smooth strain histories which satisfy $\varepsilon_{ij}(0) = 0$. This criterion differs from limits on the dynamic functions which are defined for steady-state sinusoidal loading in that it is an energy integral for an initially undeformed material. The concept of dissipative material does not imply a relaxation modulus function which is monotonically decreasing. An example is given of a (nonphysical) relaxation modulus in dimensionless variables, $E(t) = \cos t$, for $0 \leq t \leq \infty$. This can be shown to be dissipative [3.4.6] but it is nonmonotonic.

Nonlinearly viscoelastic materials also dissipate energy. Tan δ as defined above only applies to linear materials. In nonlinear materials, an energy ratio Ψ continues to have a clear physical interpretation even though the stress–strain curve is no longer elliptical.

§3.5   Resonance of structural members

This section is particularly relevant in applications of linearly viscoelastic materials and in experimental methods for their characterization, and is referred to extensively in later sections. Torsion is considered here since the distributed system for torsion is comparatively simple. In contrast, tension–compression vibration of a rod is dispersive even in the elastic case as a result of the Poisson effect.

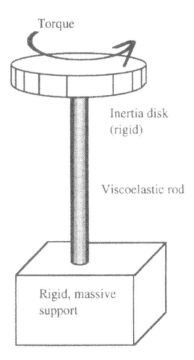

Figure 3.3   Diagram of a lumped torsional oscillator.

§3.5.1   Resonance, lumped system

Let us consider a linearly viscoelastic rod, fixed at one end and attached to a rigid disk of rotary inertia I at the other end, with the aim of determining the dynamic compliance as a function of frequency. This system is lumped in the sense that all the inertia is considered to reside in the disk (since it is large compared with the rod) and all the compliance is considered to reside in the rod (since it is slender and made of a more flexible material). An external torque,

$$M_{ext} = M_0 e^{i\omega t}, \tag{3.5.1}$$

sinusoidal in time with angular frequency ω, is applied to the disk. From Newton's second law,

$$M_{ext} - M_r = I\frac{d^2\theta}{dt^2}, \tag{3.5.2}$$

with M_r as the torque of the rod upon the disk, and θ as the angular displacement. Since the rod is viscoelastic,

$$M_r = G^* k\theta = G'(1 + i \tan \delta)k\theta, \tag{3.5.3}$$

with k as a geometrical factor which depends on the cross-sectional size and shape of the rod. For a round, straight rod, $k = \pi d^4/32L$, with d as the rod diameter and L as the rod length.

So

$$I\frac{d^2\theta}{dt^2} + G'(1+i\tan\delta)\,k\theta - M_{ext} = 0. \tag{3.5.4}$$

Since the applied torque is sinusoidal in time, so is angular displacement: $\theta = \theta_0 e^{i\omega t}$, with θ_0 as a complex amplitude to represent the possibility of a phase shift. By substituting,

$$\left[M_{ext} - G'(1+i\tan\delta)\,k\theta\right] = I\frac{d^2\theta}{dt^2},$$

$$\left[M_0 - G'(1+i\tan\delta)\,k\theta_0\right] = -I\omega^2\theta_0. \tag{3.5.5}$$

So

$$\frac{\theta_0}{M_0} = \frac{1}{G'(1+i\tan\delta)\,k - I\omega^2}. \tag{3.5.6}$$

The i tan δ term indicates the angular displacement is out of phase with the driving torque even at very low frequency for which the inertial term $I\omega^2$ is negligible.

The ratio of angular displacement amplitude to driving torque amplitude θ_0/M_0 represents a structural compliance; conversely, M_0/θ_0 is a structural rigidity. By contrast, G is a material rigidity and $J = 1/G$ is a material compliance.

Define the normalized dynamic torsional compliance

$$\Gamma \equiv \frac{G'k\theta_0}{M_0}. \tag{3.5.7}$$

Calculate Γ from Eq. 3.5.6, keeping in mind the fact that G' and $\tan\delta$ are functions of frequency,

$$\Gamma = \frac{1}{1 + i\tan\delta - (\omega/\omega_1)^2}, \tag{3.5.8}$$

in which

$$\omega_1 \equiv \sqrt{\frac{G'(\omega_1)k}{I}} \tag{3.5.9}$$

is the natural angular frequency. When $\omega = \omega_1$, Γ attains its maximum value. The dynamic compliance is a complex quantity which has a magnitude and a phase associated with it. The phase angle ϕ between torque and angular displacement may be determined from

$$\tan\phi = \mathrm{Im}(\Gamma^{-1})/\mathrm{Re}(\Gamma^{-1}), \tag{3.5.10}$$

$$\tan\phi = \frac{\tan\delta}{1 - (\omega/\omega_1)^2}. \tag{3.5.11}$$

This equation shows that the phase angle ϕ is approximately the same as δ for sufficiently low frequency $\omega \ll \omega_1$, is $\pi/2$ radians at ω_1, and is π radians for high frequency $\omega \gg \omega_1$. The magnitude of Γ is

$$|\Gamma| = \sqrt{\Gamma\Gamma^*}, \tag{3.5.12}$$

with Γ^* as the complex conjugate of Γ:

$$\Gamma\Gamma^* = \frac{1}{\left[1 - (\omega/\omega_1)^2\right]^2 + \tan^2\delta}. \tag{3.5.13}$$

The dependence of $|\Gamma|$ on frequency is referred to as a Lorentzian function, sometimes called a Lorentzian line shape in view of the form of the spectral lines associated with electromagnetic radiation (light) from atomic resonances. At the natural frequency $\omega = \omega_1$, the maximum value of $|\Gamma|$ is

$$|\Gamma| = \frac{1}{\tan\delta},$$

so the compliance at resonance diverges in the elastic case ($\delta \to 0$). Observe that the structural compliance of the resonance of the lumped torsional system (for small loss) is a factor $1/\tan\delta$ larger than the compliance under static conditions. This may be expressed as the "magnification factor" associated with resonance to be $1/\tan\delta$.

Consider now the width of the resonance curve $|\Gamma|$ vs. frequency. When $|\Gamma|$ is at half its maximum value, $\Gamma\Gamma^*$ is at $1/4$ maximum. So

$$[1 - (\omega/\omega_1)^2]^2 = 3\tan^2\delta. \tag{3.5.14}$$

Now an approximation is used to obtain an expression for the width,

$$\left[1-\left(\omega/\omega_1\right)^2\right] = \frac{\left(\omega_1^2 - \omega^2\right)}{\omega_1^2} = \frac{\left(\omega_1 - \omega\right)\left(\omega_1 + \omega\right)}{\omega_1^2} \approx \frac{2\left(\omega_1 - \omega\right)}{\omega_1}$$
$$\equiv \frac{\Delta\omega}{\omega_1} \tag{3.5.15}$$

with the approximation $\omega_1 + \omega \approx 2\omega_1$ valid if the resonance curve is narrow. So the relative width of the resonance curve is

$$\frac{\Delta\omega}{\omega_1} \approx \sqrt{3}\tan\delta. \tag{3.5.16}$$

This $\Delta\omega$ represents the *full* width of the resonance curve at half maximum amplitude, or twice the half excursion represented by $(\omega_1 - \omega)$. The relative width of the resonance curve is referred to as Q^{-1} with Q as the "*quality factor*", a term also used in describing electronic resonant circuits. At times one considers the frequency at which the response is $1/\sqrt{2}$ of maximum, rather than half maximum; in that case $Q^{-1} = (\Delta\omega)/(\omega_1) \approx \tan\delta$. Since $20\log_{10}\sqrt{2} = 3.01$, the corresponding frequency is referred to as a "3 dB point" in connection with the logarithmic decibel (dB) scale. Derivation of exact relations between Q and $\tan\delta$ generally requires a model of the viscoelastic response of the material. For example, the exact expression [3.4.1] for a material with a complex modulus and damping independent of frequency is

$$Q^{-1} = \sqrt{1 + \tan \delta} - \sqrt{1 - \tan \delta}.$$

Observe that the width of the resonance curve for small loss is independent of the details of the inertia disk and rod stiffness and dimensions; it depends only on the loss tangent. Since an approximation was made that the resonance curve is narrow, this expression is valid only for materials with a small loss tangent. For that reason, resonant vibration methods are commonly used in experimental procedures to determine the loss tangent of low-loss materials. One can develop an exact relationship, valid for large loss, between $\tan \delta$ and the relative width Q^{-1} of the resonance curve. To do so requires knowledge of the functional form of the viscoelastic properties as they depend on frequency [3.4.1].

Observe also that the response near resonance depends sensitively on $\tan \delta$. For low-loss materials (e.g., $\tan \delta < 10^{-2}$), a subresonant test gives a Lissajous figure which is difficult to distinguish from a straight line. In such a test, materials with $\tan \delta$ between about 10^{-3} and 10^{-9} cannot be easily distinguished (unless the signals are rather free of noise). Yet the resonant behavior of such materials differs dramatically. Applications are discussed in §10.6 and §10.8.

§3.5.2 Resonance, distributed system

Consider torsional oscillation of a circular cylindrical rod of a linearly viscoelastic material, fixed at one end and free of constraint, with an applied moment—sinusoidal in time—at the other end. This is referred to as a "fixed-free" configuration in the experimental literature. The stiffness and inertia are distributed throughout the rod. There are an infinite number of degrees of freedom. The dynamic structural rigidity is the ratio of applied external moment M to the angular displacement θ; the ratio is complex since M and θ need not be in phase. The rod length is L and its radius is R. End inertia is not considered here. The dynamic structural rigidity is [1.6.1, 3.5.1],

$$\frac{M}{\theta} = \left[\frac{\pi}{2} R^4 \right] \left[\rho \omega^2 L \right] \frac{\operatorname{ctn} \Omega^*}{\Omega^*} \qquad (3.5.17)$$

in which

$$\Omega^*(\omega) = \sqrt{\frac{\rho \omega^2 L^2}{G^*(\omega)}}. \qquad (3.5.18)$$

Methods for the solution of boundary value problems such as this are treated in Chapter 5; the derivation of the above expression is presented in §7.2. The normalized compliance based on Eq. 3.5.17 is plotted in Fig. 3.5

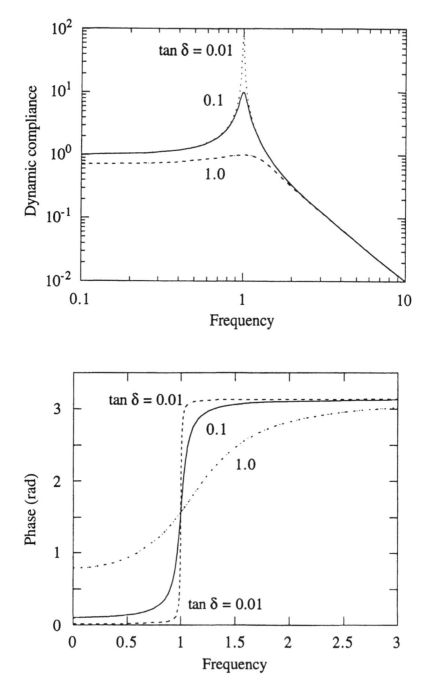

Figure 3.4   Behavior of a lumped torsional oscillator, for several loss tangents. G is assumed constant. Frequency is normalized to the natural frequency. (top) Normalized dynamic compliance vs. normalized frequency. (bottom) Phase vs. normalized frequency.

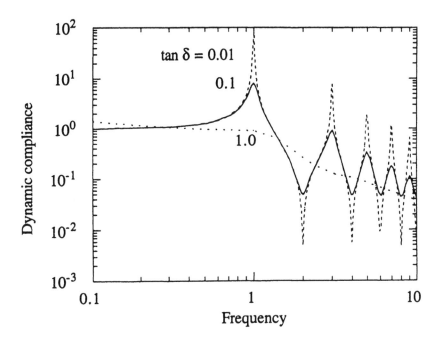

Figure 3.5 Normalized dynamic compliance for a distributed torsional oscillator (a cylindrical rod fixed at one end and driven at the free end) for several loss tangents vs. normalized frequency. Plot is based on Eq. 3.5.17, without appeal to small loss approximation.

as a function of normalized frequency for several values of tan δ. Observe that the resonance peaks are in the ratio 1, 3, 5, 7, 9.

We develop in the following, simple expressions for the rigidity at resonance and determine the width of the resonance curve under the assumption of small loss. The aim is to cultivate physical insight and to provide simple expressions for the interpretation of experiments.

Case 1. Let us find the rigidity at the first resonance, which has the lowest frequency ω_1. Assume the loss is small, tan δ << 1. Expand the cotangent as a Taylor series about $\pi/2$, which corresponds to the lowest resonance frequency:

$$f(x) = f(a) + f'(a)(x - a) + f''(a)(x - a)^2/2! + \dots$$

Observe that $d[\text{ctn } x]/dx = -\csc^2 x$. So

$$\text{ctn } \Omega^* \approx \text{ctn}\left[\frac{\pi}{2}\right] + \left[-\csc^2\left(\frac{\pi}{2}\right)\right]\left[\Omega^* - \frac{\pi}{2}\right] + \dots \approx \left[\frac{\pi}{2} - \Omega^*\right]$$

so

$$\mathrm{ctn}\,\Omega^* \approx \left[\frac{\pi}{2}\right] - \sqrt{\frac{\rho\omega^2 L^2}{G'(1+i\tan\delta)}} \, ,$$

but at the first resonance, the real part of Ω^* is $\pi/2$ to minimize the structural rigidity.

$$\sqrt{\frac{\rho\omega_1^2 L^2}{G'}} = \left[\frac{\pi}{2}\right],$$

so, by writing a Taylor expansion of the square root and of the inverse, and by retaining the lowest order terms to obtain a simple form for low loss,

$$\mathrm{ctn}\,\Omega^* \approx \left[\frac{\pi}{2}\right]\left[1 - \left(1 - \frac{1}{2}i\tan\delta\right)\right] = \left[\frac{\pi}{4}\right]i\tan\delta.$$

So the dynamic rigidity at the first resonance is

$$\frac{M}{\theta} = \left[\frac{\pi}{2}R^4\right]\left[\rho\omega_1^2 L\right]\left[\frac{1}{2}\right]i\tan\delta, \tag{3.5.19}$$

in which i indicates a 90° phase shift at resonance. ω_1 here is the angular frequency at the first resonance. At resonance, the dynamic rigidity is proportional to the loss tangent so the dynamic compliance becomes large as $\tan\delta$ becomes small.

Case 2. Let us find the rigidity at the next highest resonance of the viscoelastic rod. For an elastic rod fixed at one end and free to rotate at the other end (fixed-free condition) the next resonance occurs at an angular frequency ω_3, about three times that of the lowest resonance. Assume the loss is small, $\tan\delta \ll 1$. Expand the cotangent about $3\pi/2$ which corresponds to the next highest resonance frequency.

$$\mathrm{ctn}\,\Omega^* \approx \mathrm{ctn}\left[\frac{3\pi}{2}\right] + \left[-\csc^2\left(\frac{3\pi}{2}\right)\right]\left[\Omega^* - \frac{3\pi}{2}\right] + \ldots$$

$$= 0 - 1\left\{\sqrt{\frac{\rho\omega^2 L^2}{G'(1+i\tan\delta)}} - \frac{3\pi}{2}\right\}$$

$$= \frac{3\pi}{2}\left[1 - (1+i\tan\delta)^{-1/2}\right],$$

since

$$\sqrt{\frac{\rho\omega_3^2 L^2}{G'}} = \frac{3\pi}{2}.$$

so

$$\operatorname{ctn} \Omega^* \approx \frac{3\pi}{4} i \tan \delta,$$

so

$$\frac{M}{\theta} = \left[\frac{\pi}{2}R^4\right]\left[\rho\omega_3^2 L\right]\left[\frac{1}{2}\right] i \tan \delta, \tag{3.5.20}$$

which has the same form as in the case of the lowest resonance. For small loss, G' is nearly independent of frequency, so the angular frequency ω of the next highest resonance is about three times that of the first, and the dynamic rigidity M/θ at this resonance is nine times greater than that of the first resonance. The higher resonant modes, therefore, will appear at much smaller amplitudes than the first mode, assuming a constant drive torque amplitude M. These higher modes occur at frequencies in the ratio 1, 3, 5, 7, 9 ... provided that the loss tangent is small. This resonance structure in the fixed-free configuration is also observed in musical instruments such as the clarinet, in which one end is effectively closed and the other end is open. By contrast, a rod in a free-free configuration has both ends unconstrained. The resonant frequencies are then in the ratio 1, 2, 3, 4, 5 ... . This resonance structure is observed in the flute, which is effectively open at both ends.

Case 3. Consider the region near the first resonance with the aim of determining the width of the resonance curve. Let ω_1 be the angular frequency of the first resonance:

$$\omega_1 = \left[\frac{\pi}{2}\right]\left[\frac{G'}{\rho L^2}\right]^{1/2}$$

so

$$\operatorname{ctn} \Omega^* \approx \left[\frac{\pi}{2}\right]\left[1 - \{\omega/\omega_1\}\left(1 - \frac{1}{2}i\tan\delta\right)\right].$$

Evaluate the magnitude of the rigidity:

$$\left|\operatorname{ctn}\Omega^*\right|^2 \approx \left[\frac{\pi}{2}\right]^2\left[(1-\omega/\omega_1)^2 + \{\omega/2\omega_1\}^2 \tan^2\delta\right]. \tag{3.5.21}$$

Find ω such that the dynamic compliance is half its value at resonance. For $\omega = \omega_1$,

$$\left|\text{ctn}\,\Omega^*\right|^2 \approx \left[\frac{\pi}{2}\right]^2 \left[\frac{1}{4}\right]\tan^2\delta, \text{ so } 4\left[(\omega_1 - \omega)/\omega_1\right]^2 \approx 3\tan^2\delta. \qquad (3.5.22)$$

So the full width of the resonance curve at half maximum is

$$\frac{\Delta\omega}{\omega_1} \approx \sqrt{3}\tan\delta. \qquad (3.5.23)$$

This is the same result obtained in the case of lumped inertia, even though the exact expression for the shape of the resonance curve is different. Neither the rod diameter nor the density appear in Eq. 3.5.23; the loss is obtained from the *width* of the resonance curve (dynamic rigidity or compliance vs. frequency). We remark that Eq. 3.5.23 applies to the case of small damping, $\tan\delta \ll 1$.

§3.6   *Decay of resonant vibration*

Suppose that in the lumped torsional system considered in §3.5, the external sinusoidal torque is suddenly removed, so $M_{ext} = 0$ in Eq. 3.5.4. Suppose moreover that the viscoelastic material is passive. So

$$I\frac{d^2\theta}{dt^2} + G'(1 + i\tan\delta)\,k\theta = 0. \qquad (3.6.1)$$

Consider a solution for the angular displacement of the form

$$\theta(t) = \theta_0 e^{i\lambda t}. \qquad (3.6.2)$$

Then

$$\frac{d^2\theta}{dt^2} = -\theta_0\lambda^2 e^{i\lambda t} = -\theta(t)\lambda^2. \qquad (3.6.3)$$

So

$$G'(1 + i\tan\delta)k = I\lambda^2, \qquad (3.6.4)$$

and

$$\lambda = \omega_1\sqrt{1 + i\tan\delta}, \qquad (3.6.5)$$

with

$$\omega_1 \equiv \sqrt{\frac{G'k}{I}}.$$

So

$$\theta(t) = \theta_0 \exp\left\{i\omega_1 t\sqrt{1 + i\tan\delta}\right\}. \qquad (3.6.6)$$

If we again assume the loss to be small, $\tan\delta \ll 1$; then expand the square root as a power series about 1; and retain the lowest order term,

$$\lambda \approx \omega_1\left(1 + \frac{1}{2}i\tan\delta\right),$$

so

$$\theta(t) \approx \theta_0 e^{i\omega_1 t} e^{-(\omega_1 t/2)\tan\delta}. \qquad (3.6.7)$$

This form shows the effect of the natural frequency and the mechanical damping separately; it represents a damped sinusoid, which decays in amplitude with time (Fig. 3.6). The sign in the square root solution is chosen to correspond to a passive material, which damps the vibrations rather than amplifying them.

When $(\omega_1 t/2)\tan\delta = 1$, the amplitude has decayed to $1/e$ of its value at time zero; the corresponding time is $t_{1/e} = 2/\omega_1\tan\delta$. The number n of cycles required for the amplitude to decay by a factor of $1/e$ is

$$n = t_{1/e}\frac{\omega_1}{2\pi} =$$

$$t_{1/e}\nu_1 = n = \frac{1}{\pi\tan\delta}. \qquad (3.6.8)$$

The loss tangent may be expressed in terms of the time $t_{1/e}$ for decay of vibration and the period T of oscillation:

$$\tan\delta = \frac{1}{\pi}\frac{T}{t_{1/e}}. \qquad (3.6.9)$$

To evaluate the rate of decay, consider adjacent cycles, at time $t_1 = 0$ and $t_2 = 1/\nu = 2\pi/\omega$. The corresponding angular displacement amplitudes are $A_1 = \theta_0$ and

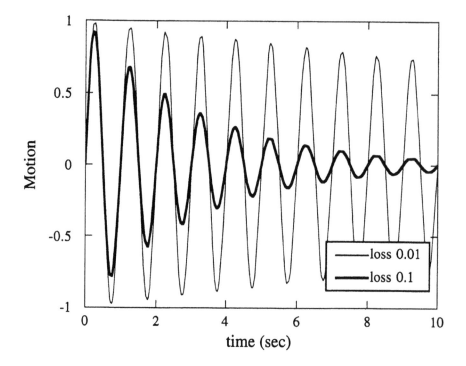

Figure 3.6   Free decay of resonant vibration motion with time for materials with tan δ = 0.1 (thick line) and 0.01 (thin line).

$$A_2 = \theta_0 \exp\{(\omega_1 2\pi \tan \delta)/2\omega_1\}.$$

Define the *log decrement* as

$$\Lambda \equiv \ln \frac{A_1}{A_2}, \qquad\qquad (3.6.10)$$

then

$$\Lambda = \pi \tan \delta. \qquad\qquad (3.6.11)$$

Experimentally (see Fig. 3.7 and §6.8), free decay of vibration is often used to determine the loss tangent of low-loss materials. As discussed in §1.5, viscoelasticity can be demonstrated by observing the free decay of vibration of tuning forks or cantilevers of various materials.

In the study of differential equations, the equation

$$m\frac{d^2u}{dt^2} + b\frac{du}{dt} + C\,u = 0 \qquad\qquad (3.6.12)$$

Figure 3.7 Forced vibration of an elastic string fixed to a viscoelastic support. (After S. Harris. With permission.)

is often solved [3.6.1]. Here u is a variable; and m, b, and C are positive constants. This equation is seen as equivalent to Eq. 3.6.1 (if tan δ is constant), in which $i = \sqrt{(-1)}$ shifts the phase of a sinusoid by 90° and so for sinusoids the i represents a first derivative. The differential equation may be rewritten:

$$\frac{d^2u}{dt^2} + \frac{b}{m}\frac{du}{dt} + \omega_1^2 u = 0 \qquad (3.6.13)$$

with

$$\omega_1 = \sqrt{C/m}.$$

Substitution of the following trial solution (with U_0 as a constant):

$$u(t) = U_0 e^{pt} \qquad (3.6.14)$$

into the above differential equation leads to the following algebraic equation:

$$p^2 + \frac{b}{m}p + \omega_1^2 = 0 \qquad (3.6.15)$$

So

$$p = -\frac{1}{2}\frac{b}{m} \pm \sqrt{\frac{1}{4}(b/m)^2 - \omega_1^2}. \tag{3.6.16}$$

The case $(b/m) < 2\omega_1$ gives rise to a complex exponential which represents a damped sinusoid. This is the case considered above and is referred to as underdamped. If $(b/m) = 2\omega_1$, then $u(t) = U_0 e^{-\omega_1 t}$, and there are no vibrations, just a gradual return to equilibrium. This case is called critical damping. If $(b/m) > 2\omega_1$, the two roots for p are both real and negative, and $u(t)$ becomes a sum of two decaying real exponentials. Again there are no vibrations. This case is called overdamping or heavy damping. For $(b/m) \gg \omega_1$, the response approximates a single exponential. This condition is considered very heavy damping.

If the damping coefficient b has its origin in viscoelastic behavior rather than simple viscous drag, we must recognize that the loss tangent in general depends on frequency. Also, in a general viscoelastic material, the stiffness C must depend on frequency by virtue of the Kramers–Kronig relations. In the study of a system near resonance, it may be expedient to ignore the frequency dependence of tan δ provided that its magnitude is small. To understand this, let us examine the Fourier transform [3.6.2] of a damped sinusoid:

$$\mathcal{F}\left\{\mathcal{H}(t)e^{-at}\sin\omega_1 t\right\} = \frac{\omega_1}{(a + i\omega_1)^2 + \omega_1^2}. \tag{3.6.17}$$

Since $a = (\tfrac{1}{2}\omega_1)\tan\delta$, a small loss corresponds to a narrow peak in the frequency content. So even if tan δ varies with frequency, only a small range of frequencies occurs in the signal, and tan δ is effectively constant over that range. Free decay for small tan δ approximates a sinusoid at one frequency. The case of heavy damping with nonconstant loss is more complicated and can be handled with transform methods [1.6.8].

§3.7 Wave propagation and attenuation

Consider the following simple one-dimensional problem of waves in a thin rod of linearly viscoelastic material. Waves [3.7.1, 3.7.2] are transmitted down a long uniform rod of length L, cross-sectional area A, dynamic modulus E^*, and density ρ. In the following, a one-dimensional analysis is performed on the rod; neglect of the Poisson effects is warrantable if the wavelength of stress waves in the rod is much greater than the rod diameter. The stress–strain relation is taken as

$$\sigma = E^*\varepsilon = E^*\frac{\partial u}{\partial z}, \tag{3.7.1}$$

in which z is a coordinate along the rod axis and u is a displacement in that direction. Consider now a differential element of length dz along the rod. By applying Newton's second law,

$$\rho A dz \frac{\partial^2 u}{\partial t^2} = A\sigma\left(z+dz, t\right) - A\sigma\left(z, t\right) + f\left(z, t\right) dz, \qquad (3.7.2)$$

in which f is the externally applied force per unit length. By dividing by dz,

$$\rho A \frac{\partial^2 u}{\partial t^2} = \frac{\partial A\sigma}{\partial z} + f\left(z, t\right). \qquad (3.7.3)$$

By incorporating the stress–strain relation,

$$\frac{\partial^2 u}{\partial t^2} = \frac{E^*}{\rho} \frac{\partial^2 u}{\partial z^2} + \frac{f\left(z, t\right)}{A\rho}. \qquad (3.7.4)$$

This is a *wave equation*. Consider a trial solution of the form

$$u(z, t) = u_0 \, e^{i(kz-\omega t)}, \qquad (3.7.5)$$

and let the forcing function f(z, t) vanish in the region of interest. Then

$$\omega^2 u_0 = k^2 u_0 \frac{E^*}{\rho}, \qquad (3.7.6)$$

$$k \sqrt{\frac{E'}{\rho}(1+i\tan\delta)} = \omega, \qquad (3.7.7)$$

$$k \sqrt{\frac{E'}{\rho}} \sqrt{1+i\tan\delta} = \omega. \qquad (3.7.8)$$

Suppose that the loss is small, expand the square root containing the loss tangent, and retain the lowest order terms. Moreover, define

$$c = \sqrt{\frac{E'}{\rho}}, \qquad (3.7.9)$$

$$k \approx \frac{\omega}{c}\left(1+\frac{1}{2}i\tan\delta\right). \qquad (3.7.10)$$

Then Eq. 3.7.3 has a solution,

$$u\left(z, t\right) = u_0 \exp\left\{i\omega\left(\frac{z}{c} - t\right)\right\} \exp\left\{-z \frac{\omega}{2c} \tan\delta\right\}. \qquad (3.7.11)$$

In the first complex exponential, the argument $\{(z/c) - t\}$ is constant if $dz/c = dt$, or $dz/dt = c$. So the wave speed c is interpreted as the *phase velocity* of the wave. As time t increases, z must increase to keep the phase argument constant, so the wave moves to the right (direction of increasing z). A left-moving wave would be written $u(z, t) = u_0 \exp i\omega\{(z/c) + t\}$.

As for the interpretation of k (as a real number, for zero loss), observe that a sinusoid goes through a complete cycle when its argument goes through 2π. So $kz = 2\pi$ corresponds to one spatial cycle in z, or one *wavelength*, given the name λ. So $k = 2\pi/\lambda$. Similarly, at a given point in space, one cycle occurs when $\omega t = 2\pi$. The time for one cycle is called the *period* T of the wave; and the inverse of the period is the number of cycles per unit time, the *frequency* $\nu = 1/T$. So $\omega = 2\pi\nu$.

The second exponential in Eq. 3.7.11 is a real exponential, and signifies a decrease of wave amplitude with distance z. This decrease is known as *attenuation*. The attenuation coefficient (per unit length) is defined for small loss as

$$\alpha \approx \frac{\omega}{2c} \tan\delta. \qquad (3.7.12)$$

α, the attenuation per unit length, is commonly given in units of nepers per length, in which "neper" is dimensionless. One neper is a decrease in amplitude of a factor of $1/e$. This corresponds to $\alpha z = 1$ in Eq. 3.7.11 with the second term taken as $e^{-\alpha z}$. The word "neper" is a degradation of the name of the inventor of the natural logarithm, Napier. The attenuation can also be expressed [3.7.1] in decibels per unit length by the following [1.6.23]:

$$20 \log_{10}(e^\alpha) = 8.686\ \alpha,$$

so

$$\alpha\ (dB/cm) = 8.68\ \alpha\ (neper/cm).$$

If the rod is elastic ($\delta = 0$), the stress waves propagate without dispersion; that is, the wave speed is independent of frequency, provided the one-dimensional assumption is justified by the wavelength being much larger than the rod diameter. Moreover, the waves propagate without attenuation. If the rod is viscoelastic, wave speed depends on frequency since E' depends on frequency, and there is attenuation.

Exact relations for wave speed c and attenuation α can be extracted with further effort [1.6.1],

$$c = \sqrt{\frac{|E^*|}{\rho}} \sec\frac{\delta}{2}. \qquad (3.7.13)$$

$$\alpha = \frac{\omega}{c}\tan\frac{\delta}{2}. \qquad (3.7.14)$$

Some authors use units of nepers per wavelength; since $\lambda v = c$ such attenuations give the loss without any frequency term, as follows:

$$\alpha\lambda = 2\pi\tan\frac{\delta}{2}. \qquad (3.7.15)$$

The attenuation per unit wavelength λ is also expressed as

$$\alpha\lambda = \pi/Q \text{ nepers} \approx 27.3/Q \text{ dB}.$$

Here Q can be considered in terms of energy dissipated in a wave, as the number of wavelengths for an attenuation of π nepers, or as the number of cycles required for the amplitude to drop to $e^{-\pi}$ of its original value [3.7.1].

§3.8 *Measures of damping*

The loss angle δ, or the loss tangent $\tan\delta$, may be considered as the fundamental measure of damping in a linear material. Other measures, such as those developed above, are often cited in analyses of viscoelastic behavior. For the purpose of comparison, these measures of damping and their relationship to the loss angle are presented in Table 3.1.

The approximations are valid for the case of small damping, $\tan\delta \ll 1$. They are within 1% of the exact value up to a loss tangent of about 0.2 [3.4.1]. Derivation of exact relations generally requires a model of the viscoelastic response of the material.

§3.9 *Nonlinear materials*

Nonlinearly viscoelastic materials (see §2.12 for constitutive equations and §5.10 for examples of analysis) can also be excited by sinusoidal loads. The simplest dynamic example of the effect of nonlinearity is nonlinear elasticity in one dimension. Suppose

$$\sigma = f(\varepsilon). \qquad (3.9.1)$$

Table 3.1 Measures of Damping

Measure	Relation to δ	Phenomenon
Loss angle	δ	Phase, stress, strain
Loss tangent	$\tan \delta$	Tangent of the phase angle
$[E''(\omega)]/[E'(\omega)]$	$\tan \delta(\omega)$	Ratio of imaginary part to real part of modulus
Quality factor	$Q \approx (\tan \delta)^{-1}$	Resonance peak width
Log decrement	$\Lambda \approx \pi \tan \delta$	Free decay
Decay time $t_{1/e}$	$\tan \delta \approx (1/\pi) \times (T/t_{1/e})$	Free decay
Specific damping capacity	$\Psi = 2\pi \tan \delta$	Ratio of energy dissipated to energy stored
Attenuation	$\alpha\lambda = 2\pi \tan (\delta/2)$	Wave attenuation (neper/λ)
Attenuation	$\alpha = (\omega/c) \tan (\delta/2)$	Wave attenuation (neper/m)

Write f as a power series, so

$$\sigma = a_1\varepsilon^1 + a_2\varepsilon^2 + a_3\varepsilon^3 + \dots . \tag{3.9.2}$$

If the strain is sinusoidal,

$$\varepsilon = \cos \omega t, \tag{3.9.3}$$

trigonometric identities, e.g.,

$$\cos^2 \omega t = \frac{1}{2}(1 + \cos 2\omega t), \tag{3.9.4}$$

$$\cos^3 \omega t = \frac{1}{4}\cos 3\omega t + \frac{3}{4}\cos \omega t, \tag{3.9.5}$$

may be used [3.9.1] to express the stress as a sum of sinusoids.

$$\sigma(t) = \sum_{n=1}^{N} a_n \sin \omega_n t. \tag{3.9.6}$$

Consequently, the effect of the nonlinearity is to generate higher harmonics (integer multiples) of the driving frequency. If the driving signal contains several frequencies, the material responds at new frequencies corresponding to sums and differences of the drive frequencies [3.9.1].

As for a viscoelastic material obeying nonlinear superposition, analysis (§5.10) shows that higher harmonics are generated as well [3.9.2]. In the electrical engineering community, this sort of response is called *harmonic distortion*. In a Green–Rivlin solid which obeys the Green–Rivlin multiple integral expansion (§2.12), the response contains higher harmonics as in the material obeying nonlinear superposition. These two types of material can be distinguished by applying a strain history containing a superposition of several frequencies [3.9.3, 3.9.4]. In a Green–Rivlin solid, oscillations occur at new frequencies not originally present in the original excitation (§5.10). In the electrical engineering community, this sort of response is called *intermodulation distortion*.

Response of a nonlinear material to cyclic load may also be visualized via a plot of stress vs. strain, as shown in Fig. 3.8. In a nonlinear material, the plot is no longer elliptical in shape, as it is in a linear material. Harmonic distortion due to the material gives rise to a nonelliptical plot. Many types of curves are observed in real materials. The area within the closed curve, as in the case of linear materials, represents energy per volume dissipated per cycle.

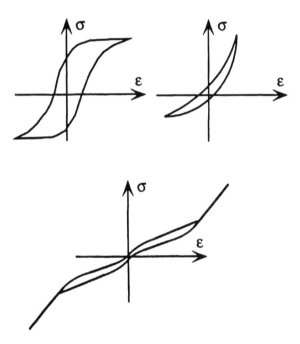

Figure 3.8   Response (stress σ vs. strain ε) of several nonlinear materials to cyclic load. The loop is no longer elliptical as it is in linear materials.

In the dynamic response of nonlinearly viscoelastic materials, the loss angle δ can still be defined as the phase angle between the drive signal and the fundamental component of the response at that frequency. However, the interpretation of the loss angle in terms of energy dissipation is no longer the same as in linear materials. In nonlinear solids it is more sensible to speak directly of energy dissipation per cycle.

§3.10 *Summary*

The dynamic properties of linearly viscoelastic materials have been developed in terms of the transient functions. In dynamic (sinusoidal) loading the stress–strain ratio or dynamic modulus has a magnitude and phase and can be represented as a complex number. The real and imaginary parts are associated with storage and dissipation of mechanical energy, respectively. The real and imaginary parts of the dynamic modulus and compliance are not independent but are related by the Kramers–Kronig relations. In structural resonance, the material loss tangent manifests itself in a broadening and lowering of the resonance peak, and in the damping of free vibration of structural members. As for low-loss materials, we remark that in a subresonant test using conventional mechanical testing equipment, materials with tan δ between about 10^{-3} and 10^{-9} cannot be easily distinguished. Yet the resonant behavior of such materials differs dramatically. In wave propagation, the loss tangent is associated with attenuation of waves. The examples considered are directly relevant to experimental methods for determining material properties and to applications of viscoelastic materials.

Examples

Example 3.1

Express the dynamic modulus $E^* = E' + iE''$ in terms of the loss tangent.

Solution

$$E^* = E' + iE'' = E'\left(1 + i\frac{E''}{E'}\right) = E'(1 + i\tan\delta).$$

Example 3.2

Express the dynamic modulus $E^* = E' + iE'$ in terms of $|E^*|$ and the phase δ. Also express $|E^*|$ in terms of E' and δ.

Solution

$$E^* = |E^*|\frac{(E' + iE'')}{|E^*|} = |E^*|(\cos\delta + i\sin\delta).$$

This form may be visualized by expressing the complex modulus as a point in the complex plane.

$$|E^*| = \sqrt{(E' + iE'')(E' - iE'')} = \sqrt{E'^2 + E''^2} = E'\sqrt{1 + \tan^2 \delta} = \frac{E'}{\cos \delta}.$$

Example 3.3

A viscoelastic ball is dropped from a height h upon a rigid floor. To what fraction f of that height will it rebound? An approximate value is satisfactory.

Solution

As an initial approximation, represent the impact event as half a cycle of harmonic loading. Proceed by equating the energy dissipated in the material by viscoelastic damping with the reduction in potential energy associated with the difference in height. This procedure neglects any energy loss due to air resistance, radiation of sound energy during the impact, and friction. For a linearly viscoelastic material, following Fig. E3.1, the stress–strain curve for harmonic loading is an ellipse. The area of the ellipse represents energy dissipated per unit volume per cycle, and the impact is considered as half a cycle. This is an approximation since the actual impact begins with zero stress and zero strain; a transient term is neglected. So the energy density W_d dissipated in half a cycle is proportional to the area within the ellipse,

$$W_d = \frac{1}{2}\pi ab, \tag{E3.3.1}$$

with a as the semiminor axis and b as the semimajor axis of the ellipse. Consider the stress to be normalized to Young's modulus so that the ellipse appears on dimensionless axes for simpler interpretation. We will consider a ratio of energies so that the modulus will not appear in the final result in any case. By referring to Fig. 3.2, $b = B\sqrt{2}$ and $a = A/\sqrt{2}$, so from Eq. 3.2.21,

$$W_d = \frac{1}{2}\pi AB. \tag{E3.3.2}$$

But the normalized stored energy is

$$W_s = \frac{1}{2}B^2 \cos \delta,$$

so

$$\frac{W_d}{W_s} = \pi \frac{A}{B} = \pi \tan \delta. \tag{E3.3.3}$$

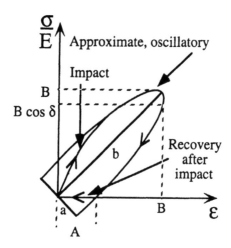

Figure E3.1   Stress σ vs. strain ε for a half cycle of harmonic loading compared
with an impact.

Consider now the stored and dissipated energies in terms of ratio f of
the drop height h_{drop} to the rebound height h_{reb}, the ball mass m, and the
acceleration due to gravity g:

$$f = \frac{h_{reb}}{h_{drop}}.$$ (E3.3.4)

$$W_s = mgh_{drop}, \quad W_d = mg\left(h_{drop} - f \cdot h_{drop}\right) = mgh_{drop}(1-f),$$ (E3.3.5)

so

$$\frac{W_d}{W_s} = 1 - f.$$ (E3.3.6)

By equating the expressions for energy ratio,

$$1 - f \approx \pi \tan \delta,$$ (E3.3.7)

so

$$f \approx 1 - \pi \tan \delta.$$ (E3.3.8)

This approximation may be improved somewhat by considering the two
small "triangles" at the base of the half ellipse in Fig. E3.1. Each of these has
area $\frac{1}{2} a^2$. But a = A/√2 = (1/√2)B sin δ so the total area to be subtracted
from the half ellipse area is

$$\frac{1}{2} B^2 \sin^2 \delta.$$

The dissipated energy is

$$W_d = \frac{1}{2} \pi AB - \frac{1}{2} B^2 \sin^2 \delta = \frac{1}{2} B^2 \left[\pi \sin \delta - \frac{1}{2} \sin^2 \delta \right]$$

Again, the normalized stored energy is

$$W_s = \frac{1}{2} B^2 \cos \delta \quad \text{and} \quad \frac{W_d}{W_s} = 1 - f.$$

So

$$f = \frac{h_{reb}}{h_{drop}} \approx 1 - \left(\pi \tan \delta - \frac{1}{2} \sin^2 \delta \right).$$

This result still does not take into account the fact that the load history starts at zero and therefore neglects an additional area in the diagram. For large δ the approximation breaks down since large δ gives rise to a negative height ratio f which cannot occur.

A better approximation to rebound resilience may be had by considering free decay of vibration following Eq. 3.6.7, for small $\tan \delta$ with ω as the natural angular frequency.

$$x(t) \approx x_0 \, e^{-(\omega t/2) \tan \delta} \sin \omega t.$$

The moving mass is considered to come in contact with a one-dimensional massless viscoelastic spring at time zero, to execute one-half cycle of damped vibration, and then to rebound, losing contact at time $t = \pi/\omega$.

The velocity is

$$v = \frac{dx}{dt} = x_0 \omega e^{-(\omega t/2) \tan \delta} \left[\cos \omega t - \frac{1}{2} \tan \delta \sin \omega t \right].$$

The impact velocity v_0 is calculated at time $t = 0$:

$$v_0 = x_0 \omega.$$

The rebound velocity is calculated at time $t = \pi/\omega$, when the displacement is again zero.

$$v_1 = -x_0 \omega e^{-(\pi/2)\tan \delta}.$$

The height ratio is related to the velocity ratio via the relation between kinetic and potential energy.

$$f = \frac{h_{reb}}{h_{drop}} = \frac{v_1^2}{v_0^2} = \left\{ e^{-(\pi/2)\tan \delta} \right\}^2 = e^{-\pi \tan \delta}.$$

So,

$$\tan \delta \approx \frac{1}{\pi} \ln \left\{ \frac{h_{drop}}{h_{reb}} \right\}. \qquad (E3.3.9)$$

This approximation makes physical sense at the extremes of damping: the height ratio f is unity for zero damping and tends to zero for very high damping. The result is approximate since a simplified form of the vibration equation is used, and since the treatment is one dimensional. Moreover, energy loss due to radiated sound and stress waves in the substrate is not considered. The actual impact force is a pulsed function of time which may be Fourier analyzed as a distribution of frequencies, so tan δ over a range of frequency contributes to the observed rebound resilience. A more detailed treatment is in §10.10.

Physically, the fraction f may be interpreted as a resilience: 100% for a perfectly elastic material and zero for the case of zero rebound. This may be contrasted with the *coefficient of restitution* which is defined as the ratio of *speed* v after an impact to the speed before the impact. Since gravity converts potential energy corresponding to the height h to kinetic energy as the ball is dropped according to mgh = ½ mv², it is evident that there is a quadratic relationship between the height ratio and the coefficient of restitution.

Example 3.4

Show that the change in storage compliance over the full frequency range is proportional to the area under the loss compliance curve on a log frequency scale.

Solution

Use the Kramers–Kronig relation, Eq. 3.3.8, with ω = 0 to obtain the full frequency range.

$$J'(0) - J'(\infty) = \int_0^\infty \frac{J''(\omega)}{\omega} d\omega = \int_0^\infty J''(\omega) \, d \ln \omega. \qquad (E3.4.1)$$

Example 3.5

Determine the dynamic behavior associated with an exponential relaxation function

$$E(t) = (\Delta E)e^{-t/\tau}. \qquad (E3.5.1)$$

Solution

Consider the standard integrals

$$\int_0^\infty e^{-ax}\cos bx\, dx = \frac{a}{a^2+b^2}, \qquad (E3.5.2)$$

$$\int_0^\infty e^{-ax}\sin bx\, dx = \frac{b}{a^2+b^2}. \qquad (E3.5.3)$$

Then, with Eqs. 3.2.5 and 3.2.6,

$$E''(\omega) = (\Delta E)\,\omega\int_0^\infty e^{-t/\tau}\cos\omega t\, dt = (\Delta E)\frac{\omega\tau}{1+\omega^2\tau^2}, \qquad (E3.5.4)$$

$$E'(\omega) = E'(0) + (\Delta E)\,\omega\int_0^\infty e^{-t/\tau}\sin\omega t\, dt = E'(0) + (\Delta E)\frac{\omega^2\tau^2}{1+\omega^2\tau^2}. \qquad (E3.5.5)$$

Example 3.6

Show that

$$J' = \frac{1}{E'}\frac{1}{1+\tan^2\delta}.$$

Observe that $J^* = J' - iJ''$. Discuss the physical interpretation of this in terms of the stress–strain diagram.

Solution

$$J^* = \frac{1}{E^*} = \frac{1}{E'+iE''} = \frac{1}{E'+iE''}\frac{E'-iE''}{E'-iE''} = \frac{E'}{E'^2+E''^2} - i\frac{E''}{E'^2+E''^2}. \qquad (E3.6.1)$$

Since

$$\tan \delta = \frac{E''}{E'},$$

$$\frac{E'}{E'^2 + E''^2} = \frac{1}{E'} \frac{1}{1 + \tan^2 \delta}, \tag{E3.6.2}$$

so

$$J' = \frac{1}{E'} \frac{1}{1 + \tan^2 \delta}. \tag{E3.6.3}$$

The physical interpretation of this in terms of the stress–strain diagram is that in the elliptical Lissajous figure, the slope $1/J'$ corresponding to the stress–strain ratio at maximum stress is greater than the slope E' corresponding to the stress–strain ratio at maximum strain.

Example 3.7

Observe in §3.5 that the structural compliance of the resonance of the lumped torsional system (for small loss) is a factor $1/\tan \delta$ larger than the compliance under static conditions. The "magnification factor" is therefore $1/\tan \delta$. What is the corresponding magnification factor for the first resonance of the distributed rod in torsion?

Solution

At the first resonance of the distributed torsion case,

$$\omega_1 = \frac{\pi}{2} \sqrt{\frac{G'}{\rho L^2}}, \tag{E3.7.1}$$

so

$$\frac{\pi}{2} = \sqrt{\frac{\rho \omega_1^2 L^2}{G'}}, \tag{E3.7.2}$$

so

$$\omega_1^2 = \frac{G' \pi^2}{4 \rho L^2}. \tag{E3.7.3}$$

Substitute in the relation for dynamic rigidity at the first resonance,

$$\frac{M}{\theta} = \left[\frac{1}{2}\pi R^4\right]\left[\rho\omega_1^2 L\right]\frac{1}{2}i\tan\delta, \qquad (E3.7.4)$$

$$\frac{M}{\theta} = G'\frac{\pi R^4 \pi^2}{2L\ 8}i\tan\delta, \qquad (E3.7.5)$$

but the static rigidity of the rod in torsion is

$$\frac{M}{\theta} = G'\frac{\pi R^4}{2L}. \qquad (E3.7.6)$$

But $8/\pi^2 \approx 0.81057$, so the magnification factor for the first resonance of the distributed system is not $1/\tan\delta$, but instead is $0.81/\tan\delta$.

Example 3.8

In §3.4, the principles of nonnegative stored energy and nonnegative rate of dissipation of energy were discussed in connection with inequalities for the dynamic functions. Are these real physical laws? Do they follow from physical laws? Would it be possible to create a material for which $\tan\delta < 0$? If so, what would be required? Is it possible to have negative stiffness?

Solution

Some of the mathematical principles actually entail assumptions about the material or the nature of the system.

Negative stiffness in dynamic systems is certainly possible. As shown in §3.5, a phase difference of 180° occurs between load and deformation above resonance in an elastic material, or far above resonance in a viscoelastic material. This phase shift corresponds to a negative *structural* stiffness since motion occurs in the opposite direction as force. In static deformation, negative stiffness entails a condition of unstable equilibrium. Examples include a bead rolling on a smooth bowl which is concave down, as well as a buckled column in response to a lateral load. When the buckled column is straight, it is in unstable equilibrium and it exhibits negative stiffness for lateral perturbations. Systems with negative stiffness must contain stored energy or they must receive energy from some external source. Such systems are certainly possible, but once prepared, they will diverge from the condition of unstable equilibrium unless restrained.

Negative damping (or negative attenuation) corresponds to a *gain* of energy in vibration or wave motion. Such gain is possible in light of the principle of energy conservation, if there is an external source of energy. In passive materials, there is no such energy source by definition, so $E'' \geq 0$. Mechanical examples are given in §7.9. Some more familiar examples in electrical engineering and optics are as follows. An electrical amplifier exhibits gain, and the energy is supplied by the electrical power supply. A laser (the word means light amplification by stimulated emission of radiation) amplifies light, and as with other amplifiers it can serve as an oscillator. The laser is provided with an external source of power.

Example 3.9

What is the shape of the load–deformation diagram for a lumped system under harmonic load as the natural frequency is approached?

Solution

Under quasistatic conditions, well below the lowest resonance, the load–deformation diagram follows the stress–strain behavior and is elliptical; the width of the ellipse is related to tan δ by Eq. 3.2.26. At resonance, the phase ϕ between load and deformation is 90°, following Eq. 3.5.11. This diagram is a circle or an ellipse with principal axes aligned with the coordinate axes.

Example 3.10

Suppose $E(t) = E_2 + E_1 e^{-t/\tau}$. What is the relaxation strength Δ? Show that

$$\tan \delta = \frac{\Delta}{\sqrt{1+\Delta}} \frac{\omega \tau_m}{1+\omega^2 \tau_m^2},$$

with

$$\tau_m = \tau \sqrt{1+\Delta}.$$

What is the peak tan δ?

Solution

The relaxation strength Δ is, by definition, the ratio of change in stiffness through the relaxation to the relaxed stiffness, so by considering limits at short time and long time, $\Delta = E_1/E_2$. By definition tan δ = E''/E'; by substituting from Example 3.5,

$$\tan\delta = \frac{E_1\dfrac{\omega\tau}{1+\omega^2\tau^2}}{E_2+E_1\dfrac{\omega^2\tau^2}{1+\omega^2\tau^2}} = \frac{E_1\omega\tau}{E_2(1+\omega^2\tau^2)+E_1\omega^2\tau^2}$$

$$= \frac{\dfrac{E_1}{E_2}\omega\tau}{1+\left(1+\dfrac{E_1}{E_2}\right)\omega^2\tau^2} \tag{E3.10.1}$$

$$= \frac{\dfrac{E_1}{E_2}}{\sqrt{1+\dfrac{E_1}{E_2}}}\cdot\frac{\omega\tau\sqrt{1+\dfrac{E_1}{E_2}}}{1+\omega^2\left(\tau\sqrt{1+\dfrac{E_1}{E_2}}\right)^2}.$$

Since $\Delta = E_1/E_2$,

$$\tan\delta = \frac{\Delta}{\sqrt{1+\Delta}}\frac{\omega\tau_m}{1+\omega^2\tau_m^2}, \tag{E3.10.2}$$

with

$$\tau_m = \tau\sqrt{1+\Delta}.$$

So, if E'' is a Debye peak, $\tan\delta$ also is a Debye peak, but with a shift in frequency.

Example 3.11

Suppose the loss compliance is constant over a frequency range, so $J''(\omega) = J_0''$ for $0 \le \omega \le \omega_c$, with ω_c as a cutoff frequency, and $J'' = 0$ for $\omega > \omega_c$. Find the storage compliance and discuss the physical interpretation.

Solution

By integrating the Kramers–Kronig relation Eq. 3.3.8 with the aid of some identities,

$$J'(\omega) - J'(\infty) = \frac{J_0}{\pi}\left[\ln(\omega_c^2 - \omega^2) - \ln\omega^2\right]. \tag{E3.11.1}$$

Now if we have $\omega << \omega_c$,

$$J'(\omega) - J'(\infty) \approx \frac{2J_0}{\pi} \left[\ln(\omega_c) - \ln \omega \right]. \qquad (E3.11.2)$$

So for this case a constant loss compliance is associated with a linear dispersion in the storage compliance vs. log frequency. Not uncommonly, the behavior of real materials is approximated by this example. In real materials, sharp cutoffs in the mechanical loss are not observed; a cutoff was incorporated in this example to achieve convergence and it was made sharp for simplicity in calculation. The singular resonant behavior in the storage compliance at the cutoff frequency is not representative of physically observed mechanical behavior of *materials*.

Example 3.12

Suppose the loss is concentrated at only one frequency: $J''(\omega) = k\delta(\omega - \omega_c)$ in which δ is the Dirac delta function. Find the storage compliance and discuss the physical interpretation.

Solution

Again by integrating the appropriate Kramers–Kronig relation,

$$J'(\omega) - J'(\infty) = \frac{2k\omega_c}{\pi(\omega_c^2 - \omega^2)}. \qquad (E3.12.1)$$

The storage compliance here exhibits an undamped resonant behavior in which the compliance tends to infinity as ω_c is approached from below. The negative value of dynamic compliance above ω_c refers to the 180° phase shift between force and displacement of a lumped system. Structural members exhibit such resonances as described in §3.5; the Kramers–Kronig relations make no distinction between continuum properties and structural ones. Homogeneous materials do not exhibit resonances in their mechanical properties. In composite materials (which are heterogeneous) there is the possibility of microresonance of the structural elements; given the typical size scales involved such behavior could only occur at ultrasonic frequencies above 1 MHz.

Optical properties of materials (specifically the electric polarizability) are also governed by the Kramers–Kronig relations. Materials can exhibit resonant absorption of light due to microresonance on the atomic scale. For example, transparent materials such as glass and water exhibit resonant absorption of ultraviolet light due to resonance of electrons in the material. These materials are transparent (they have a small loss term) to visible light, which has lower frequencies than that of the resonant absorption. The dispersion of visible light manifests itself as the rainbow and in the spectral colors seen in light refracted through diamonds.

Problems

3.1 Find $\tan \delta$ if the relaxation function is $E(t) = At^{-n}$.

3.2 Why do tires become warm after you have driven on the highway?

3.3 Estimate roughly the loss tangent of your ear.

3.4 Show that

$$J'' = \frac{1}{E'} \frac{\tan \delta}{1 + \tan^2 \delta}.$$

Show that

$$E' = \frac{1}{J'} \frac{1}{1 + \tan^2 \delta}.$$

3.5 What would happen if engineering materials were in fact perfectly elastic ($\tan \delta = 0$)? What would happen if the earth and all the solid matter on it were perfectly elastic?

3.6 Why are marching soldiers instructed to "break step" when crossing bridges?

3.7 Suppose two sinusoids have a phase difference ξ. What is the phase difference between their Fourier transforms?

3.8 It is said that some opera singers are able to shatter a glass by singing the correct note. Is this possible? How? Does it matter what kind of glass is used?

3.9 Why are there tuning forks but no tuning knives or tuning spoons?

3.10 Compare the detailed shape of the resonance peak for a lumped system and for a distributed system. This can be done by plotting both curves for a given loss tangent or by making a perturbation analysis.

3.11 Consider an underdamped lumped resonating system. The envelope of the displacement amplitude is a smooth exponential. For comparison, calculate the rate of decay of the total mechanical energy, considered as the sum of kinetic and potential energies [P3.11.1]. Must the decay of mechanical energy be uniform? Discuss.

3.12 Show that

$$|E^*| = E'\sqrt{1 + \tan^2 \delta}.$$

3.13 Show that

$$\sigma(t) = |E^*(\omega)|\varepsilon_0 e^{i(\omega t + \delta)} \text{ with } E^* \equiv E' + i E''.$$

3.14 What is the shape of the stress–strain diagram for a material under harmonic load as a natural frequency is approached?

3.15 Show that a single relaxation time process gives a semicircle in a Cole–Cole plot.

3.16 In §3.4, the principles of nonnegative stored energy and nonnegative rate of dissipation of energy were discussed in connection with inequalities for the dynamic functions. Must we in fact have $dE'(\omega)/d\omega \geq 0$? Hint: consider the *structural* compliance of a system which exhibits resonance. Can materials exhibit resonance on a microstructural scale?

3.17 Consider a viscoelastic spacecraft spinning in outer space and neglect any gravitational pull from other objects. Is angular momentum conserved? Is kinetic energy conserved? Suppose that the spin axis does not exactly coincide with any principal axis of inertia, so that there is some wobble.

3.18 What conclusions can be drawn about a system if oscillations of decreasing magnitude are observed following a step load to a system?

3.19 Determine tan δ from the stress–strain loop in Fig. 3.2, (a) from the width of the loop, and (b) from the ratio of dissipated to stored energy density. Compare the results.

References

3.3.1 Kronig, R., "On the theory of dispersion of X-rays", *J. Opt. Soc. Am.* 12, 547, 1926.

3.3.2 Kronig, von R. de L. and Kramers, H. A., "Absorption and dispersion in X-ray spectra" (Zur Theorie der Absorption und dispersion in den Röntgenspektren, in German), *Z. Phys.* 48, 174–179, 1928.

3.3.3 Kubo, R. and Ichimura, M., "Kramers Kronig relations and sum rules", *J. Math Phys.* 13, 1454–1461, 1972.

3.3.4 Post, E. J., *Formal Structure of Electromagnetics*, North-Holland, Amsterdam, 1962.

3.3.5 Hu, B., "Kramers Kronig in two lines", *Am. J. Phys.* 57, 821, 1989.

3.4.1 Graesser, E. J. and Wong, C. R., "The relationship of traditional damping measures for materials with high damping capacity: a review", in *M³D: Mechanics and Mechanisms of Material Damping*, ed. V. K. Kinra and A. Wolfenden, ASTM, Philadelphia, PA, 1992, ASTM STP 1169.

3.4.2 Nashif, A. D., Jones, D. I. G., and Henderson, J. P., *Vibration Damping*, J. Wiley, NY, 1985.

3.4.3 Lazan, B. L., *Damping of Materials and Members in Structural Mechanics*, Pergamon, NY, 1968.

3.4.4 Ungar, E. E. and Kerwin, E. M., "Loss factors of viscoelastic systems in terms of energy concepts", *J. Acoust. Soc. Am.* 34, 954–957, 1962.

3.4.5 Christensen, R. M., "Restrictions upon viscoelastic relaxation functions and complex moduli", *Trans. Soc. Rheol.* 16, 603–614, 1972.

3.4.6 Gurtin, M. E. and Herrera, I., "On dissipation inequalities and linear viscoelasticity", *Q. Appl. Math* 23, 235–245, 1965.

3.5.1 Gottenberg, W. G. and Christensen, R. M. "An experiment for determination of the material property in shear for a linear isotropic viscoelastic solid", *Int. J. Eng. Sci.* 2, 45–56, 1964.

3.6.1 Main, I. G., *Vibrations and Waves in Physics*, Cambridge University Press, England, 1978.

3.6.2 Lathi, B. P., *Linear Systems and Signals*, Berkeley-Cambridge Press, Carmichael, CA, 1992.

3.7.1 Thurston, R. N., "Wave propagation in fluids and normal solids", in *Physical Acoustics*, ed. E.P. Mason, Vol. 1A, Academic, NY, 1964, pp. 1–110.

3.7.2 Bohn, E., *The Transform Analysis of Linear Systems*, Addison Wesley, Reading, MA, 1963.

3.9.1 Main, I. G., *Vibrations and Waves in Physics*, Cambridge, 1978.

3.9.2 Lakes, R. S. and Katz, J. L., "Viscoelastic properties of wet cortical bone III. A non-linear constitutive equation", *J. Biomechanics* 12, 689–698, 1979.

3.9.3 Lockett, F. J. and Gurtin, M. E., "Frequency response of nonlinear viscoelastic solids", Brown University Technical Report, NONR 562(10), NONR 562(30), 1964.

3.9.4 Lockett, F. J., *Nonlinear Viscoelastic Solids*, Academic, NY, 1972.

P3.11.1 Karlow, E. A., "Ripples in the energy of a damped harmonic oscillator", *Am. J. Phys.* 62, 634–636, 1994.

chapter four

Conceptual structure of linear viscoelasticity

§4.1   Introduction

The purpose of this chapter is to introduce the spectra, to derive approximate interrelations between the viscoelastic functions, and to assemble all the viscoelastic functions with their interrelations, with the aim of developing physical insight regarding the structure of viscoelasticity theory.

Thus far, we have considered the creep compliance $J(t)$ and the relaxation modulus $E(t)$ as the transient functions; and the storage modulus $E'(\omega)$, the loss modulus $E''(\omega)$, and the loss tangent $\tan \delta(\omega)$ as the dynamic functions. These viscoelastic functions are directly measurable, but different types of experiment, and even different types of apparatus, are used to determine each one.

The viscoelastic functions are interrelated. Let us recapitulate the salient interrelations obtained in previous sections. We have, in §2.4, found the exact relationship, a convolution, between the relaxation modulus $E(t)$ and the creep compliance $J(t)$:

$$\int_0^t J(t-\tau)\, E(\tau)\, d\tau = \int_0^t E(t-\tau)\, J(\tau)\, d\tau = t.$$

The complex dynamic modulus $E^* = E' + iE''$ is related in a simpler way (§3.2) to the complex dynamic compliance $J^* = J' - iJ''$: by an inverse.

$$J^* = \frac{1}{E^*}.$$

The dynamic moduli $E'(\omega)$ and $E''(\omega)$ are related to the relaxation modulus $E(t)$ by one-sided Fourier transforms, as developed in §3.2:

$$E'(\omega) \equiv E(\infty) + \omega \int_0^\infty \{E(t') - E(\infty)\} \sin \omega t' dt'.$$

$$E''(\omega) \equiv \omega \int_0^\infty \{E(t') - E(\infty)\} \cos \omega t' dt'.$$

The real part and the imaginary part of a complex viscoelastic function are related by the Kramers–Kronig relations. For the compliance, they are as follows:

$$J'(\omega) - J'(\infty) = \frac{2}{\pi} \left[\wp \int_0^\infty \frac{\varpi J''(\varpi)}{\varpi^2 - \omega^2} d\varpi \right],$$

$$J''(\omega) = \frac{2\omega}{\pi} \left[\wp \int_0^\infty \frac{J'(\varpi) - J'(\infty)}{\omega^2 - \varpi^2} d\varpi \right].$$

Interrelations among the viscoelastic functions are of use in converting experimental results from one form to another. The integral forms can be evaluated with the aid of a computer; however, the approximate relations developed below continue to be useful for a quick examination of data, for some special situations, and particularly in the development of physical insight.

§4.2   Spectra in linear viscoelasticity

§4.2.1   Definition: exact interrelations

Materials governed by a simple relaxation mechanism may be approximated by a single exponential or a unique sum of exponentials. However, such cases are rare; most real materials exhibit behavior which covers many decades of time scale and which cannot be described in this way. If the governing mechanisms are nevertheless considered to give rise to exponential components, one may envisage a relaxation function E(t) composed of a distribution of exponentials, as follows:

$$E(t) - E_e = \int_{-\infty}^\infty H(\tau) e^{-t/\tau} d\ln\tau = \int_0^\infty \frac{H(\tau)}{\tau} e^{-t/\tau} d\tau, \qquad (4.2.1)$$

in which $H(\tau)$ is called the relaxation spectrum and $E_e = \lim_{t\to\infty} E(t)$ is the "equilibrium" modulus, the stiffness after an infinite time of relaxation. The above equation *defines* the spectrum $H(\tau)$. Similarly, the creep compliance $J(t)$ may be written:

$$J(t) = J(0) + \frac{t}{\eta} + \int_0^\infty \frac{L(\tau)}{\tau} \left(1 - e^{-t/\tau}\right) d\tau, \qquad (4.2.2)$$

in which η is the asymptotic viscosity (which is infinite in solids) and $L(\tau)$ is the retardation spectrum. These definitions follow those of Ferry [1.6.18]; some authors, e.g., Christensen [1.6.1] define the spectra which differ from the above by a factor of τ in the integrand.

The dynamic functions are given in terms of the relaxation spectrum as follows, as presented in the classic works of Ferry [1.6.18] and Nowick and Berry [1.6.23]:

$$E'(\omega) = E_e + \int_0^\infty \frac{H(\tau)}{\tau} \frac{\omega^2 \tau^2}{1 + \omega^2 \tau^2} d\tau, \qquad (4.2.3)$$

$$E''(\omega) = \int_0^\infty \frac{H(\tau)}{\tau} \frac{\omega \tau}{1 + \omega^2 \tau^2} d\tau. \qquad (4.2.4)$$

The spectra are related to each other as follows [1.6.18, 1.6.23], with the integrals considered as principal parts. For a viscoelastic liquid [1.6.18], η represents an asymptotic viscosity; for solids, the inverse of this viscosity vanishes.

$$L(\tau) = \frac{H(\tau)}{\left[E_e - \int_{-\infty}^\infty \frac{H(u)}{\frac{\tau}{u} - 1} d\ln u \right]^2 + \pi^2 H^2(\tau)}. \qquad (4.2.5)$$

$$H(\tau) = \frac{L(\tau)}{\left[J(0) + \int_{-\infty}^\infty \frac{L(u)}{1 - \frac{u}{\tau}} d\ln u - \frac{\tau}{\eta} \right]^2 + \pi^2 L^2(\tau)}. \qquad (4.2.6)$$

If $H(\tau)$ can be written as a sum of delta functions, then $L(\tau)$ also consists of a sum of delta functions, albeit with different time constants. This can be demonstrated by applying a delta function method to the exact transformation equations [4.2.1]; however, it is more straightforward to convert H(τ) to a sum of exponentials in E(t), convert E(t) to J(t) and then to L(τ).

§4.2.2 *Particular spectra*

A variety of particular spectra have been developed for the description of various materials. One which has been useful is the *lognormal spectrum* [4.2.2]

$$H(\tau) = \frac{b}{\sqrt{\pi}} e^{-b^2 z^2}, \qquad (4.2.7)$$

in which z = log τ/τ_m. It describes materials in which a molecular relaxation process operates in a range of atomic environments distributed in a Gaussian fashion about a mean value. The damping peak due to this spectrum is broader than a Debye peak, and the breadth depends on the value of the parameter b. Measurable functions for this spectrum are calculated numerically rather than analytically.

The "*box*" spectrum [4.2.3] is constant over a range of time values:

$$H(\tau) = H_0 \qquad \text{for } \tau_1 \le \tau \le \tau_2, \qquad (4.2.8)$$

$$H(\tau) = 0, \qquad \text{for } \tau > \tau_2, \tau < \tau_1.$$

The box spectrum describes certain polymers and is useful in that analytical expressions are available for both the transient and dynamic functions associated with it. Specifically,

$$E(t) = E_e + H_0[Ei(-t/\tau_1) - Ei(-t/\tau_2)], \qquad (4.2.9)$$

in which

$$Ei(y) \equiv -\int_y^\infty e^{-x}/x \, dx$$

is the exponential integral function.

$$E'(\omega) = E_e + \frac{H_0}{2} \ln \frac{1+\omega^2\tau_2^2}{1+\omega^2\tau_1^2}, \qquad (4.2.10)$$

$$E''(\omega) = H_0\left(\tan^{-1}\omega\tau_2 - \tan^{-1}\omega\tau_1\right). \qquad (4.2.11)$$

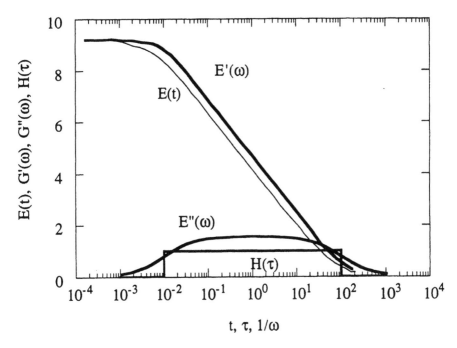

Figure 4.1   Box spectrum covering four decades of time t and measurable modulus functions E(t), E'(ω), and E''(ω). Arbitrary units of stiffness on the ordinate; arbitrary units of time on the abscissa.

A box spectrum covering four decades of time and measurable modulus functions are plotted in Fig. 4.1. E'' is nearly constant over a wide range of frequency; E' and E(t) vary as the log of frequency and of time, respectively, over a wide range.

The *"wedge"* spectrum [4.2.3] is wedge-shaped in a log–log plot. It has been used by Tobolsky [4.2.3] to model the glass–rubber transition of poly-isobutylene.

$$H(\tau) = k\tau^{-1/2} \qquad \text{for } a \le \tau \le b,$$

$$H(\tau) = 0, \qquad \text{for } \tau > b, \tau < a.$$

For the wedge relaxation spectrum

$$H(\tau) = k\tau^{-1/2} \qquad \text{for } a \le \tau \le b,$$

integration of the exact interrelation gives the following retardation spectrum:

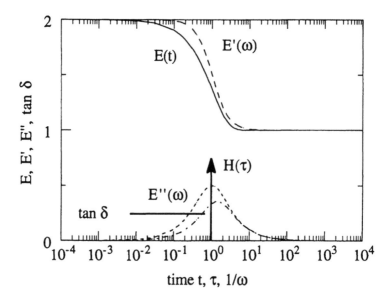

Figure 4.2   Viscoelastic functions for a delta function relaxation spectrum at time $\tau = 1$. Arbitrary units of time. These functions correspond to a standard linear solid. E'' and tan δ have Debye peak functional dependence.

$$L(\tau) = \cfrac{1}{H(\tau)\left\{\left(\ln\dfrac{\sqrt{b}+\sqrt{\tau}}{\sqrt{b}-\sqrt{\tau}}\dfrac{\sqrt{\tau}-\sqrt{a}}{\sqrt{\tau}+\sqrt{a}}\right)^2 + \pi^2\right\}} \qquad \text{for } a \le \tau \le b,$$

$$L(\tau) = 0 \qquad\qquad\qquad\qquad\qquad\qquad \text{for } \tau \le a \text{ and } \tau \le b.$$

If the original wedge spectrum covers many decades,

$$L(\tau) \approx \frac{1}{\pi^2 H} = \frac{\tau^{1/2}}{\pi^2 k},$$

except over about half a decade of time scale near the spectrum edges. Here $L(\tau)$ increases with time over most of the range in contrast to $H(\tau)$ which decreases, again illustrating the emphasis on longer times in the retardation spectrum in comparison with the relaxation spectrum.

We may also consider a *delta function* spectrum

$$\frac{H(\tau)}{\tau} = A\delta(\tau - \tau_1), \qquad\qquad\qquad\qquad (4.2.12)$$

in which A is a constant.

The measurable functions are then those of the standard linear solid. We have

$$E(t) = E_e + Ae^{-t/\tau_1},$$
(4.2.13)

$$E' = E_e + A\frac{\omega^2\tau_1^2}{1+\omega^2\tau_1^2},$$
(4.2.14)

$$E'' = A\frac{\omega\tau_1}{1+\omega^2\tau_1^2}.$$
(4.2.15)

By plotting these, assuming $E_e = 1$ and $A = 1$ in units of stress vs. log time (and log $1/\omega$ to view the dynamic functions on the same scale) for $\tau_1 = 1$ unit of time, we obtain Fig. 4.2. Compare with Fig. 4.1. Observe also that most of the change in the measurable functions for a standard linear solid occurs within one decade of time or frequency.

§4.3   Approximate interrelations among viscoelastic functions

Many approximate interrelations have been developed, [4.3.1, 4.3.2]; they display the connection between properties in an accessible way.

§4.3.1   Interrelations involving the spectra

§4.3.1.1   Relaxation spectrum from relaxation modulus
The exact relation is the definition of the spectrum,

$$E(t) - E_e = \int_0^\infty \frac{H(\tau)}{\tau}e^{-t/\tau}\,d\tau,$$

but one may desire to calculate the spectrum from the relaxation modulus since the latter is measurable. Since no exact analytical inversion formula is available, approximate analytical forms or numerical methods are used.

The box spectrum may be used to obtain useful approximations. By differentiating $E(t)$ with respect to t in Eq. 4.2.9 and making use of Leibnitz' rule,

$$t\frac{dE(t)}{dt} = \frac{dE(t)}{d\ln t} = H_0\left(e^{-t/\tau_1} - e^{-t/\tau_2}\right).$$
(4.3.1)

If the spectrum is so broad that $\tau_1 \ll \tau_2$ and if $\tau_1 \ll t \ll \tau_2$, then $e^{-t/\tau_1} \approx 0$ and $e^{-t/\tau_2} \approx 1$. Then

$$H_0 \approx -\frac{dE(t)}{d\ln t},\tag{4.3.2}$$

with the approximation improving as the spectrum becomes arbitrarily broad. If the spectrum is slowly varying, we have

$$H(\tau) \approx -\frac{dE(t)}{d\ln t}\Big|_{t=\tau}.\tag{4.3.3}$$

To obtain an approximation by a different approach, plot $e^{-t/\tau}$ vs. τ on a logarithmic scale. Observe that $e^{-t/\tau} \approx 1$ for $\tau \gg t$ and $e^{-t/\tau} \approx 0$ for $\tau \ll t$. The transition in $e^{-t/\tau}$ covers about one logarithmic decade; if the spectrum $H(\tau)$ varies much more slowly than $e^{-t/\tau}$, the latter may be approximated by a step function. The symbol H is used for the relaxation spectrum; the Heaviside step function is distinguished by use of the script letter $\mathcal{H}$. They also differ by the fact that the spectrum $H(\tau)$ is always a function of one time quantity τ while the step function $\mathcal{H}$ in the following is a function of the difference between two quantities. So [1.6.18],

$$E(t) - E_e = \int_0^\infty \frac{H(\tau)}{\tau}\mathcal{H}(\tau - t)\,d\tau = \int_t^\infty \frac{H(\tau)}{\tau}\,d\tau,\tag{4.3.4}$$

so

$$H(\tau) \approx -t\frac{dE(t)}{dt} = -\frac{dE(t)}{d\ln t}\Big|_{t=\tau}.\tag{4.3.5}$$

This is known as the Alfrey approximation [4.3.3].

§4.3.1.2 *Relaxation spectrum from dynamic storage modulus*
The box spectrum may be used to approximately relate E' to $H(\tau)$. Let us calculate $\omega\, dE'/d\omega = dE'/d\ln\omega$ since a semilog plot of E' vs. ω shows a straight line far from the spectrum edges in Fig. 4.1. The condition $\tau_1 \ll t \ll \tau_2$ corresponds to $\omega\tau_1 \ll 1 \ll \omega\tau_2$, so $\omega\, dE'/d\omega \approx H_0$. Again we recognize that the assumption of $t = 1/\omega$ far from the spectrum edges corresponds to a slowly varying spectrum. So the final approximation is

$$H(\tau) \approx \omega\frac{dE'}{d\omega} = \frac{dE'}{d\ln\omega}\Big|_{1/\omega=\tau}.\tag{4.3.6}$$

To develop this approximation by direct construction, the portion $(\omega^2\tau^2)/(1 + \omega^2\tau^2)$ of the integrand in Eq. 4.2.3 is approximated as a step function in that it varies vs. log time much more rapidly than does the spectrum which is assumed to vary slowly. To visualize this, plot $(\omega^2\tau^2)/(1 + \omega^2\tau^2)$ on a log scale covering many decades; compare Fig. 4.2 with Fig. 4.1.

$$E'(\omega) - E_e = \int_0^\infty \frac{H(\tau)}{\tau}\frac{\omega^2\tau^2}{1+\omega^2\tau^2}\,d\tau \approx \int_0^\infty \frac{H(\tau)}{\tau}\mathcal{H}\left(\tau - \frac{1}{\omega}\right)d\tau$$

$$= \int_{1/\omega}^\infty \frac{H(\tau)}{\tau}\,d\tau.$$

(4.3.7)

$$H(\tau) \approx \omega\frac{dE'(\omega)}{d\omega} = \frac{dE'(\omega)}{d\ln\omega}\Big|_{\omega=1/\tau}.$$

(4.3.8)

§4.3.1.3   Relaxation spectrum from dynamic loss modulus

The box spectrum may be used to approximately relate E'' to $H(\tau)$. The condition $\tau_1 \ll t \ll \tau_2$ gives $E''(\omega) \approx \pi H_0/2$. For a slowly varying spectrum we make the correspondence $\tau = 1/\omega$, and the approximation

$$H(\tau) \approx \frac{2}{\pi}E''(\omega)\Big|_{1/\omega=\tau}.$$

(4.3.9)

E'' may be related to $H(\tau)$ by direct construction. The kernel in Eq. 4.2.4 has the form of a hump covering about one decade. If the spectrum varies sufficiently slowly, the hump may be regarded in comparison as a delta function [4.3.1]. The "area" of the hump, i.e., the integral of $[(\omega\tau)/(1 + \omega^2\tau^2)]$, is $\pi/2$.

Hence

$$E''(\omega) = \int_0^\infty \frac{H(\tau)}{\tau}\frac{\omega\tau}{1+\omega^2\tau^2}\,d\tau \approx \int_0^\infty \frac{H(\tau)}{\tau}\tau\frac{\pi}{2}\delta\left(\tau - \frac{1}{\omega}\right)d\tau$$

$$= \frac{\pi}{2}H\left(\tau = \frac{1}{\omega}\right),$$

(4.3.10)

so

$$H(\tau) \approx \frac{2}{\pi}E''(\omega)\Big|_{\omega=1/\tau}.$$

(4.3.11)

By referring to Fig. 4.1, which shows the viscoelastic functions derived from exact interrelations for a box spectrum, this approximation is seen to be accurate away from the spectrum edges. If the spectrum contains a sharp peak, as in Fig. 4.2, E'' is a poor approximation for the spectrum $H(\tau)$. Since a slowly varying spectrum was assumed in obtaining the approximation, there should be no surprise.

§4.3.1.4   *Retardation spectrum from creep or dynamic compliance*
Procedures similar to the above give approximate relations involving the creep behavior of solids [1.6.23].

$$L(\tau) \approx \frac{d\,J(t)}{d\ln t}\bigg|_{t=\tau}. \tag{4.3.12}$$

$$L(\tau) \approx -\frac{d\,J'(\omega)}{d\ln\omega}\bigg|_{\omega=1/\tau}. \tag{4.3.13}$$

$$L(\tau) \approx \frac{2}{\pi}J''(\omega)\bigg|_{\omega=1/\tau}. \tag{4.3.14}$$

§4.3.2   *Interrelations involving measurable functions*

§4.3.2.1   *Transient and dynamic moduli*
By setting the expressions Eqs. 4.3.2, 4.3.8, and 4.3.11 for the spectra equal, one obtains approximate interrelations among the moduli.

$$\frac{2}{\pi}E''(\omega)\bigg|_{\omega=1/\tau} \approx -\frac{d\,E'(\omega)}{d\ln\omega}\bigg|_{\omega=1/\tau} \approx -\frac{d\,E(t)}{d\ln t}. \tag{4.3.15}$$

The first approximate equality provides an approximation to the Kramers–Kronig relations. Moreover, from the last two of Eq. 4.3.15, to lowest order,

$$E'(\omega)\big|_{\omega=1/\tau} \approx E(t). \tag{4.3.16}$$

Similarly [1.6.23]

$$J'(\omega)\big|_{\omega=1/\tau} \approx J(t).$$

This approximation (Eq. 4.3.16) may be compared with the curves in Fig. 4.1, which neglects E_e that would represent a baseline. The percentage of error associated with this approximation becomes smaller as E_e becomes larger in relation to the total dispersion. The approximation is therefore most accurate for weakly viscoelastic materials, those in which the dispersion in stiffness is small compared with the stiffness.

Approximations of improved accuracy and no greater complexity can be derived with a little additional effort from the interrelations between transient and dynamic properties [1.6.1]. For example,

$$E(t) \approx E'(\omega)\big|_{\omega=2/\pi t}, \tag{4.3.17}$$

as considered in Example 4.2.

The transient and dynamic moduli and the spectrum are linked by the following exact equation [4.3.1]:

$$E'(\omega)_{\omega=1/t} - E(t) = \int_{-\infty}^{\infty} H(\tau)\left[\frac{1}{1+(t/\tau)^2} - e^{-t/\tau}\right] d\ln\tau.$$

The quantity in the [ ] brackets forms a hump covering about one decade of time scale. One can approximate the relation by assuming $H(\tau) = k\tau^{-m}$ over the range of time in which this peak differs significantly from zero:

$$E(t) \approx E'(\omega)_{\omega=1/t} - H\left\{\frac{\pi}{2}\csc\frac{m\pi}{2} - \Gamma(m)\right\}, \text{ for } -1 < m < 2.$$

For $m = 0$ the quantity in the { } brackets converges to the Euler–Mascheroni constant ≈ 0.5772. If the spectrum is known approximately, the transient and dynamic moduli can be related. Moreover, an approximate relation for H in terms of G'' can be obtained by substituting

$$H(\tau) = k\tau^{-m}$$

in

$$E''(\omega) = \int_{-\infty}^{\infty} H(\tau)\frac{\omega\tau}{1+\omega^2\tau^2} d\ln\tau.$$

Integration, again approximating the hump as a delta function [4.3.1] gives

$$E''(\omega) \approx H\frac{\pi}{2}\sec\frac{m\pi}{2}, \text{ for } -1 < m < 1.$$

The zero-order approximation corresponds to the case $m = 0$,

$$H \approx \frac{2}{\pi}E'',$$

as seen in Eq. 4.3.11.

§4.3.2.2   Creep compliance and relaxation modulus

We have considered previously in §2.6.4 an exact interrelation for a material with a relaxation modulus of power law form $E(t) = At^{-n}$. The value n is the slope of the relaxation curve plotted on a log–log scale. The relation is (Eq. 2.6.51)

$$J(t) = \frac{1}{A\Gamma(1-n)\Gamma(n+1)}t^{n}. \qquad (4.3.18)$$

The gamma function $\Gamma(x)$ is defined for $n > 0$ as

$$\Gamma(x) = \int_{0}^{\infty} t^{x-1}e^{-t}\, dt. \qquad (4.3.19)$$

The identity $\Gamma(x)\Gamma(1-x) = \pi/(\sin \pi x)$ may be used to obtain the following [1.6.18, 4.3.1]:

$$E(t) = \frac{\sin n\pi}{n\pi}\frac{1}{J(t)}, \qquad (4.3.20)$$

with $n < 1$. The expression is exact for a power law material. For an arbitrary material, one can fit a power law to the behavior for a particular time; the fit will be approximate over a limited interval of time scale, and Eq. 4.3.20 becomes approximate. Moreover, for small n, the expression converges to the elastic relation $E = 1/J$.

§4.3.2.3   The loss tangent

The loss tangent is related to the slope of the log of moduli or compliances on log frequency or log time scales. The approximations are based on the *exact* relationship for a power law relaxation, $E(t) = At^{-n}$, for which creep is also a power law and the loss angle is (Example 4.1)

$$\delta = \frac{n\pi}{2}. \qquad (4.3.21)$$

For an arbitrary relaxation function a power law may be fitted for a particular value of time. The loss tangent obtained is then an approximation. The approximations are most satisfactory if the slope is small. Moreover,

$$\tan \delta = \frac{E''}{E'} \approx \frac{\pi}{2}\frac{dE'}{d\ln\omega}\frac{1}{E'}. \qquad (4.3.22)$$

from Eq. 4.3.15, but

$$\frac{d\ln E'}{d\ln\omega} = \frac{1}{E'}\frac{dE'}{d\ln\omega},$$

and so [1.6.23]

$$\tan\delta \approx \frac{\pi}{2}\frac{d\ln E'}{d\ln\omega}. \tag{4.3.23}$$

Similarly, the loss tangent is related to the slope of the relaxation [1.6.23] curve by the following:

$$\tan\delta \approx -\frac{\pi}{2}\frac{d\ln E(t)}{d\ln t}\Big|_{t=1/\omega}. \tag{4.3.24}$$

A different approach to this derivation is given by Zener [1.6.22].
   The loss tangent is related to the slope of the creep curve by the following:

$$\tan\delta \approx \frac{\pi}{2}\frac{d\ln J(t)}{d\ln t}\Big|_{t=1/\omega}. \tag{4.3.25}$$

The loss tangent is related to the slope of the dynamic compliance [1.6.23] by the following:

$$\tan\delta \approx -\frac{\pi}{2}\frac{d\ln J'}{d\ln\omega}. \tag{4.3.26}$$

§4.3.2.4   *Approximate interrelations among spectra*
Smith [4.3.1] obtains the following approximate relations between the spectra with $\tau = \omega^{-1}$.

$$L(\tau) \approx \frac{H(\tau)}{\{G'(\omega)-G''(\omega)+1.37H(\tau)\}^2 + \pi^2 H^2(\tau)},$$

$$H(\tau) \approx \frac{L(\tau)}{\{J'(\omega)-J''(\omega)+1.37L(\tau)\}^2 + \pi^2 L^2(\tau)}.$$

These relations require knowledge of the dynamic properties as well as a spectrum.

Tschoegl [1.6.13] obtains the following, based on the exact form (Example 4.3) for the box spectrum

$$L(\tau) \approx \frac{1}{H(\tau)} \frac{1}{\left[\dfrac{G_e}{H(\tau)} + \ln\dfrac{b-\tau}{\tau-a}\right]^2 + \pi^2}, \, a \leq \tau \leq b,$$

$$L(\tau) = 0, \, \tau < a, \text{ and } \tau > b.$$

§4.3.2.5   Higher order approximations

Further approximate interrelationships have been given [1.6.18, 4.3.4]. It is possible to achieve improved accuracy by the use of higher order derivatives [4.3.4]. Specifically, the spectrum is written as a series expansion involving derivatives of a measurable function, and then substituted into the defining equation for the spectrum.

Results include the following [4.3.5]:

$$H(\tau) \approx \frac{2}{\pi} \left\{ E''(\omega) - \frac{d^2 E''}{d(\ln\omega)^2} \right\} \Big|_{\omega=1/\tau}. \qquad (4.3.27)$$

$$H(\tau) \approx -\left\{ \frac{dE(t)}{d\ln t} - \frac{d^2 E}{d(\ln t)^2} \right\} \Big|_{t=2\tau}. \qquad (4.3.28)$$

Caution is required with these, since all experimental data contain some noise, and differentiation of the data increases the noise. Spectra which vary abruptly with time may be extracted via numerical methods [4.3.6].

§4.3.2.6   Summary

The approximate interrelations are useful for developing our intuition concerning the connection between the viscoelastic functions, as well as for calculations. Here we see that viscoelastic loss is always associated with creep, relaxation, and dispersion of the dynamic stiffness; the slope of the modulus or compliance curves vs. log time or log frequency is approximately related to the loss.

§4.4   Conceptual organization of viscoelastic functions

In this section we collect and discuss the viscoelastic functions studied so far according to the conceptual view of Gross [1.6.9].

The dynamic functions $E^* = E' + i\,E''$ and $J^* = J' - i\,J''$ are considered to be the ones most closely related to direct experience since they are directly connected with observable phenomena (Fig. 4.3). The rate of decay of

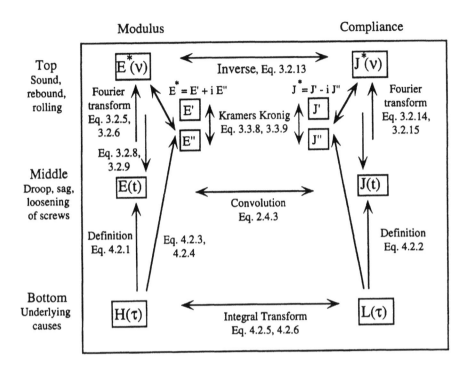

Figure 4.3 Conceptual organization of the viscoelastic functions.

vibration of a bell or bar after it is struck, acoustic "deadness" or ringing (Table 4.1), and the degree of rebound of a ball are related to the phase angle δ between stress and strain. The fundamental resonant frequency of a struck bell or bar depends on the dynamic modulus as well as the object's size and shape. Indeed, some of the first experiments done in childhood in exploring the degree to which round objects can be bounced or rolled represent an informal measure of tan δ; the formal relationship is developed later in Chapter 10. The dynamic functions are therefore considered to be at the top level in this conceptual scheme. The relation between the dynamic modulus and the dynamic compliance is also the most easily grasped mathematically: it is simply an inverse as in the case of elasticity. The difference is that the viscoelastic dynamic functions are complex numbers, while the elastic moduli are real numbers.

Dynamic functions of frequency also describe several optical phenomena which are directly experienced. The phase angle between electric field and electric displacement governs optical phenomena such as color (related to absorption of light at different frequencies). Dispersion of light (the rainbow; the rainbow effect in gemstones, referred to as "fire") arises from dependence of the refractive index or optical compliance of the material on frequency.

The transient properties, relaxation and creep, are considered one level down in terms of direct perception (Fig. 4.3). The associated phenomena of sag and droop can be directly perceived, but such motion is only perceived

Table 4.1 Perceived Sound of an Object with a Resonant
Frequency in the Acoustic Range, Following an Impact
Giving Rise to Free Decay of Vibration

tan δ	Cycles for decay to one tenth of initial amplitude	Sound
≤0.001	≥730	"Clang", approaching pure tone
0.01	73	"Bong": clear pitch
0.1	7.3	"Bunk": discernible pitch
0.5	1.5	"Thud": almost without pitch

Source: Adapted from Kerwin and Ungar [4.4.1].

if it lies within our attention span in the time domain. Rapid creep, faster than 1 sec, is not perceived by us as droop. Slow processes, such as the flow of glaciers and of rock, and even relaxations which occur in hours or days are not immediately apparent. The transient functions are therefore considered to be at the middle level. The interrelation between transient modulus and compliance is more complicated than in the dynamic case: a convolution. The transient functions were introduced first as fundamental, primarily because most engineers or scientists interested in viscoelasticity initially prefer the time domain to the frequency domain.

The spectra, which relate to the underlying causes for the phenomena, represent the bottom level since they are not directly observable; they are even difficult to extract accurately from experimental data. The interrelation between the spectra is a complicated integral transform. The spectra have a significance related to causes only if the fundamental relaxation process is an exponential one in time, $e^{-t/\tau}$. As we shall see in Chapter 8, exponentials occur naturally in a variety of processes. Some relaxation mechanisms, however, give rise naturally to a power law relaxation function; a spectrum of exponentials would be less appropriate for materials governed by such mechanisms. The spectra are nonetheless useful for visualization since in the absence of local microinertial effects, no relaxation can be more abrupt than a Debye relaxation or single relaxation time process; this corresponds to a spectrum in the form of a delta function.

Viscoelastic response is a single physical reality which can be manifest in many ways, some of which are conceptualized in terms of the viscoelastic functions. A thorough understanding of viscoelasticity entails a simultaneous appreciation of all these manifestations and their analytical representations. Each viscoelastic function, if known over the full range of time or frequency, contains complete information regarding the material's linear behavior. Nevertheless, each viscoelastic function emphasizes a different aspect of the behavior. For example, in §2.6.1, a material with an exponential relaxation response was seen to exhibit a creep response with a *longer* time constant than that for relaxation. Moreover, as seen in Fig. 4.4, the creep response emphasizes longer term processes than does the relaxation response; the corresponding spectra contain the same difference in emphasis (see Example 4.3).

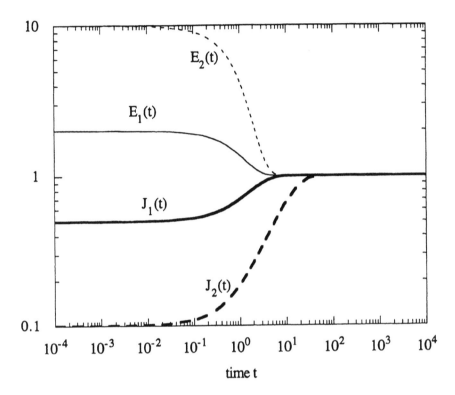

Figure 4.4 Equivalent relaxation and creep curves corresponding to a single relaxation time and a single retardation time, respectively, calculated via Eqs. 2.6.17, 2.6.24, and 2.6.25. $E_1(t)$ and $J_1(t)$ represent a factor of two change in stiffness in relaxation and creep, respectively. $E_2(t)$ and $J_2(t)$, represent a factor of ten change in stiffness in relaxation and creep, respectively. The transition in the creep curve appears to occur at a longer time than in the relaxation curve according to Eq. 2.6.25. Units of time, stiffness, and compliance are arbitrary.

There is a complementary relation between time and frequency as embodied in the Fourier transform. As with the other integral equations considered, the Fourier transform is nonlocal in that the value of a function at a particular frequency depends on the value of its transform at *all* values of time. All values of time are not equally weighted; we have considered approximations based on locality or on "weak" nonlocality associated with derivatives. In that vein, we may consider frequencies ν to be complementary to time according to $\nu = 1/2\pi t$.

The range of time and frequency permitted by mathematics is the full range from zero to infinity. The range of time and frequency permitted by physics is bounded. Mechanical disturbances in real solids are limited to frequencies below about 10^{13} Hz since the wavelength of a sound wave in a material cannot be shorter than the spacing between atoms. A portion of the range of times and frequencies of interest in science and engineering is shown in Fig. 4.5, with some of the salient phenomena. Under some circumstances

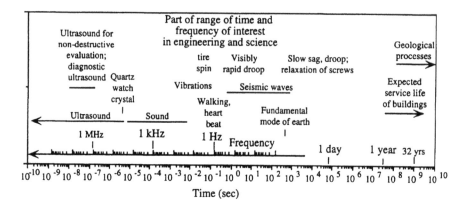

Figure 4.5   A portion of the range of time and frequency of interest in engineering and science: 20 decades.

material behavior over a range of 20 decades or more may be of interest. In geological problems and in the study of historical buildings, longer time scales than those shown (320 years) are relevant. Ultrasonic waves of frequency up to 10 or 20 MHz are commonly used in nondestructive testing of engineering materials and in medical diagnosis; higher frequencies are used in scientific research. It is therefore sensible to think in a logarithmic sense to appreciate the range of times and frequencies of interest.

§4.5   Summary

The physical reality of viscoelastic behavior may be represented mathematically by a variety of functions of time or frequency. Some of these, the dynamic functions, are regarded as most closely related to direct perception. All these functions are interrelated; most of the exact relationships are integral transforms. Simpler approximate interrelations are obtained under the assumption of slowly varying viscoelastic functions. In considering the full range of viscoelastic behavior we observe that 20 decades or more (a factor of 10^{20}) of time and frequency may be of interest in scientific study and engineering applications of viscoelastic materials.

Examples

Example 4.1

Suppose the relaxation function follows a power law: $E(t) = At^{-p}$. Determine the loss tangent.

Solution

Substitute $E(t)$ in the transform integrals, with $E_e = 0$, and make the substitution $\omega t' = z$.

$$E'(\omega) = \omega \int_0^\infty \frac{\sin \omega t'}{At'^p} dt' = \omega \int_0^\infty \frac{\sin z}{Az^p} \omega^p \frac{1}{\omega} dz.$$

This is a standard integral form, so

$$E'(\omega) = \omega^p \frac{\pi}{2\Gamma(p)\sin\frac{p\pi}{2}}.$$

Similarly

$$E''(\omega) = \omega^p \frac{\pi}{2\Gamma(p)\cos\frac{p\pi}{2}}.$$

$$\tan \delta(\omega) = \frac{E''(\omega)}{E'(\omega)} = \tan\frac{p\pi}{2}.$$

$$\delta = \frac{p\pi}{2}.$$

So the loss angle is proportional to the slope of the power law relaxation function on a log–log plot. The relationship is exact for power law relaxation, and may be used as an approximation if a portion of a relaxation curve can be fitted by a power law.

Example 4.2

Show that

$$E(t) \approx E'(\omega)\big|_{\omega=2/\pi t}.$$

Solution

The exact interrelation is an inverse one-sided Fourier transform [1.6.18]. In the absence of an equilibrium modulus,

$$E(t) = \frac{2}{\pi} \int_0^\infty E'(\omega) \frac{\sin \omega t}{\omega} d\omega. \tag{E4.2.1}$$

The quantity $\omega^{-1} \sin \omega t$ is a peaked function (Fig. E4.2), the integral of which is $\pi/2$. If $E'(\omega)$ varies sufficiently slowly, the peak will seem sharp by comparison, so the approximation

$$\omega^{-1} \sin \omega t \approx \frac{\pi}{2} \delta(\omega - \omega_0) \qquad \text{(E4.2.2)}$$

is sensible [1.6.1]. To find ω_0 in terms of t, recognize that ω_0 specifies the position of the delta function on the angular frequency axis. The function $\omega^{-1} \sin \omega t$ is peaked but not sharply peaked, so the "position" is diffuse. An effective position may be defined in terms of the first moment of this function, so

$$\omega_0 = \frac{\displaystyle\int_0^\infty \omega \frac{\sin \omega t}{\omega} \, d\omega}{\displaystyle\int_0^\infty \frac{\sin \omega t}{\omega} \, d\omega} \qquad \text{(E4.2.3)}$$

The integral in the numerator does not converge. However, multiply the integrand by $e^{-p\omega}$ which gives

$$\int_0^\infty e^{-p\omega} \sin \omega t \, d\omega = \frac{t}{p^2 + t^2}. \qquad \text{(E4.2.4)}$$

Taking the limit as $p \to 0$, gives $\omega_0 = (1/t)/(\pi/2)$ so $\omega_0 = 2/\pi t$. The sifting property of the delta function refers to the fact that $\delta(x - a)$ in an integral with respect to x sifts out the value of the multiplied function at $x = a$, as discussed in Appendix 1. Using the sifting property in the exact equation gives

$$E(t) \approx E'(\omega)\big|_{\omega = 2/\pi t} \qquad \text{(E4.2.5)}$$

as desired.

This is a better approximation than Eq. 4.3.16 as can be seen by examining Figs. 4.1 and 4.2.

Example 4.3

Determine the retardation spectrum $L(\tau)$ corresponding to the box spectrum of relaxation times $H(\tau) = \{H_0, a \leq \tau \leq b, \text{ and } 0, \tau < a, \tau > b\}$. Discuss the physical implications.

Solution

Use the exact interrelation between the spectra, Eq. 4.2.5.

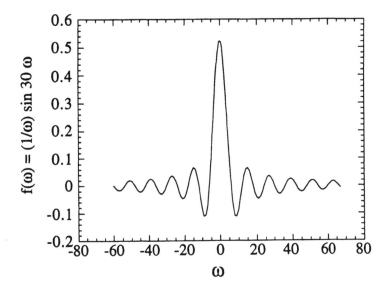

Figure E4.2 Plot of $\omega^{-1}\sin\omega t$.

$$L(\tau) = \frac{1}{H_0} \frac{1}{\left[\dfrac{G_e}{H_0} + \ln\dfrac{b-\tau}{\tau-a}\right]^2 + \pi^2}, a \le \tau \le b, \qquad (E4.3.1)$$

$$L(\tau) = 0, \tau < a, \text{ and } \tau > b.$$

Observe in plotting this function (Fig. E4.3) that while the assumed relaxation spectrum is flat, the retardation spectrum increases sharply with time [4.3.1]. Consequently, the slower viscoelastic processes are emphasized more in a creep test than in a relaxation test.

Example 4.4

Suppose that for a material the relaxation strength is 0.1. What is the maximum tan δ (a) if a Debye relaxation is assumed, and (b) if it is desired that tan δ be uniform over the acoustic range 20 Hz to 20 kHz? Discuss.

Solution

For the Debye form, maximum tan δ is

$$\tan\delta = \frac{1}{2}\frac{\Delta}{\sqrt{1+\Delta}} = 0.05.$$

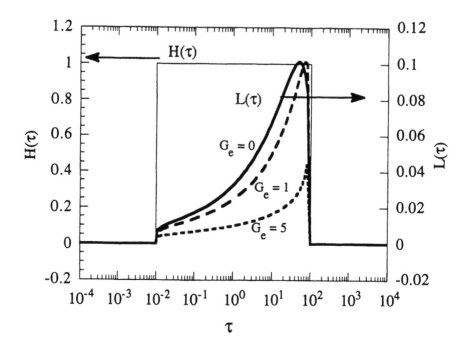

Figure E4.3   Plot of a box function relaxation spectrum $H(\tau)$ and the corresponding retardation spectrum $L(\tau)$.

For a constant loss from angular frequency ω_1 to ω_2, from Example 3.4,

$$J'(0) - J'(\infty) = \int_0^\infty J''(\omega)\, d\ln\omega \approx J''(\ln\omega_1 - \ln\omega_2). \qquad \text{(E4.4.1)}$$

This result is approximate since if tan δ is constant, J' must vary with frequency so J'' cannot be constant. Divide both sides by J' to obtain the relaxation strength $\delta J'/J'$.

$$\tan\delta \approx \frac{1}{6.91}\frac{\delta J'}{J'} = 0.0145. \qquad \text{(E4.4.2)}$$

For a particular given relaxation strength, the wider the range of frequency the smaller the magnitude of the damping.

Example 4.5

To what degree do viscoelastic materials remember the path by which they arrived at the present state of stress and strain? Does path memory depend on the details of the relaxation function?

Solution

Consider the stress which results from a superposition of two strain step functions of amplitude a and b at times t_1 and t_2.

$$\varepsilon(t) = \varepsilon_0[a\mathcal{H}(t - t_1) + b\mathcal{H}(t - t_2)]. \tag{E4.5.1}$$

The resulting stress is, by the principle of superposition and the definition of the relaxation modulus,

$$\sigma(t) = \varepsilon_0[aE(t - t_1) + bE(t - t_2)]. \tag{E4.5.2}$$

Suppose the relaxation function is a simple exponential, $E(t) = E_0 e^{-t/\tau}$, with τ as the relaxation time. Then the stress is

$$\sigma(t) = \varepsilon_0 E_0\{a \ e^{-(t-t_1)/\tau} + b \ e^{-(t-t_2)/\tau}\} \tag{E4.5.3}$$

$$\sigma(t) = \varepsilon_0 E_0\{a \ e^{-t/\tau}e^{t_1/\tau} + b \ e^{-t/\tau}e^{t_2/\tau}\} \tag{E4.5.4}$$

$$\sigma(t) = \varepsilon_0 E_0\{a \ e^{t_1/\tau} + b \ e^{t_2/\tau}\} \ e^{-t/\tau} \tag{E4.5.5}$$

$$\sigma(t) = \varepsilon_0 E_0\{a \ e^{t_1/\tau} e^{-t_2/\tau} + b \ e^{t_2/\tau} e^{-t_2/\tau}\} \ e^{-t/\tau} e^{t_2/\tau} \tag{E4.5.6}$$

$$\sigma(t) = \varepsilon_0 E_0\{a \ e^{-(t_2-t_1)/\tau} + b\} \ e^{-(t-t_2)/\tau} \tag{E4.5.7}$$

In the last expression, the quantity in the { } brackets is constant independent of time t. Therefore [E4.5.1], the response is the same as if a single step strain had been applied at time t_2. The material which follows a single exponential exhibits memory of transient strain, but no memory of any details of the strain history. Consider, by contrast, relaxation functions which are not exponential, such as the stretched exponential

$$E(t) = (E_0 - E_\infty) \ e^{-(t/\tau)^\beta} + E_\infty,$$

or sums of exponentials. The rate of relaxation of an exponential relaxation is exponential, but the rate of relaxation for the stretched exponential is $-(\beta/\tau^\beta t^{1-\beta})e^{-(t/\tau)^\beta}$, which depends on the time following perturbation.

Example 4.6

Consider path memory for a history in which strain is maintained constant for time t_1 to t_2, then the stress is forced to zero at time t_2 by a second step in strain. How does the response to such a history depend on the relaxation function? Is there anything special about the exponential function?

Solution

For time $t_1 < t < t_2$, the strain is $\varepsilon(t) = \varepsilon_0 \mathcal{H}(t - t_1)$, so the stress is $\sigma(t) = \varepsilon_0 E(t - t_1)$. The stress at time t_2 is $\varepsilon_0 E(t_2 - t_1)$, and the corresponding strain increment is $\varepsilon_0 E(t_2 - t_1)/E(0)$ since we are dealing with the immediate response to the second step function. To null the stress for time $t > t_2$, the strain history must be

$$\varepsilon(t) = \varepsilon_0 \left[\mathcal{H}(t - t_1) - \frac{E(t_2 - t_1)}{E(0)} \mathcal{H}(t - t_2) \right], \tag{E4.6.1}$$

so

$$\sigma(t) = \varepsilon_0 \left[E(t - t_1) - \frac{E(t_2 - t_1)}{E(0)} E(t - t_2) \right]. \tag{E4.6.2}$$

Suppose again that the relaxation function is an exponential, $E(t) = E_0 e^{-t/\tau}$,

$$\sigma(t) = \varepsilon_0 E_0 \left[e^{-(t-t_1)/\tau} - e^{-(t_2-t_1)/\tau} e^{-(t-t_2)/\tau} \right]$$

$$= \varepsilon_0 E_0 \left[e^{-t/\tau} e^{+t_1/\tau} - e^{-t_2/\tau} e^{+t_1/\tau} e^{-t/\tau} e^{+t_2/\tau} \right] = 0. \tag{E4.6.3}$$

So for an exponential relaxation function, the action of interrupting the relaxation by applying a second step strain to momentarily force the stress to zero, results in zero stress for all time following. For other relaxation functions there is a rebound effect ($\sigma(t) > 0$) which arises from the fact that the momentarily enforced equilibrium is actually a mixture of nonequilibrium states [E4.5.1].

Example 4.7

Show that

$$J''(\omega) \approx -\frac{\pi}{2} \frac{dJ'(\omega)}{d\ln\omega} = -\frac{\pi}{2} \omega \frac{dJ'(\omega)}{d\omega}.$$

Hint: use the Kramers–Kronig relation for J', differentiate within the integral, and assume that J'' is slowly varying.

Solution

The Kramers–Kronig relation is

$$J'(\omega) - J'(\infty) = \frac{2}{\pi} \left[\wp \int_0^\infty \frac{\omega J''(\omega)}{\omega^2 - \omega^2} d\omega \right].$$

By differentiating and assuming that J'' varies sufficiently slowly that it can be regarded as constant compared with the kernel,

$$\frac{dJ'(\omega)}{d\omega} \approx \frac{2}{\pi} 2\omega J'' \int_0^\infty \frac{\omega}{\left(\omega^2 - \omega^2\right)^2} d\omega.$$

This is a standard integral of the following form (recall that ω is a variable of integration which can be renamed):

$$\int \frac{x}{\left(x^2 - a^2\right)^2} dx = -\frac{1}{2\left(x^2 - a^2\right)}.$$

$$\frac{dJ'(\omega)}{d\omega} \approx \frac{2}{\pi} 2\omega J'' \left\{ -\frac{1}{2\left(\omega^2 - \omega^2\right)} \right\}_{\omega=0}^\infty.$$

$$\frac{dJ'(\omega)}{d\omega} \approx -\frac{\pi}{2} \omega \frac{dJ'(\omega)}{d\omega}.$$

Moreover,

$$\frac{d\ln\omega}{d\omega} = \frac{1}{\omega},$$

so

$$J''(\omega) \approx -\frac{\pi}{2} \frac{dJ'(\omega)}{d\ln\omega}.$$

Problems

4.1 Consider the stretched exponential form $E(t) = (E_0 - E_\infty)e^{-(t/\tau_r)^\beta} + E_\infty$, with $0 < \beta \leq 1$. Determine the tan δ associated with this relaxation. Use an analytical approximation scheme or use a numerical approach to evaluate the transformation integrals. Discuss how the loss peak differs from a Debye peak.

4.2 For a glassy polymer, suppose that G = 1 GPa at 1 Hz and tan δ ≈ 0.08 with the loss approximately independent of frequency. Estimate the stiffness at 30 kHz, at 1 MHz, and after 2 weeks of creep.

4.3 If you have done an informal creep test in an earlier assignment, transform your results to $|E^*|$ (or $|G^*|$) and tan δ vs. frequency and discuss.

4.4 The approximate interrelations convert nonlocal integral equations into local expressions or weakly nonlocal ones involving derivatives. Some of the approximate interrelations are exact for power law relaxation. Analytically determine the effect at time t of a superposed exponential relaxation of time τ as a function of t/τ.

4.5 For $H(\tau)/\tau = E_0(\delta(\tau - \tau_1) + \delta(\tau - 10\tau_1))$ and for $H(\tau)/\tau = E_0(\delta(\tau - \tau_1) + \delta(\tau - 100\tau_1))$ determine E(t), J(t), and L(τ) and plot them vs. log time. For simplicity let $E_e = E_0$. Discuss.

4.6 Show that

$$H(\tau) \approx \frac{2}{\pi}\left(E''(\omega) - \frac{d^2E''}{d(\ln\omega)^2}\right)\Big|_{\omega=1/\tau}.$$

4.7 If $J(t) = A\,t^n$, what is the retardation spectrum? Hint [1.6.10]: consider the definition of the gamma function as an integral representation. Write J(t) in terms of the definition of the retardation spectrum and perform a change of variables to make contact with the gamma integral representation.

4.8 To further illustrate path memory effects calculate the stress from Eq. E4.5.2; by assuming a relaxation function consisting of a sum of two exponentials of the same amplitude but different time constants, plot the response and discuss.

4.9 Sketch as a *space curve* the dependence of stress and strain as they depend on time during dynamic loading of a viscoelastic material. Illustrate that the projection of this curve on the stress–strain plane is an ellipse, and that the projection on the stress–time plane is a sinusoid. What does the space curve look like in the case of an elastic material?

References

4.2.1 Gross, B., "On creep and relaxation", *J. Appl. Phys.* 18, 212–221, 1947.

4.2.2 Nowick, A. S. and Berry, B. S., "Lognormal distribution function for describing anelastic and other relaxation processes", *IBM J. Res. Dev.* 5, 297–312, 1961.

4.2.3 Tobolsky, A. B., *Properties and Structure of Polymers*, J. Wiley, NY, 1960.

4.3.1 Smith, T. L., "Approximate equations for interconverting the various mechanical properties of linear viscoelastic materials", *Trans. Soc. Rheol.* 2, 131–151, 1958.

4.3.2 Marvin, R. S., "The linear viscoelastic behavior of rubberlike polymers and its molecular interpretation", in *Viscoelasticity, Phenomenological Aspects*, Academic, NY, 1960.

4.3.3 Alfrey, T., *Mechanical Behaviour of High Polymers*, Interscience, NY, 1948.

4.3.4 Schwarzl, F. R. and Struik, L. C. E., "Analysis of relaxation measurements", in *Advances in Molecular Relaxation Processes*, Vol. 1, Elsevier, Amsterdam, 1967, pp. 201–255.

4.3.5 Schwarzl, F. and Staverman, A. J., "Higher approximation methods for the relaxation spectrum from static and dynamic measurements of visco-elastic materials", *Appl. Sci. Res.* A4, 127–141, 1953.

4.3.6 Wiff, D. R., "RQP method of inferring a mechanical relaxation spectrum", *J. Rheol.* 22, 589–597, 1978.

4.4.1 Kerwin, E. M., Jr. and Ungar, E. E., "Requirements imposed on polymeric materials in structural damping applications", in *Sound and Vibration Damping with Polymers*, ed. R. D. Corsaro and L. H. Sperling, American Chemical Society, Washington, DC, 1990.

E4.5.1 Rekhson, S. M., "Viscoelasticity of glass", in *Glass: Science and Technology*, Vol. 3, Viscosity and Relaxation, ed. D. R. Uhlmann and N. J. Kreidl, Academic, NY, 1986, Chapter 1.

chapter five

Viscoelastic stress and deformation analysis

§5.1   Introduction

In this chapter we consider problems of determining stress, strain, or displacement fields in viscoelastic bodies. First the constitutive equations are generalized from one dimension to three dimensions. Then a problem in simple beam theory is solved for a viscoelastic material, illustrating the approach of direct construction. It is simpler in this case to use the correspondence principle which simplifies solution for certain boundary value problems in viscoelastic materials. As input to the correspondence principle one uses a corresponding solution for an elastic material. Next, several examples involving use of the correspondence principle are developed. Finally, problems not amenable to the correspondence principle are examined.

§5.2   Three-dimensional constitutive equation

In three dimensions, Hooke's law of linear *elasticity* is given by

$$\sigma_{ij} = C_{ijkl}\varepsilon_{kl}, \tag{5.2.1}$$

with C_{ijkl} as the elastic modulus tensor. The usual Einstein summation convention is used in which we sum over repeated indices [5.2.1, 5.2.2]. Each component of strain can have an independent time dependence. There are 81 components of C_{ijkl}. As discussed in §2.8, the number of independent constants is reduced to 21 by considering symmetry in the stress and strain, and a strain energy density function. Isotropic materials, which have properties independent of direction, are describable by two independent elastic constants.

To obtain the constitutive equation for *viscoelastic* materials, repeat for each component of strain and stress arguments identical to those given in §2.2 for one dimension. The following is the result:

$$\sigma_{ij}(t) = \int_0^t C_{ijkl}(t-\tau)\frac{d\varepsilon_{kl}}{d\tau}\,d\tau. \tag{5.2.2}$$

This constitutive equation can accommodate any degree of anisotropy. Each independent component of the modulus tensor can have a different time dependence.

Many practical materials are approximately isotropic (their properties are independent of direction). For isotropic elastic materials, the constitutive equation is, in Lamé form [5.2.1],

$$\sigma_{ij} = \lambda\varepsilon_{kk}\delta_{ij} + 2\mu\varepsilon_{ij}, \tag{5.2.3}$$

in which λ and μ are the two independent Lamé elastic constants, δ_{ij} is the Kronecker delta (1 if i = j, 0 if i ≠ j), and $\varepsilon_{kk} = \varepsilon_{11} + \varepsilon_{22} + \varepsilon_{33}$. Engineering constants such as Young's modulus E, shear modulus G, and Poisson's ratio ν can be extracted from these tensorial constants. Specifically [5.2.3],

$$G = \mu,$$

$$B = \frac{G(3\lambda + 2G)}{(\lambda + G)},$$

$$\nu = \frac{\lambda}{2(\lambda + G)}.$$

In isotropic viscoelastic solids, the corresponding relation between strain history and stress is

$$\sigma_{ij}(t) = \int_0^t \lambda(t-\tau)\delta_{ij}\frac{d\varepsilon_{kk}}{d\tau}\,d\tau + \int_0^t 2\mu(t-\tau)\frac{d\varepsilon_{ij}}{d\tau}\,d\tau. \tag{5.2.4}$$

There are two independent viscoelastic functions which can have different time dependence. As in the case of elasticity, we may consider engineering functions rather than tensorial ones. The viscoelastic time-dependent functions are Young's modulus E(t), shear modulus G(t) = μ(t), Poisson's ratio ν(t), and bulk modulus B(t).

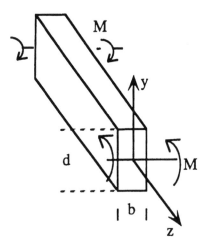

Figure 5.1   Pure bending of a bar.

§5.3   Pure bending by direct construction

The problem addressed here is determination of the relation between the moment and the curvature under the assumptions of elementary beam theory, for a linearly viscoelastic beam. The assumptions are (i) plane sections remain plane, (ii) effects of shear are neglected, and (iii) the material has identical properties in tension and compression. Consider a beam of isotropic viscoelastic material of rectangular cross-sectional dimensions b and d, that is uniform along the beam axis (the z direction), and bent by moments M(t) applied at each end (Fig. 5.1).

The relation between curvature Ω and strain ε (in the z direction) is obtained from (a).

$$\varepsilon(y,t) = -\Omega(t)\big(y - y_n(t)\big) = -\Omega(t)y', \qquad (5.3.1)$$

in which y_n is the position of the neutral axis. Substitute Eq. 5.3.1 in the Boltzmann superposition integral which by virtue of linearity entails assumption (c). Stress is in the z direction: $\sigma(y, t) = \sigma_{zz}(y, t)$ and the absence of other stresses follows from the above assumption that we neglect shear.

$$\sigma(y,t) = -y' \int_0^t E(t-\tau)\frac{d\Omega}{d\tau}\,d\tau. \qquad (5.3.2)$$

The equilibrium conditions $\Sigma F = 0$, $\Sigma M = 0$ give

$$\int \sigma(y, t) dx\, dy = 0, \tag{5.3.3}$$

$$\int \sigma(y, t) y\, dx\, dy + M(t) = 0. \tag{5.3.4}$$

By substituting Eq. 5.3.2 in Eq. 5.3.3,

$$b \int_{-d/2}^{d/2} -y' \int_0^t E(t - \tau) \frac{d\Omega}{d\tau}\, d\tau\, dy = 0. \tag{5.3.5}$$

By interchanging the order of integration with the aim of locating the neutral axis,

$$\int_0^t E(t - \tau) \frac{d\Omega}{d\tau}\, d\tau \left(-b \int_{-d/2}^{d/2} (y - y_n(t)) dy \right) = 0, \tag{5.3.6}$$

$$-b \int_{-d/2}^{d/2} (y - y_n(t)) dy = -b \left[\frac{y^2}{2} - y_n y \right]_{-d/2}^{d/2} = 0. \tag{5.3.7}$$

So $y_n = 0$ so the neutral axis coincides with the centroid of the cross section. To examine the moments, substitute Eq. 5.3.2 in Eq. 5.3.4.

$$b \int_{-d/2}^{d/2} -y^2 \int_0^t E(t - \tau) \frac{d\Omega}{d\tau}\, d\tau\, dy = M(t), \tag{5.3.8}$$

$$\frac{bd^3}{12} \int_0^t E(t - \tau) \frac{d\Omega}{d\tau} = M(t). \tag{5.3.9}$$

$$\frac{M(t)}{I} = \int_0^t E(t - \tau) \frac{d\Omega}{d\tau}\, d\tau. \tag{5.3.10}$$

This is the desired relationship between the curvature history $\Omega(t)$ and moment in the bending of a viscoelastic beam.

In Eq. 5.3.10, I is the area moment of inertia which is $\frac{1}{12} bd^3$ for a rectangular section. This last result gives the moment history in terms of the curvature history. If we know the moment and wish to calculate the curvature, an appropriate relation can be obtained by taking a Laplace transform of Eq. 5.3.10,

$$sE(s)\Omega(s) = \frac{M(s)}{I}, \qquad (5.3.11)$$

but from Laplace transformation of the Boltzmann integrals in the compliance and modulus formulations,

$$sJ(s) = \frac{1}{sE(s)}, \qquad (5.3.12)$$

so

$$\Omega(s) = \frac{1}{I}M(s)sJ(s). \qquad (5.3.13)$$

Thus by taking the inverse transform,

$$\Omega(t) = \frac{1}{I}\int_0^t J(t-\tau)\frac{dM}{d\tau}d\tau. \qquad (5.3.14)$$

This last expression gives the curvature history in terms of the moment history for the bent viscoelastic beam. Observe that most of the effort in obtaining Eq. 5.3.10 is identical to that which was already expended (in earlier studies of elastic response of bent beams) in obtaining the corresponding solution for an elastic material. One may surmise from the similarity in the elastic and viscoelastic analyses that there should be an easier way to solve boundary value problems for viscoelastic materials. For a substantial class of such problems this is the case, as is discussed below.

§5.4   Correspondence principle

The correspondence principle [5.4.1, 5.4.2, 5.4.3] states that if a solution to a linear elasticity problem is known, the solution to the corresponding problem for a linearly viscoelastic material can be obtained by replacing each quantity which can depend on time by its Laplace transform multiplied by the transform variable (p or s), and then by transforming back to the time domain. There is the restriction that the interface between boundaries under prescribed load and boundaries under prescribed displacement may not change with time, although the loads and displacements can be time dependent.

The correspondence principle is demonstrated as follows. In solving problems in mechanics, one uses the equilibrium equations,

$$\frac{\partial \sigma_{ij}}{\partial x_i} + F_j = 0, \tag{5.4.1}$$

and the strain–displacement relations,

$$\varepsilon_{ij} = \frac{1}{2}\left(\frac{\partial u_j}{\partial x_i} + \frac{\partial u_i}{\partial x_j}\right). \tag{5.4.2}$$

One also has the boundary conditions on part of the surface of the object Γ_T, for which the surface tractions are prescribed,

$$T_j = \sigma_{ij} n_i, \tag{5.4.3}$$

with n as the unit normal. On another part of the surface the surface displacements are prescribed,

$$u_j = U_j. \tag{5.4.4}$$

Finally, one makes use of the constitutive equation which relates stress to strain. In some elasticity problems, compatibility conditions for the strains must be used. Here the standard notation is used in which indices can have values from 1 to 3; repeated indices are summed over. The relations given are for the three-dimensional theory of elasticity; simplified counterparts may be used in analyses at the level of elementary mechanics as was done in the prior section. Among the above equations, only the viscoelastic constitutive equation contains a product of time-dependent quantities. Therefore, by following a Laplace transform, only the viscoelastic constitutive equation will differ from its elastic counterpart, and the difference is a factor of the transform variable s (sometimes called p). In one dimension, the Boltzmann integral in the Laplace plane is

$$\sigma(s) = sE(s)\varepsilon(s).$$

Thus, as the governing equations are carried through to obtain a solution, the distinction between the elastic and viscoelastic cases in the Laplace domain consists of merely the extra factors of s, which then appear in the final solution. So if we have an elasticity solution, the corresponding viscoelasticity solution may be obtained by replacing each quantity which can depend on time by its s (or p) multiplied the Laplace transform, and by transforming back to the time domain. In order that the elastic and viscoelastic solutions correspond, the interface between boundary regions under specified displacement and under specified stress must not change with time.

There is also a correspondence principle involving Fourier transforms [5.4.3, 1.6.1]. If a solution to a linear elasticity problem is known, the solution to the corresponding problem for a linearly viscoelastic material can be obtained by replacing each quantity which can depend on frequency by its Fourier transform. This principle is simpler conceptually than the above Laplace principle since the Fourier transform of a transient viscoelastic response function is a dynamic response function which is itself *measurable experimentally*. The Fourier transform of the stress history is given by

$$\sigma(\omega) = \int_{-\infty}^{\infty} \sigma(t) e^{-i\omega t} dt = \int_{-\infty}^{\infty} \int_{-\infty}^{t} E(t-\tau) \frac{d\varepsilon(\tau)}{d\tau} d\tau \, e^{-i\omega t} \, dt. \qquad (5.4.5)$$

By using the shift, convolution, and derivative theorems for Fourier transforms discussed in Appendix 2, the constitutive equation becomes

$$\sigma(\omega) = E^*(\omega)\varepsilon(\omega). \qquad (5.4.6)$$

There is no extra multiple of the transform variable in this case since the ω is absorbed in the definition of the dynamic functions. As in the case of the Laplace correspondence principle described above, Fourier transformation of the other equations (strain displacement, equilibrium, boundary conditions, etc.) used in obtaining an elasticity solution leads to no change in the appearance of these equations. To apply this correspondence principle, begin with a known elasticity solution. Wherever an elastic constant (such as G) appears replace it with the corresponding complex dynamic viscoelastic function (such as G*). The result will be valid for linearly viscoelastic materials. The same restrictions associated with the Laplace type of correspondence principle apply here.

§5.5 Pure bending reconsidered

We have the solution relating curvature Ω to moment M for an elastic beam,

$$\Omega(t) = \frac{M(t)}{EI}. \qquad (5.5.1)$$

Apply the correspondence principle to obtain the solution of the bending problem for a linearly viscoelastic material. In the Laplace domain, with $E(s)J(s) = 1/s^2$ from the Boltzmann integral,

$$\Omega(s) = \frac{M(s)}{sE(s)I} = \frac{1}{I} sM(s)J(s). \qquad (5.5.2)$$

By performing the inverse transform,

$$\Omega(t) = \frac{1}{I} \int_0^t J(t-\tau) \frac{dM}{d\tau} d\tau, \qquad (5.5.3)$$

which is identical to the relation obtained earlier by direct construction, but is obtained with less effort.

We may also consider the relation

$$\sigma(y) = \frac{My}{I}.$$

No elastic constants are present. By the correspondence principle,

$$\sigma(y,s) = \frac{M(s)y}{I}.$$

Therefore, in the viscoelastic case, by an inverse transform,

$$\sigma(y,t) = \frac{M(t)y}{I}.$$

Relaxation has no effect on the relationship between stress and the applied bending moment. If this is counterintuitive, consider that in the elastic case a bar of steel and a bar the same shape of rubber have the same relationship between stress and moment; consider the reason why.

§5.6   Further examples with the correspondence principle

In this section we consider several examples in which stress distributions are taken as given from classical elasticity. In several cases of practical interest, such as stress concentrations and Saint Venant's principle, viscoelasticity can have a significant effect on stress distribution. The correspondence principle is also of use in the study of composite materials as examined in Chapter 9.

§5.6.1   Constitutive equations

One may obtain the viscoelastic constitutive equations in three dimensions by using the elastic versions and the correspondence principle rather than repeating the direct demonstrations as suggested above. For example, in the elastic case, $\sigma_{ij} = C_{ijkl}\varepsilon_{kl}$; in the Laplace plane, $\sigma_{ij}(s) = sC_{ijkl}(s)\varepsilon_{kl}(s)$; by transforming back,

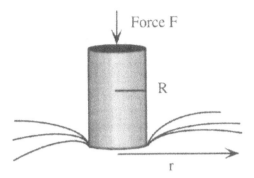

Figure 5.2   Indentation of a semi-infinite solid by a cylindrical indenter.

$$\sigma_{ij}(t) = \int_0^t C_{ijkl}(t-\tau)\frac{d\epsilon_{kl}}{d\tau}\,d\tau,$$

as was argued above in §5.2.

§5.6.2   Rigid indenter on a semi-infinite solid

Consider a rigid circular cylindrical indenter of radius R pressed with load F on a semi-infinite viscoelastic solid substrate (Fig. 5.2). This could represent a building erected upon the compliant earth, or an industrial press operation. A solution for an elastic solid of Young's modulus E and Poisson's ratio v is available. The indenter displacement is [5.2.2]

$$w = \frac{F(1-v^2)}{2RE}. \qquad (5.6.1)$$

The pressure distribution q(r) as a function of radial coordinate r is

$$q(r) = \frac{F}{2\pi R\sqrt{R^2-r^2}}. \qquad (5.6.2)$$

These elastic solutions are valid if F, q, and w are slowly varying functions of time; the time scale of variation is much longer than the period associated with any resonance. If rapid variations occurred, a new solution containing inertial terms would be required. In the viscoelastic case which allows similar slow time dependence, the stiffness E and Poisson's ratio v depend on time as well. In the Laplace domain,

$$q(r,s) = \frac{F(s)}{2\pi R\sqrt{R^2 - r^2}}; \qquad (5.6.3)$$

by transforming back, the time-dependent pressure follows the time dependence of the force F,

$$q(r,t) = \frac{F(t)}{2\pi R\sqrt{R^2 - r^2}}. \qquad (5.6.4)$$

This case is simple since there are no elastic constants present. As for
the displacement w,

$$w(s) = \frac{F(s)}{2RsE(s)}\left(1 - s^2 v(s)v(s)\right) = \frac{1}{2R}F(s)\left(sJ(s) - s^3 J(s)v(s)v(s)\right). \qquad (5.6.5)$$

By transforming back, the time-dependent displacement is

$$w(t) = \frac{1}{2R}\int_0^t J(t-\tau)\frac{dF}{d\tau}d\tau$$

$$-\frac{1}{2R}\left\{\int_0^t J(t-\tau)\left[\int_0^\tau v(\tau-\eta)\left[\int_0^\eta v(\eta-\xi)\frac{d^3F(\xi)}{d\xi^3}d\xi\right]d\eta\right]d\tau\right\}. \qquad (5.6.6)$$

The second term is obtained by repeatedly applying the convolution
theorem to $s^3 J(s)v(s)v(s)$, peeling off one expression at a time. These are
cascaded convolutions. We remark that if Poisson's ratio is independent of
time, the expression becomes much simpler since

$$w(s) = \frac{1}{2R}F(s)sJ(s)\left(1 - v^2\right), \qquad (5.6.7)$$

so

$$w(t) = \frac{1}{2R}\left(1 - v^2\right)\int_0^t J(t-\tau)\frac{dF(\tau)}{d\tau}d\tau. \qquad (5.6.8)$$

If the force F is a step function in time, the indentation displacement is
proportional to the creep compliance. Therefore, for the case of constant
Poisson's ratio, a time-dependent indentation test can be used to determine
the creep compliance.

§5.6.3 Viscoelastic rod held at constant extension

Does a stretched viscoelastic rod held at constant extension get fatter or thinner with time [1.6.8]? Consider first polymer materials, and then arbitrary materials. When an elastic rod is stretched, it gets thinner provided it is made of an "ordinary" material for which $v > 0$. We do not consider here materials for which $v < 0$ [5.6.1]. As for a viscoelastic rod, consider the interrelations between Poisson's ratio and the stiffnesses. For an isotropic elastic material, the elastic constants are related [5.2.1] as follows (see also Appendix 7).

$$E = 2G(1+v), \ G = \mu,$$

$$B = \lambda + \frac{2}{3}\mu, \ B = \frac{2G(1+v)}{3(1-2v)},$$

$$B = \frac{E}{3(1-2v)}, \ B = \frac{GE}{3(3G-E)},$$

$$v = \frac{E}{2G} - 1, \ v = \frac{3B-2G}{6B+2G}, \ v = \frac{1}{2} - \frac{E}{6B},$$

in which E is Young's modulus, G is the shear modulus, B is the bulk modulus, v is Poisson's ratio, and λ and μ are the Lamé constants given above. The last equation for Poisson's ratio is the easiest to convert into viscoelastic form in the context of the stated problem for polymers. Consider the bulk compliance $\kappa = 1/B$, then the Poisson's ratio may be written as follows:

$$v = \frac{1}{2} - \kappa \frac{E}{6}. \tag{5.6.9}$$

If we consider a polymer material, there is much more relaxation in the shear and axial properties (typically a factor of 1000 or more through the glass–rubber transition) than in the bulk properties (possibly a factor of two [1.6.18]). For this case approximate B(t) as a constant. By using the correspondence principle and passing to the Laplace domain,

$$sv(s) = \frac{1}{2} - \frac{1}{6}s\kappa E(s), \tag{5.6.10}$$

since the bulk compliance κ is assumed to be constant. By transforming back,

$$v(t) = \frac{1}{2} - \kappa \frac{E(t)}{6}. \tag{5.6.11}$$

Since E(t) decreases monotonically, v(t) increases for this case (in a typical polymer $v \approx 1/3$ in the glassy region, $v \approx 1/2$ in the rubbery region) so the viscoelastic polymer rod under constant tensile strain becomes *thinner* with time.

As for a general material for which both Young's and bulk moduli are time dependent, again by applying the correspondence principle,

$$sv(s) = \frac{1}{2} - \frac{1}{6}s^2\kappa(s)E(s).$$

By dividing by s and transforming back,

$$v(t) = \frac{1}{2} - \frac{1}{6}\int_0^t E(t-\tau)\frac{d\kappa(\tau)}{d\tau}\,d\tau. \qquad (5.6.12)$$

It is not immediately obvious from this expression if there is now any restriction on whether Poisson's ratio increases or decreases; see Problem 5.3.

§5.6.4   Stress concentration

The stress concentration factor is the ratio of the maximum stress near an heterogeneity to the stress far from it. In three-dimensional elasticity theory, expressions for stress concentration factors usually contain Poisson's ratio. We may consider the effect of relaxation in Poisson's ratio of viscoelastic materials on the stress concentration factor (SCF). For example, for a spherical cavity in biaxial tension [5.6.2],

$$\text{SCF} = \frac{\sigma_{max}}{\sigma_0} = \frac{12}{(7-5v)}. \qquad (5.6.13)$$

So

$$\sigma_{max}(7-5v) = 12\sigma_0.$$

By applying the correspondence principle,

$$\sigma_{max}(s)(7-5sv(s)) = 12\sigma_0(s).$$

By transforming back,

$$7\sigma_{max}(t) - 5\int_0^t v(t-\tau)\frac{d\sigma_{max}(\tau)}{d\tau}\,d\tau = 12\sigma_0(t). \qquad (5.6.14)$$

An explicit relation can be obtained given an explicit functional form for $v(t)$. However, suppose for the sake of simplicity that for asymptotically short times the material behaves elastically with $v = 1/3$, and at asymptotically long times also behaves elastically with $v \approx 1/2$. Under such asymptotic conditions, the material is considered to behave elastically. The stress concentration factor goes from 2.25 to 2.667, a 19% increase. The functional form of the time-dependent stress concentration factor is not obtained by an asymptotic argument; it can only be obtained by solving the convolution equation. In a polymer, such a change in Poisson's ratio would likely be accompanied by a relaxation of several orders of magnitude in shear modulus. In many cases, relaxation in Poisson's ratio is neglected since its effect is often much less than that of the relaxation in stiffness. A counterexample is given in the next section.

§5.6.5   Saint Venant's principle

Saint Venant's principle is important in the application of elasticity solutions in many practical situations in which boundary conditions are satisfied in the sense of resultants rather than pointwise. For example, a bending moment may be applied to a beam via a complex array of bolted joints, which generate a locally complex stress pattern. In view of Saint Venant's principle, one expects to observe bending-type stresses far from the ends.

Saint Venant's principle states that a localized self-equilibrated load system produces stresses which decay with distance more rapidly than stresses due to forces and moments. It is applicable in many situations of interest in engineering. There are some counterexamples (see, e.g., Fung [5.2.3]). Consider a sandwich panel with rigid face sheets and an elastic material of Poisson's ratio v sandwiched between them. For Poisson's ratios in the vicinity of 0.5, stresses applied to the end will decay [5.6.3] with distance z as $\sigma(z) \propto e^{-\gamma z}$. The decay rate is

$$\gamma \propto \sqrt{\frac{3(1-2v)}{3-4v}}. \tag{5.6.15}$$

The form of $\gamma(v)$ makes it difficult to proceed with a transform solution unless a specific functional form of $v(t)$ is chosen. We may assume the material is a viscoelastic polymer in which the Poisson's ratio changes from $1/3$ at asymptotically short times to nearly $1/2$ at asymptotically long times. The decay rate γ for end stresses tends to zero as relaxation proceeds. Therefore, the distance $1/\gamma$, over which there is significant stress, diverges as Poisson's ratio relaxes to $1/2$. As relaxation progresses in this example, stress propagates farther and farther into the core material. Practical sandwich panels incorporate as core anisotropic honeycombs or relatively stiff foam (with a Poisson's ratio of about 0.3) rather than soft polymers. So this

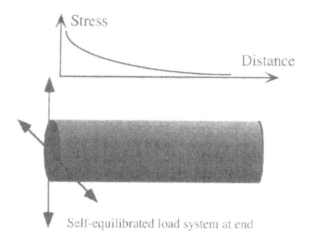

Figure 5.3   Decay of end stress with distance in a rod according to Saint Venant's principle.

example serves to show that dramatic effects can result from relaxation in Poisson's ratio, rather than as a practical case study.

As for unconstrained rods, a self-equilibrated load system applied to a rod (Fig. 5.3) with free lateral surfaces gives rise to stress which decays with distance according to Saint Venant's principle. The distance required for substantial decay is comparable to the rod radius for normal values of Poisson's ratio, but decay can be slow [5.6.4] for Poisson's ratio approaching −1. If Poisson's ratio changes with time in a viscoelastic material with negative Poisson's ratio, the spatial rate of stress decay can also change with time.

§5.7   Dynamic problems

It is usually expedient to treat dynamic problems in the frequency domain. One may proceed ab initio or make use of the dynamic correspondence principle to convert an elastic solution [5.7.1–5.7.3] into a viscoelastic one.

§5.7.1   Longitudinal vibration and waves in a rod

A uniform rod of length L, cross-sectional area A, dynamic modulus E*, and density ρ is rigidly fixed at one end and has the other end free. A one-dimensional analysis is performed on the rod; neglect of the Poisson effects is warrantable if the wavelength of stress waves in the rod is much greater than the rod diameter. In §3.7, the stress–strain relation was combined with Newton's second law and the viscoelastic stress–strain relation to obtain the following *wave equation:*

$$\frac{\partial^2 u}{\partial t^2} = \frac{E^*}{\rho} \frac{\partial^2 u}{\partial x^2}.$$

(5.7.1)

This admits traveling wave solutions such as

$$u(x,t) = u_0 \exp\left\{ i\omega\left(\frac{x}{c} - t \right) \right\} \exp\left\{ -x \frac{\omega}{2c} \tan \delta \right\}, \qquad (5.7.2)$$

for small loss, with the wave speed

$$c = \sqrt{\frac{E'}{\rho}}. \qquad (5.7.3)$$

The real exponential in Eq. 5.7.2 represents wave attenuation.
    If we are given a wave equation for elastic materials,

$$\frac{\partial^2 u}{\partial t^2} = \frac{E}{\rho} \frac{\partial^2 u}{\partial x^2}, \qquad (5.7.4)$$

the corresponding wave equation for viscoelastic materials is obtained by replacing the elastic modulus E with the complex modulus E^*. The wave speed and attenuation from this wave equation are then determined as in §3.7.
    Standing wave solutions corresponding to vibrations are also admissible. A standing wave can be formed as a superposition of two traveling waves, one going right and the other going left. For example, for an elastic material [5.7.1],

$$u_n = A_n \cos \frac{\omega_n x}{c} + B_n \sin \frac{\omega_n x}{c}. \qquad (5.7.5)$$

The coefficients A_n are determined from the boundary condition that displacement is zero at one end, $x = 0$. By substituting in the wave equation, $A_n = 0$. At the other end there is no force, so $\partial u / \partial x = 0$ at $x = L$. So (ω_n / c) $B_n \cos(\omega_n L / c) = 0$, and

$$\omega_n = \frac{c\pi n}{2L}, \quad n = 1, 3, 5 \ldots .$$

Here n, an integer, is the mode number. So for fixed-free vibration of a rod, the odd modes are present. If both ends are free, all modes are present. Several experimental methods make use of vibrating rods as presented in §6.8. For viscoelastic materials, the dynamic compliance at the resonant frequencies becomes finite, and the resonance peaks acquire finite width proportional to $\tan \delta$, as presented in §3.5.

§5.7.2 Torsional waves and vibration in a rod

In this section a detailed solution is obtained for an elastic material; then the correspondence principle is applied to obtain a viscoelastic solution.

The equation of motion in linear elasticity is

$$\frac{\partial \sigma_{ij}(x_i, t)}{\partial x_i} + F_j(x_i, t) = \rho \frac{\partial^2 u_j(x_i, t)}{\partial t^2} \tag{5.7.6}$$

in which ρ is the material density and u_i is the particle displacement [5.2.3]; the Einstein summation convention for repeated indices is used. For torsional oscillation, there is only one component of displacement (in the circumferential direction), u_θ, in cylindrical coordinates. The constitutive equation is

$$\sigma = 2G\varepsilon, \tag{5.7.7}$$

with ε as the tensorial shear strain. So, by using the strain displacement relation from elasticity theory,

$$\text{div grad } Gu_\theta = \rho \frac{\partial^2 u_i}{\partial t^2}. \tag{5.7.8}$$

In cylindrical coordinates,

$$\nabla^2 \mathbf{A} = \mathbf{a}_r \left[\nabla^2 A_r - \frac{A_r}{r^2} - \frac{2}{r^2} \frac{\partial A_\theta}{\partial \theta} \right] + \mathbf{a}_\theta \left[\nabla^2 A_\theta + \frac{A_\theta}{r^2} + \frac{2}{r^2} \frac{\partial A_r}{\partial \theta} \right] + \mathbf{a}_z \nabla^2 A_z, \tag{5.7.9}$$

in which $\mathbf{A}$ is a vector, $\mathbf{a}_r$, $\mathbf{a}_\theta$, and $\mathbf{a}_z$ are unit vectors; z is along the rod axis; r is radial; and θ is in the circumferential direction.

Also

$$\nabla^2 A_\theta = \frac{1}{r} \frac{\partial}{\partial r} \left\{ r \frac{\partial A_\theta}{\partial r} \right\} + \frac{1}{r^2} \frac{\partial^2 A_\theta}{\partial \theta^2} + \frac{\partial^2 A_z}{\partial z^2}. \tag{5.7.10}$$

So

$$\frac{\partial^2 u_\theta}{\partial r^2} + \frac{1}{r} \frac{\partial u_\theta}{\partial r} + \frac{\partial^2 u_\theta}{\partial z^2} - \frac{u_\theta}{r^2} = \frac{\rho}{G} \frac{\partial^2 u_\theta}{\partial t^2}. \tag{5.7.11}$$

This was the starting point of the analysis of Christensen [1.6.1]. This is also a *wave equation*. Consider a solution of the form

$$u_\theta = rf(z)\, e^{i\omega t}. \tag{5.7.12}$$

This represents torsional waves in that planar cross sections rotate with respect to each other but do not deform. The first term vanishes and the second and fourth terms of the wave equation cancel. The following remains. Other dynamic motions are possible in shear. Radial functions containing Bessel functions are also admissible.

$$\frac{\partial^2 u_\theta}{\partial t^2} = \frac{G}{\rho}\frac{\partial^2 u_\theta}{\partial z^2}. \tag{5.7.13}$$

This *wave equation* is similar to Eq. 5.7.4 for longitudinal waves in one dimension. The following constitutes a solution for traveling waves. Here the angular frequency is $\omega = 2\pi\nu$ with ν as frequency, and the wave number is $k = 2\pi/\lambda$ with λ as the wavelength.

$$u_\theta(x,\, t) = u_{\theta 0} e^{i(kx-\omega t)}.$$

Substitute in the wave equation to obtain the *dispersion relation*.

$$\omega^2 = k^2 \frac{G}{\rho}.$$

Again, $\omega/k = \lambda\nu$ represents the phase velocity or wave speed c_T, which for torsional waves depends on the shear modulus G rather than on Young's modulus.

$$c_T = \sqrt{\frac{G}{\rho}}.$$

The treatment is purely elastic so far. The wave speed for torsional waves is independent of frequency in the elastic case, so these waves propagate non-dispersively. The wave propagation solution for a viscoelastic material is obtained by using the correspondence principle to replace G by G^* in the dispersion relation. Then the dispersion relation becomes

$$k\sqrt{\frac{G'}{\rho}}\sqrt{1+i\tan\delta} = \omega.$$

Substitute with $\tan\delta \ll 1$, in the wave equation,

$$u_\theta(x,t) = u_{\theta0}\exp i\omega\left\{\frac{x}{c_T} - t\right\}\exp\left\{-x\frac{\omega}{2c_T}\tan\delta\right\},$$

which shows wave propagation and attenuation.

Consider now standing waves, which correspond to vibrations. Return to Eq. 5.7.12 in the elastic case:

$$f(z) = A\sin\frac{\Omega z}{L} + B\cos\frac{\Omega z}{L}, \qquad (5.7.14)$$

with

$$\Omega = \omega L\sqrt{\frac{\rho}{G}}. \qquad (5.7.15)$$

To calculate the moment necessary to drive these vibrations, the shear stress is obtained from the displacement field by the strain–displacement relation and stress–strain relation and is integrated over the cross section. The cross section is assumed here to be a solid circle of radius R.

$$M(z,t) = \frac{\pi}{2}R^4\frac{\rho\omega^2 L}{\Omega}\left\{A\sin\frac{\Omega z}{L} - BA\cos\frac{\Omega z}{L}\right\}e^{i\omega t}. \qquad (5.7.16)$$

The values of A and B come from the end conditions. Consider a rod free at one end and fixed at the other end. The dynamic rigidity (torque M divided by angular displacement θ) in torsional vibration of an elastic right circular cylinder of radius R, density ρ, length L, and shear modulus G fixed at one end and dynamically driven at the other end is given by the following [1.6.1].

$$\frac{M}{\theta} = \frac{\pi R^4}{2}\left[\rho\omega^2 L\right]\frac{\operatorname{ctn}\Omega}{\Omega}. \qquad (5.7.17)$$

The development is elastic so far.

For a viscoelastic cylinder, the Fourier correspondence principle yields

$$\frac{M^*}{\theta} = \frac{\pi R^4}{2}\left[\rho\omega^2 L\right]\frac{\operatorname{ctn}\Omega^*}{\Omega^*}, \qquad (5.7.18)$$

in which

$$\Omega^* = \omega L\sqrt{\frac{\rho}{G^*}}, \qquad (5.7.19)$$

and G* is the complex shear modulus. These vibrations represent standing waves. If the loss tangent is small, considerable simplification is possible, as has been done in §3.5.

Results for torsional vibration of a circular cylindrical rod for materials with several values of tan δ are plotted in Fig. 3.5 (see also [5.7.4]). For low loss, resonance peaks occur in the structural compliance as anticipated by the analysis in §3.5; the higher harmonics occur at progressively smaller amplitude. The phase angle passes through 90° at the resonances. When tan δ = 1, the dynamic structural compliance displays no resonance peak due to the heavy damping. The dynamic compliance θ/M differs from the normalized compliance $\Gamma = G'k\theta/\tau$ considered in §3.5. Since the dynamic storage modulus G' depends on frequency by virtue of the Kramers–Kronig relations, the compliance θ/M rises at low frequency.

Kolsky [5.7.5] obtains the torsional wave equation, Eq. 5.7.13, by more elementary means as follows, by considering the torque M on a thin differential slice of length dz of the bar. The slice experiences a twisting moment M on one side and M + dM on the other side, since the moment varies with position z in the dynamic case. Newton's second law for the slice is written in terms of its angle of rotation θ and its mass moment of inertia I = ½ πρr⁴dz, with ρ as density and r as radius.

$$\frac{\partial M}{\partial z} dz = I \frac{\partial^2 \theta}{\partial t^2}.$$

The moment and angular displacement are related by the shear modulus G of the bar material.

$$M = \frac{1}{2} \pi G r^4 \frac{\partial \theta}{\partial z}.$$

By substitution,

$$\frac{G}{\rho} \frac{\partial^2 \theta}{\partial z^2} = \frac{\partial^2 \theta}{\partial t^2},$$

which is the wave equation.

Torsional vibration is simpler than longitudinal vibration since in the latter there is the following effect of Poisson's ratio. For long wavelength, the Poisson expansion and contraction occur freely. This is the case considered above. As the wavelength approaches the rod diameter, the Poisson contraction, since it varies with axial position, is accompanied by shear. For sufficiently high frequency, the wavelength of longitudinal waves becomes smaller than the cylinder size. In that case, the Poisson expansion and contraction are suppressed, leading to dispersion in which wave speed depends on frequency.

§5.7.3 Bending waves and vibration

For bending, analyze wave motion by the correspondence principle. For bending of an elastic rod, the relationship between moment and curvature is given by

$$M(z,t) = -EI\frac{\partial^2 u}{\partial x^2}, \tag{5.7.20}$$

in which z is a coordinate along the rod axis, u is a displacement in that direction, and I is the area moment of inertia of the cross section. Consider now a differential element of length dz along the rod. By applying Newton's second law, and considering that for small displacements the angular acceleration of the element may be neglected, the sum of moments [5.7.1] on the elements vanishes.

$$M(z+dz,t) - M(z,t) = Vdz, \tag{5.7.21}$$

so

$$\frac{\partial M}{\partial z} = V,$$

with V as the shear force.

As for the lateral motion,

$$\rho A dz \frac{\partial^2 u}{\partial t^2} = V(z+dz,t) - V(z,t) + f(z,t)dz. \tag{5.7.22}$$

By dividing by dz and assuming the forcing function f(z, t) to vanish,

$$\rho A \frac{\partial^2 u}{\partial t^2} = \frac{\partial V}{\partial z}.$$

By using the moment shear relation,

$$\frac{\partial^2 u}{\partial t^2} = \frac{E I}{\rho A}\frac{\partial^4 u}{\partial z^4}. \tag{5.7.23}$$

This is also a wave equation; however, the wave speed depends on frequency or wavelength even in the elastic case. To demonstrate this, substitute a trial wave solution of the form

$$u(z,t) = \sin(kz - \omega t).$$

Then, the dispersion relation is

$$\omega^2 = \frac{E}{\rho} \frac{I}{A} k^4. \tag{5.7.24}$$

The wave speed, specifically the phase velocity, for bending waves in an elastic rod is

$$c_B = \lambda v = \frac{\omega}{k} = k \sqrt{\frac{E}{\rho} \frac{I}{A}}. \tag{5.7.25}$$

Recall that for the angular frequency, $\omega = 2\pi v$ and for the wave number $k = 2\pi/\lambda$. The wave speed for bending waves is inversely proportional to the wavelength, so there is *dispersion*, in contrast to the case of torsional waves.

The wave speed c_B is the *phase velocity*, the speed at which the wave crests, or other features of constant phase, propagate [5.7.5]. In a dispersive medium such as this one, the energy of a pulse travels at a different speed, called the *group velocity*. For bending waves both the phase and group velocity given by the elementary formula for a slender bar tend to infinity for a short wavelength. This is not realistic. Therefore, when the wavelength becomes comparable to the bar thickness, more sophistication must be incorporated in the analysis. Specifically, the rotary motion and shear deformations of segments of the bar must be taken into account when the wavelength of flexural waves is small.

The wave propagation solution for a viscoelastic material is obtained as was done above for torsion by using the correspondence principle to replace E by E* in the dispersion relation. The complex modulus depends on frequency; therefore, the wave speed depends on frequency. The imaginary part of the complex modulus gives rise to wave attenuation. Then the dispersion relation becomes

$$\omega^2 = \frac{E^*}{\rho} \frac{I}{A} k^4. \tag{5.7.26}$$

By substituting in the wave equation, one finds that the loss tangent gives rise to attenuation, and the frequency-dependent real part of the modulus alters the pre-existing dispersion of waves.

Standing waves can be excited in a bar of finite length. Such waves are useful in experimental settings [1.6.23]. For a bar free at both ends, the natural frequencies go approximately as 3^2, 5^2, and 7^2. For a bar fixed at one end and free at the other, the natural frequencies go approximately as $(1.194)^2$, $(2.988)^2$, 5^2, and 7^2.

As in the case of torsion, the loss tangent can be extracted from the width of the resonance curve under driven conditions or from the time of decay of vibrations.

§5.7.4   Waves in three dimensions

Waves in one dimension have been considered in §3.7. In three dimensions, the field equation for classical elasticity is Navier's equation, with G (also called μ) as the shear modulus, λ as the second Lamé elastic constant (no relationship to the wavelength λ), F_i as the body force, ρ as the material density, and u as the particle displacement:

$$G\frac{\partial^2 u_i}{\partial x_j^{\,2}} + (\lambda + G)\frac{\partial^2 u_j}{\partial x_j \partial x_i} + F_i = \rho\frac{\partial^2 u_i}{\partial t^2}. \qquad (5.7.27)$$

Navier's equation is obtained from the elastic constitutive equation for *isotropic* materials,

$$\sigma_{ij} = \lambda\varepsilon_{kk}\delta_{ij} + 2\mu\varepsilon_{ij}, \qquad (5.7.28)$$

in which λ and μ are the two independent Lamé elastic constants, δ_{ij} is the Kronecker delta (1 if i = j, 0 if i ≠ j) and $\varepsilon_{kk} = \varepsilon_{11} + \varepsilon_{22} + \varepsilon_{33}$; and from the equation of motion

$$\rho\frac{\partial^2 u_i}{\partial t^2} = \frac{\partial\sigma_{ij}}{\partial x_j} + F_i. \qquad (5.7.29)$$

The equation of motion is the continuum representation of Newton's second law.

In three dimensions one can have *longitudinal* waves, also called dilatational or irrotational waves [5.2.2], in which the particle displacement is in the same direction as the wave propagation, as follows:

$$u_x = A\sin\frac{2\pi}{\lambda}(x - c_L t),\ u_y = 0,\ u_z = 0, \qquad (5.7.30)$$

with λ as the wavelength. This is a plane wave solution in which surfaces of constant phase, e.g., the wave crests, form parallel planes. The solution satisfies Navier's equation provided the wave speed is taken as [5.2.2]

$$c_L = \sqrt{\frac{\lambda + 2G}{\rho}}. \qquad (5.7.31)$$

Here, λ is the Lamé elastic constant. An alternative form is [5.2.3]

$$c_L = \sqrt{\frac{E(1-v)}{\rho(1+v)(1-2v)}}. \tag{5.7.32}$$

As for interpretation, the modulus tensor element C_{1111} is given for isotropic materials by [5.2.2]

$$C_{1111} = E\frac{1-v}{(1+v)(1-2v)}, \tag{5.7.33}$$

as shown in Example 5.9. This tensor element represents stiffness in the 1 direction, when there is no strain in the 2 or 3 directions; it is a constrained modulus. Deformation in longitudinal plane waves is constrained in that there is no room for lateral motion.

In three dimensions it is also possible to have *transverse* (or shear) waves in which the particle displacement is orthogonal to the direction of wave propagation, as follows. Transverse waves are characterized by *polarization*, or the plane containing the propagation direction and the direction in which the particles move. For example, for particle motion in the y direction,

$$u_y = A\sin\frac{2\pi}{\lambda}(x-c_T t), \; u_x = 0, \; u_z = 0, \tag{5.7.34}$$

with λ as the wavelength.

The solution satisfies Navier's equation provided the wave speed is taken as

$$c_T = \sqrt{\frac{G}{\rho}}. \tag{5.7.35}$$

The ratio of wave speeds depends on Poisson's ratio. For $v = 0.25$, $c_L = c_T\sqrt{3}$; for Poisson's ratio approaching 0.5, the longitudinal wave speed is much greater than the transverse wave speed, $C_L \gg C_T$.

The above development is for elastic solids. In three-dimensional viscoelastic bodies, we may apply the dynamic correspondence principle and replace the elastic constants which appear in the particle displacement equations via the wave speeds, with the corresponding complex viscoelastic functions. Attenuations are obtained as in the one-dimensional case considered in §3.7. The attenuations for longitudinal and transverse waves can differ, depending on the material, since an isotropic material has two independent stiffness values and two independent loss tangents.

§5.8   Noncorrespondence problems

There are problems of practical interest which cannot be treated by the correspondence principle. These include extending cracks, moving loads upon a surface, and ablation (or evaporation) of a solid. The correspondence principle cannot be applied in these cases because the interface between boundary regions under specified displacement and under specified stress moves with time, contradicting one of the assumptions used in deriving the correspondence principle. Solutions can nonetheless be obtained albeit with more difficulty [1.6.5].

§5.8.1   Solution by direct construction: example

In the direct construction approach, one proceeds ab initio, as was done in §5.3. The stress or strain history of each differential element of the material is examined explicitly with no appeal to any elastic solutions.

   As a simple example consider a weight w hanging from a viscoelastic rope of cross-sectional area A, length L, and weight which is negligible compared with w. The rope is initially supported at its upper end, and, of the total length, the length $L_1 + L_2$ is initially embedded in the support. The weight is applied at time t = 0. At a later time $t_1 > 0$, a section of depth L_1 of the support crumbles (or ablates), exposing more rope; at a later time $t_2 > t_1$, a section of depth L_2 of the support crumbles. Determine the downward displacement of the weight.

   For an elastic rope, the displacement u of the weight would be, for time $t > t_2$,

$$u = \frac{w}{EA}\left(L + L_1 + L_2\right).\tag{5.8.1}$$

The correspondence principle cannot be applied in this case since regions of the rope pass from a condition of specified displacement to a condition of specified load. We may consider the deformation of each region of the rope separately so that in the elastic case,

$$u(t) = \frac{wJ}{A}\left(L\,\mathcal{H}(t) + L_1\,\mathcal{H}(t - t_1) + L_2\,\mathcal{H}(t - t_2)\right).\tag{5.8.2}$$

In the viscoelastic case,

$$u(t) = \frac{w}{A}\left(LJ(t) + L_1J(t - t_1) + L_2J(t - t_2)\right).\tag{5.8.3}$$

This solution reflects the fact that different regions of the rope are loaded at different times.

This problem is not amenable to the correspondence principle; it is of interest to attempt a "solution" obtained by that method with the actual solution. Again consider Eq. 5.8.1 for the displacement of the end of an elastic rope for time $t > t_2$. By incorporating the s-multiplied Laplace transform of the compliance $J = E^{-1}$,

$$u(s) = sJ(s)\frac{w(s)}{A}(L + L_1 + L_2).$$

By transforming back,

$$u(t) = \frac{L + L_1 + L_2}{A} \int_0^t J(t - \tau)\frac{dw(\tau)}{d\tau}d\tau.$$

To proceed further, the load history $w(t)$ is required; however, different regions of the rope are loaded at different times. Therefore, a single functional form for the load history is not available.

§5.8.2   A generalized correspondence principle

A generalized correspondence principle was developed by Graham [5.8.1] and discussed by Christensen [1.6.1]. One assumes that an elastic solution is known in which a displacement, strain, or stress $c^{el}(r, t)$ is known which is separable into a product of a function $\xi(\mu, B)$ of the elastic constants μ and B, and a function $C^e(r, t)$ of the spatial coordinates r and time t.

$$c^{el}(r, t) = \xi(\mu, B)C^e(r, t). \tag{5.8.4}$$

Suppose further that $C^e(r, t) = 0$ upon a region B_2 (part of the boundary) and the region B_1 is monotonically decreasing. Then, with s as the Laplace transform variable,

$$c^{visc}(r, t) = K(t)C^e(r, 0) + \int_0^t K(t - \tau)\frac{dC^e(r, \tau)}{d\tau}d\tau, \tag{5.8.5}$$

with

$$K(t) = \mathcal{L}^{-1}\left\{\left[\frac{1}{s}\xi(s\mu(s), sB(s))\right]\right\}. \tag{5.8.6}$$

So, a subset of problems not amenable to the correspondence principle can be solved by the generalized correspondence principle.

§5.8.3 *Specific cases*

Many noncorrespondence type problems and their solutions are given in References [1.6.1, 1.6.5]. An interesting example is that of the winding of tape or paper onto rolls [5.8.2] in which the tape is under tension and is added one layer at a time. As each layer is added it alters the stress state in the hub and in the layers present at that time. If the tape is viscoelastic, the stress state at the end of the winding phase depends on the winding speed. Storage or pausing of the tape will result in stress redistribution.

Rolling of a rigid cylinder [5.8.3] upon a viscoelastic substrate is a noncorrespondence problem since regions of the substrate experience different boundary conditions as time progresses and they come into contact and lose contact with the roller. Viscoelasticity of the substrate gives rise to rolling friction as discussed in §10.7.

Indentation of a substrate by a curved indenter also gives rise to a noncorrespondence problem [5.8.4, 5.8.5].

§5.9 *Bending in nonlinear viscoelasticity*

Problems considered thus far have been within the linear theory. We consider here a simple problem of bending of a beam of material which undergoes steady-state (secondary) creep [5.9.1] governed by the following nonlinear relation:

$$\frac{d\varepsilon}{dt} = A\sigma^n. \tag{5.9.1}$$

Although this relation between stress and strain rate is nonlinear, it deals only with creep and offers no prediction of the response to stress which is not constant in time.

We consider z to be along the beam axis and y to be along the beam depth so that the beam is bent in the plane zy. The beam has a depth d in the y direction and a width b so that $I = bd^3/12$ is the area moment of inertia as in Fig. 5.1. The "initial" response, considered as elastic, is that the curvature Ω is

$$\Omega = \frac{M}{EI} = \frac{MJ}{I},$$

and the stress is

$$\sigma = \frac{My}{I}.$$

The bending moment is

$$b \int_{-d/2}^{d/2} y\sigma \, dy = M(t).$$

(5.9.2)

The creep strain rate is, from the strain–curvature relation,

$$\frac{d\varepsilon}{dt} = y\frac{d\Omega}{dt} = A\sigma^n.$$

(5.9.3)

By combining the above,

$$M = \left\{ \Omega \frac{b^n d^{2n+1}(n)^n}{2A(4n+2)^n} \right\}^{1/n}.$$

(5.9.4)

The stress, obtained from the above is

$$\frac{\sigma(y)}{\sigma_{max}} = \frac{2}{3}\left(1+\frac{1}{2n}\right)\left|\frac{2y}{d}\right|^{1/n}, \text{ for } 0 < \frac{2y}{d} < 1,$$

(5.9.5)

$$\frac{\sigma(y)}{\sigma_{max}} = -\frac{2}{3}\left(1+\frac{1}{2n}\right)\left|\frac{2y}{d}\right|^{1/n}, \text{ for } -1 < \frac{2y}{d} < 0.$$

(5.9.6)

The stress distribution is no longer linear across the depth of the beam as it is in linear elasticity and in linear viscoelasticity. Instead, the stress is an antisymmetrical function of the coordinate y and the creep exponent n, but not on time. The stress would redistribute itself with time during the primary phase of nonlinear creep, not considered here.

§5.10 *Dynamic response of nonlinear materials*

The stress response of nonlinear materials to dynamic strain histories is treated in the following analysis. Recall that for linear materials, an analysis of the material's response to oscillatory strain or stress histories leads to the introduction of the complex moduli or compliances as done in §3.2. In a linear material, sinusoidal strain gives rise to sinusoidal stress but with a phase shift. Nonlinear solids obeying nonlinear superposition and solids obeying the Green–Rivlin series (§2.12) are considered. The strain history may contain a single frequency component or a superposition of multiple frequencies. For both types of nonlinear material considered, the single frequency strain input gives rise to multiple frequencies in the stress output. A multiple frequency strain input generates a response at new frequencies, expressed as the sum and difference of the input frequencies, for the Green–Rivlin material only.

Consider first [5.10.1, 1.6.11] a solid describable by the nonlinear super-position integral, Eq. 2.12.3, in which the kernel $E(t, \varepsilon)$ is separable:

$$E(t, \varepsilon) = E_0(t) + A(t)B(\varepsilon). \tag{5.10.1}$$

If the strain is given by

$$\varepsilon(t) = \varepsilon_0 \sin \omega t,$$

then the integral becomes

$$\frac{\sigma(t)}{\varepsilon_0} = E'(\omega)\sin \omega \tau + E''(\omega)\cos \omega \tau + A'(\omega)B(\varepsilon_0 \sin \omega \tau) + A''(\omega)B(\varepsilon_0 \cos \omega \tau), \tag{5.10.2}$$

in which

$$E'(\omega) = \omega \int_0^\infty \left[E_0(t') - E_e \right] \sin \omega t' dt' + E_e, \tag{5.10.3}$$

$$E''(\omega) = \omega \int_0^\infty \left[E_0(t') - E_e \right] \cos \omega t' dt', \tag{5.10.4}$$

$$A'(\omega) = \omega \int_0^\infty A(t') \sin \omega t' dt', \tag{5.10.5}$$

and

$$A''(\omega) = \omega \int_0^\infty A(t') \cos \omega t' dt'. \tag{5.10.6}$$

The response of the solid to a strain history containing a single frequency component,

$$\varepsilon(t) = a_1 \sin \omega_1 t, \tag{5.10.7}$$

is obtained by expanding the strain-dependent part $B(\varepsilon)$, eliminating powers of the trigonometric functions by means of multiple-angle identities, and collecting terms:

$$\frac{\sigma(t)}{\varepsilon_o} = \sin \omega t \left\{ E'(\omega) + A'(\omega) \left[B(0) + \frac{3}{4} \frac{B^{(2)}(0)}{2!} \varepsilon_o^2 - \frac{65}{16} \frac{B^{(4)}(0)}{4!} \varepsilon_o^4 + \ldots \right] \right\}$$

$$+ \cos \omega t \left\{ E''(\omega) + A''(\omega) \left[B(0) - \frac{1}{4} \frac{B^{(2)}(0)}{2!} \varepsilon_o^2 + \frac{5}{8} \frac{B^{(4)}(0)}{4!} \varepsilon_o^4 - \ldots \right] \right\}$$

$$+ \sin 2\omega t \left\{ A''(\omega) \left[\frac{1}{2} \frac{B^{(1)}(0)}{1!} \varepsilon_o + \frac{1}{4} \frac{B^{(3)}(0)}{3!} \varepsilon_o^3 + \ldots \right] \right\}$$

$$+ \cos 2\omega t \left\{ A'(\omega) \left[-\frac{1}{2} \frac{B^{(1)}(0)}{1!} - \frac{1}{2} \frac{B^{(3)}(0)}{3!} - \ldots \right] \right\}$$

$$+ A'(\omega) \left[\frac{1}{2} \frac{B^{(1)}(0)}{1!} \varepsilon_o + \frac{3}{2} \frac{B^{(3)}(0)}{3!} \varepsilon_o^3 + \ldots \right]$$

$$+ \sin 3\omega t \{\ldots\} + \text{higher order terms,}$$

$$(5.10.8)$$

where

$$B^{(n)} = \frac{d^n B}{d\varepsilon^n}.$$

The strain input at single frequency is seen to generate harmonics (integer multiples of the original frequency ω) in stress. This sort of response is called *harmonic distortion* in the electrical engineering community.

Now if the solid obeying nonlinear superposition is subjected to a strain history containing many frequencies,

$$\varepsilon(t) = \sum_{n=1}^{N} a_n \sin \omega_n t,$$

the stress is

$$\frac{\sigma(t)}{\varepsilon_o} = \sum_{n=1}^{N} \left\{ E_e \sin \omega_n t + \omega_n \int_{-\infty}^{t} \left[E(t - \tau, \varepsilon) - E_e \right] \cos \omega_n \tau \, d\tau \, a_n \right\}. \quad (5.10.9)$$

This is the sum of the responses to the individual frequency components with no interaction between them.

Consider now a solid describable by the Green–Rivlin series, Eq. 2.12.8. The frequency response of this type of solid has been treated by Lockett and Gurtin [5.10.2]; salient portions of their analysis are presented below.

In linear viscoelasticity theory, the problem of dynamic response can be analyzed with the aid of the complex, one-sided Fourier transform of the stress–relaxation function $E_1(t)$,

$$\hat{E}_1(\omega) = \int_0^\infty E_1(s) e^{-i\omega s} ds. \tag{5.10.10}$$

For the treatment of the nonlinear solid, a multiple transform may be defined analogously

$$\hat{E}(\omega_1, \omega_2, ..., \omega_n)$$

$$= \int_0^\infty \int_0^\infty ... \int_0^\infty E_n(s_1, s_2, ..., s_n) \times e^{-i(\omega_1 s_1 + \omega_2 s_2 + ... + \omega_n s_n)} ds_1 ds_2 ... ds_n. \tag{5.10.11}$$

This may be used to obtain from the Green–Rivlin series a nonlinear constitutive equation in the frequency domain.

The problem of determining the solid's stress response to a strain history containing multiple frequency components,

$$\varepsilon(t) = \sum_{n=1}^{N} a_n \sin \omega_n t, \tag{5.10.12}$$

may be approached more directly by substituting the above history in the Green–Rivlin series and making use of trigonometric identities. The first and second terms, called $\sigma_1(t)$ and $\sigma_2(t)$, of the Green–Rivlin series, become

$$\sigma_1(t) = \sum_{n=1}^{N} a_n \left[E'(\omega_n) \sin \omega_n t + E''(\omega_n) \cos \omega_n t \right], \tag{5.10.13}$$

$$\sigma_2(t) = \frac{1}{2} \sum_{n=1}^{N} \sum_{p=1}^{N} a_n a_p \omega_n \omega_p \left[\operatorname{Re} \hat{E}_2(\omega_n, \omega_p) \cos(\omega_n + \omega_p) t \right.$$

$$- \operatorname{Im} \hat{E}(\omega_n, \omega_p) \sin(\omega_n + \omega_p) t$$

$$+ \operatorname{Re} \hat{E}_2(\omega_n, -\omega_p) \cos(\omega_n - \omega_p) t$$

$$\left. - \operatorname{Im} \hat{E}_2(\omega_n, -\omega_p) \sin(\omega_n - \omega_p) t \right]. \tag{5.10.14}$$

The sums in the term of order three and higher grow rapidly in complexity with the order of the term. So, in a Green–Rivlin solid, oscillations occur at new frequencies not originally present in the original excitation. This sort of response is called *intermodulation distortion* in the electrical engineering community.

The results of these analyses may be summarized as follows:

1. For both the Green–Rivlin solid and a solid obeying nonlinear superposition, the stress response to a sinusoidally varying strain contains the original frequency, harmonics at integer multiples of this frequency, and a constant or "DC" stress.
2. If the response $\sigma(t)/\varepsilon_o$ is invariant to the sign of ε_o, as must be the case in shear for an isotropic solid or in torsion about the symmetry axis of an axisymmetrical solid, all terms in the representation of $\sigma(t)$ containing odd powers of ε_o vanish, so that all even harmonics as well as the static DC stress also vanish.
3. The response in tension–compression is not restricted in this way and may contain all harmonics plus a constant stress.
4. In musical language, these statements may be expressed as follows: the solid vibrated in shear with strain history equivalent to the tone of the flute responds with a stress corresponding to the tone of the clarinet, while in compression it could respond with the tone of the saxophone.
5. The response of the two types of solid is indistinguishable if single frequency excitation is used. A strain excitation containing several frequencies will generate "interactions" in the Green–Rivlin solid; however, the stress response of the solid obeying nonlinear superposition will be equivalent to the sum of the responses to the individual frequencies; no interactions occur in this case.
6. The dynamic response at the driving frequency $E'(\omega)$ can be less strain dependent (for identical maximum strain levels and for $t = 1/\omega$) than the relaxation response $E(t)$. This last phenomenon has been observed in bone.

§5.11   *Summary*

Many problems in determining stress distributions in viscoelastic objects can be treated simply via the correspondence principle, provided a solution for an elastic object of the same geometry is available. In problems involving changing boundary conditions or nonlinear behavior, the correspondence principle is not applicable; direct construction or other methods may be applied in those cases.

Examples

Example 5.1

An elastic bar (of negligible damping) has a Young's modulus E_1 and dimensions b and h. A thin layer of thickness $\xi \ll h$ of a linearly viscoelastic

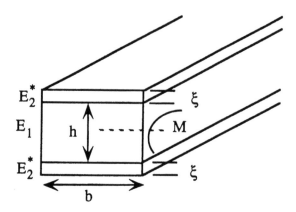

Figure E5.1   Elastic bar with viscoelastic layers.

material of stiffness E_2^* is applied to two lateral surfaces of width b, as shown in Fig. E5.1.

Show that the effective loss tangent of the composite bar in bending is

$$\tan \delta_{\text{eff}} \approx 6 \frac{\xi}{h} \frac{E_2'}{E_1} \tan \delta_2. \tag{E5.1.1}$$

Solution

Use the correspondence principle. Since the loss tangent is required, the Fourier version of the correspondence principle is appropriate. First, obtain the elastic solution for the bending rigidity. The bar's area moment of inertia is $\frac{1}{12} bh^3$. The added layers are also rectangular in cross section with area $A = b\xi$ each and are displaced from the neutral axis by a distance d. Their contribution to the total area moment is given by the parallel axis theorem:

$$Ad^2 = 2b\xi \frac{(h+\xi)^2}{2}. \tag{E5.1.2}$$

The bending rigidity, by assuming *elastic* layers, is therefore,

$$\Gamma = \frac{M}{\Omega} = E_1 \left[\frac{1}{12} bh^3 + 2b\xi \frac{E_2}{E_1} \frac{1}{4} (h+\xi)^2 \right]. \tag{E5.1.3}$$

Apply the correspondence principle,

$$\frac{M^{\bullet}}{\Omega} = E_1\left[\frac{1}{12}bh^3 + 2b\xi\frac{E_2^{\bullet}}{E_1}\frac{1}{4}(h+\xi)^2\right]. \tag{E5.1.4}$$

$$\tan\delta_{eff} = \frac{\text{Im}\{\Gamma\}}{\text{Re}\{\Gamma\}} = \frac{2b\xi\dfrac{E_2''}{E_1}\dfrac{1}{4}(h+\xi)^2}{\dfrac{1}{12}bh^3 + 2b\xi\dfrac{E_2'}{E_1}\dfrac{1}{4}(h+\xi)^2}. \tag{E5.1.5}$$

Multiply both numerator and denominator by $12/bh^3$,

$$\tan\delta_{eff} = \frac{6\dfrac{\xi}{h}\dfrac{E_2''}{E_1}\dfrac{(h+\xi)^2}{h^2}}{1+6\dfrac{\xi}{h}\dfrac{E_2'}{E_1}\dfrac{(h+\xi)^2}{h^2}}. \tag{E5.1.6}$$

For a thin layer, $\xi \ll h$, so, with the definition of $\tan\delta$,

$$\tan\delta_{eff} \approx \frac{6\dfrac{\xi}{h}\dfrac{E_2'}{E_1}\tan\delta_2}{1+6\dfrac{\xi}{h}\dfrac{E_2'}{E_1}}. \tag{E5.1.7}$$

This may be simplified further if the layer material is not too stiff:

$$\tan\delta_{eff} \approx 6\frac{\xi}{h}\frac{E_2'}{E_1}\tan\delta_2. \tag{E5.1.8}$$

Consequently, the effective damping of the composite beam is maximized if the added layer is made as thick as possible, and of a stiff material with the largest possible value of loss tangent. Polymer layers are applied to metallic structural elements to reduce vibration and noise as discussed in §10.6.

Example 5.2

A concentrated load P is suddenly applied orthogonal to the surface of a semi-infinite viscoelastic region with a horizontal boundary; then the load is held constant. What is the vertical motion of points on the surface as a function of time? Suppose Poisson's ratio is constant. Ignore inertial effects.

Solution

Use the correspondence principle. The quasistatic solution for the displacement field u of an elastic region [5.2.2] under a concentrated load P is as follows:

$$u_x = -P\frac{(1-2v)(1+v)}{2\pi Er}, \quad u_z = \frac{P(1-v^2)}{\pi Er}. \qquad (E5.2.1)$$

The coordinate z is down, perpendicular to the surface; r is the radial distance from the point of load application. By applying the correspondence principle for the time domain, and allowing the load to be time dependent,

$$u_z(s, r) = \frac{(1-v^2)}{\pi r} sP(s)J(s). \qquad (E5.2.2)$$

By transforming back,

$$u_z(r,t) = \frac{(1-v^2)}{\pi r} \int_0^t J(t-\tau)\frac{dP}{d\tau}d\tau. \qquad (E5.2.3)$$

Since the load history is a step function in time, the displacement history follows the creep function.

$$u_z(r,t) = \frac{(1-v^2)}{\pi r} P\, J(t). \qquad (E5.2.4)$$

Example 5.3

A viscoelastic rope of density ρ radius r, and length L is suddenly lowered off a cliff at time zero, and immediately a man of mass m begins to climb down at speed v. Write expressions for the stress, strain, and displacement of the rope as a function of a vertical coordinate y (in the down direction) and time t. Neglect thrashing oscillations of the rope after it is lowered.

Solution

Proceed by direct construction. The stress is

$$\sigma(y,t) = \rho g y\, \mathcal{H}(t) + \frac{mg}{\pi r^2}\, \mathcal{H}\left(t - \frac{y}{v}\right). \qquad (E5.3.1)$$

The first term represents the stress due to the weight of the rope and is simply the weight of the rope divided by its cross-sectional area. The step function $\mathcal{H}(t)$ reflects the fact that the gravitational stress begins at time zero. The second term represents the stress due to the weight of the man. The step function has a value of unity for vt > y or y < vt. Only the rope above the man experiences gravitational load due to his weight, but as he descends, more rope is stressed.

Each differential element of rope, as it receives a step stress, exhibits a strain response which follows the creep function. Therefore, the strain in the y direction is

$$\varepsilon(y,t) = \rho g y J(t) + \frac{mg}{\pi r^2} J\left(t - \frac{y}{v}\right). \tag{E5.3.2}$$

The displacement u is obtained by integrating the strain with respect to y. The upper limit in the second term is the y position of the man, $y_m = vt$.

$$u(y,t) = \rho g \frac{y^2}{2} J(t) + \frac{mg}{\pi r^2} \int_0^{vt} J\left(t - \frac{y}{v}\right) dy. \tag{E5.3.3}$$

Consider as a special case an elastic rope, for which $J(t) = J_0$ is constant and equivalent to the inverse of Young's modulus E. Then

$$u(y,t) = \rho g \frac{y^2}{2} \frac{1}{E} + \frac{mg}{\pi r^2} \frac{1}{E} \int_0^{vt} \mathcal{H}\left(t - \frac{y}{v}\right) dy. \tag{E5.3.4}$$

The integral of a step function is a slope function.

$$u(y,t) = \rho g \frac{y^2}{2} \frac{1}{E} + \frac{mg}{\pi r^2} \frac{1}{E} y \mathcal{H}\left(t - \frac{y}{v}\right). \tag{E5.3.5}$$

Example 5.4

Consider again the viscoelastic rope in Example 5.3. Is the generalized correspondence principle applicable? If so, use it to obtain a solution. Consider only the deformation due to the weight of the man.

Solution

The elastic solution in Eq. E5.3.4 can be decomposed into a product of a function ξ of the elastic constant $E = J^{-1}$, and a function $C^e(y, t)$ of the spatial coordinate y and time t, as follows:

$$\xi = J\frac{mg}{\pi r^2},$$

$$C^e(y,t) = \int_0^{vt} \mathcal{H}\left(t - \frac{y}{v}\right)dy.$$

The deformation due to the weight of the man occurs only for $y < vt$, and that deformation is zero in the monotonically decreasing region of rope below the man, as required in the statement of the generalized correspondence principle.

By following Eq. 5.8.6, evaluate K,

$$K(t) = \mathcal{L}^{-1}\left\{\frac{1}{s}\xi(sJ(s))\right\} = J(t)\frac{mg}{\pi r^2}.$$

By following Eq. 5.8.5, the viscoelastic solution for the displacement is

$$u^{visc}(y,t) = K(t)C^e(y,0) + \int_0^t K(t-\tau)\frac{dC^e(y,\tau)}{d\tau}d\tau.$$

By substituting,

$$u^{visc}(y,t) = J(t)\frac{mg}{\pi r^2}\int_0^0 \mathcal{H}\left(t - \frac{y}{v}\right)dy + \frac{mg}{\pi r^2}\int_0^t J(t-\tau)\frac{d}{d\tau}\int_0^{v\tau}\mathcal{H}\left(\tau - \frac{y}{v}\right)dy\,d\tau.$$

By interchanging the order of integration in the second term and carrying out the differentiation,

$$u^{visc}(y,t) = 0 + \frac{mg}{\pi r^2}\int_0^{vt}\int_0^t J(t-\tau)\delta\left(\tau - \frac{y}{v}\right)d\tau\,dy.$$

With the sifting property of the delta function,

$$u^{visc}(y,t) = \frac{mg}{\pi r^2}\int_0^{vt} J\left(t - \frac{y}{v}\right)dy.$$

This is identical to the second term of Eq. E5.3.3 which was obtained by direct construction. In this example, the generalized correspondence principle does not offer much simplification since most of the effort in solving this problem is in keeping track of the relationship between deformation, time, and position; and that effort is required to obtain the elastic solution.

Example 5.5

Consider bending of a circular thin plate of thickness 2h and radius a, clamped at the edge. For an elastic plate of Young's modulus E and Poisson's ratio v, the deflection w in response to a uniformly applied force F is as follows [E5.5.1], provided w << h.

$$w = \frac{F}{64\pi a^2 D}\left(a^2 - r^2\right)^2$$

with

$$D = \frac{2}{3}Eh^3 \frac{1}{1-v^2}$$

as the flexural rigidity.

How is the deflection related to the force for a viscoelastic plate, if Poisson's ratio is constant in time?

Solution

Apply the correspondence principle to the elastic solution. Poisson's ratio has been assumed constant, so it is not transformed.

$$w(s) = \frac{F(s)\left(1-v^2\right)}{\frac{2}{3}64\pi a^2 sE(s)h^3}\left(a^2 - r^2\right)^2.$$

By converting to the compliance formulation to obtain a product of transforms which can be handled with the convolution theorem,

$$w(s) = \frac{\left(1-v^2\right)}{\frac{2}{3}64\pi a^2 h^3}\left(a^2 - r^2\right)^2 F(s)sJ(s).$$

By transforming back to obtain the time dependence of the deflection,

$$w(t) = \frac{\left(1-v^2\right)}{\frac{2}{3}64\pi a^2 h^3}\left(a^2 - r^2\right)^2 \int_0^t J(t-\tau)\frac{dF}{d\tau}d\tau.$$

This result consists of the product of a geometrical factor with a Boltzmann integral as was the case with beam bending examined in §5.5. In the present example, such a simple form is obtained only for Poisson's ratio as a constant.

Example 5.6

Determine the viscoelastic Poisson's ratio of a material from its shear and tensile properties.

Solution

Referring to the interrelations among elastic constants given in §5.6.3,

$$v = \frac{E}{2G} - 1. \qquad (E5.6.1)$$

By applying the correspondence principle,

$$sv(s) = \frac{sE(s)}{2sG(s)} - 1. \qquad (E5.6.2)$$

Since the convolution theorem is to be used in evaluating the inverse transform, write this as a product in terms of the shear compliance $J_G = G^{-1}$.

$$sv(s) = \frac{1}{2} sE(s)sJ_G(s) - 1. \qquad (E5.6.3)$$

By dividing by s and transforming back,

$$v(t) = \frac{1}{2} \int_0^t E(t - \tau) \frac{dJ_G(\tau)}{d\tau} d\tau - 1. \qquad (E5.6.4)$$

Here the time-dependent Poisson's ratio is expressed in terms of a convolution of the axial relaxation modulus and the derivative of the shear creep compliance.

If *dynamic* data are available, the dynamic, frequency-dependent Poisson's ratio may be obtained by applying the Fourier version of the correspondence principle as follows:

$$v^* = \frac{E^*}{2G^*} - 1 = \frac{E' + iE''}{2(G' + iG'')} - 1. \qquad (E5.6.5)$$

$$v^* = \frac{1}{2}(E' + iE'')\left\{\frac{G'}{G'^2 + G''^2} - i\frac{G''}{G'^2 + G''^2}\right\} - 1. \qquad (E5.6.6)$$

By referring to Example 3.6, since $G^* = 1/J_G^*$,

$$v^* = \frac{1}{2}(E' + iE'')(J_G' - iJ_G'') - 1. \qquad (E5.6.7)$$

The dynamic relationship is simpler to obtain since the algebraic manipulation of complex numbers is in this case simpler than calculation of a convolution integral.

Example 5.7

Must the phase angle in Poisson's ratio be positive?

Solution

Refer to Eq. E5.6.7, and consider the following cases.

For $E = 2.6(1 + 0.1i)$ GPa and $J_G^* = 1$ GPa^{-1}, the loss tangent is 0.1 in tension and zero in torsion, and we obtain

$$v^* = 0.3 + 0.13i.$$

For $E = 2.6$ GPa, $J_G^* = 1(1 - 0.1i)$ GPa^{-1}, the loss tangent is 0.1 in torsion and zero in tension, and we obtain

$$v^* = 0.3 - 0.13i.$$

So for allowable, positive, phase angles in the axial relaxation modulus and the shear compliance, the phase angle in Poisson's ratio can be positive or negative. As for interpretation, observe that Poisson's ratio is defined as the ratio of the transverse contraction strain to the longitudinal extension strain. Therefore, there is no energy associated with Poisson's ratio or its phase angles. This is in contrast with the phase angle between stress and strain, which, as presented in §3.4, is associated with work done per cycle per unit volume.

Example 5.8

Explicitly derive Eq. 5.6.6, showing all the steps.

Solution

Consider Eq. 5.6.5,

$$w(s) = \frac{1}{2R}\left[sF(s)J(s) - J(s)v(s)v(s)s^3F(s)\right].$$

To apply the convolution theorem, which deals with a product of two functions in the Laplace plane, define $v(s)v(s)s^3F(s) = \Xi(s)$. So by following an inverse Laplace transform,

$$w(t) = \frac{1}{2R}\int_0^t J(t-\tau)\frac{dF(\tau)}{d\tau}d\tau - \frac{1}{2R}\int_0^t J(t-\tau)\frac{d\Xi(\tau)}{d\tau}d\tau.$$

Again define $\Lambda(s) = v(s)s^3F(s)$, so $\Xi(s) = \Lambda(s)v(s)$, so with another inverse Laplace transform,

$$\Xi(\tau) = \int_0^\tau v(\tau-\eta)\frac{d\Lambda(\eta)}{d\eta}d\eta.$$

Since Ξ is a function of τ, a new variable of integration is named η. Similarly, transforming back $\Lambda(s)$ gives the following. Different equivalent forms are possible depending on how one collects the multiples of s, e.g., $v(s)\{s^3F(s)\} = \{sv(s)\}\{s^2F(s)\}$.

$$\Lambda(\eta) = \int_0^\eta v(\eta-\xi)\frac{d^3F(\xi)}{d\xi^3}d\xi.$$

Combining these gives Eq. 5.6.6,

$$w(t) = \frac{1}{2R}\int_0^t J(t-\tau)\frac{dF}{d\tau}d\tau - \frac{1}{2R}\left\{\int_0^t J(t-\tau)\left[\int_0^\tau v(\tau-\eta)\left[\int_0^\eta v(\eta-\xi)\frac{d^3F(\xi)}{d\xi^3}d\xi\right]d\eta\right]d\tau\right\}.$$

Example 5.9

Analyze the behavior of a thin layer of rubbery material compressed between rigid platens. Discuss the implications both for an elastic layer and for a viscoelastic layer.

Solution

Proceed first to consider elastic materials via the tensorial version of Hooke's law, and then the elementary three-dimensional form; finally consider viscoelastic materials.

In three dimensions, Hooke's law of linear elasticity is given in tensorial form by Eq. 2.8.1,

$$\sigma_{ij} = C_{ijkl}\varepsilon_{kl}.$$

The index notation, with i, j, k, and l each able to assume values from 1 to 3 is used as is the Einstein summation convention in which repeated indices are summed over.

Consider simple tension or compression in the 1 (or x) direction: $\sigma_{11} \neq 0$, $\sigma_{22} = 0$, $\sigma_{33} = 0$. Expand the sum, assuming orthotropic symmetry.

$$\sigma_{11} = C_{1111}\varepsilon_{11} + C_{1122}\varepsilon_{22} + C_{1133}\varepsilon_{33}.$$

Use the definition of Poisson's ratio to write this in mixed tensorial and engineering symbols:

$$\sigma_{11} = C_{1111}\varepsilon_{11} + C_{1122}\left(-v_{12}\varepsilon_{11}\right) + C_{1133}\left(-v_{13}\varepsilon_{11}\right).$$

The ratio of stress to strain in simple tension in the 1 direction is Young's modulus E_1 in the 1 direction.

$$E_1 = \frac{\sigma_{11}}{\varepsilon_{11}} = \left\{C_{1111} - v_{12}C_{1122} - v_{13}C_{1133}\right\}.$$

Consider the 2 (or y) direction in which there is zero stress, and again incorporate the definition of the Poisson's ratios.

$$0 = \sigma_{22} = C_{2211}\varepsilon_{11} + C_{2222}\varepsilon_{22} + C_{2233}\varepsilon_{33} = \left\{C_{2211} - v_{12}C_{2222} - v_{13}C_{2233}\right\}.$$

Assume the material is isotropic, so both the stiffness and Poisson's ratio are independent of direction and $C_{1111} = C_{2222}$. Then

$$C_{2211} - vC_{2211} = vC_{1111}; \text{ or } C_{2211} = \frac{v}{1-v}C_{1111}.$$

By substituting the above and dropping the subscript on E since in an isotropic material there is no dependence of properties on direction,

$$E = \left\{ 1 - \frac{2v^2}{1-v} \right\} C_{1111},$$

or

$$C_{1111} = E \frac{1-v}{(1+v)(1-2v)}. \tag{E5.9.1}$$

Since $E = 2G(1 + v)$ and

$$v = \frac{3B - 2G}{6B + 2G}$$

the modulus tensor element is written

$$C_{1111} = B + \frac{4}{3}G. \tag{E5.9.2}$$

One may also work with the elementary isotropic form for Hooke's law.

$$\varepsilon_{xx} = \frac{1}{E}\left\{ \sigma_{xx} - v\sigma_{yy} - v\sigma_{zz} \right\},$$

$$\varepsilon_{yy} = \frac{1}{E}\left\{ \sigma_{yy} - v\sigma_{xx} - v\sigma_{zz} \right\},$$

$$\varepsilon_{zz} = \frac{1}{E}\left\{ \sigma_{zz} - v\sigma_{xx} - v\sigma_{yy} \right\}.$$

Consider simple tension or compression in the x direction: $\sigma_{xx} \neq 0$, $\sigma_{yy} = 0$, $\sigma_{zz} = 0$. Then

$$\frac{\sigma_{xx}}{\varepsilon_{xx}} = E.$$

Consider constrained compression, with $\varepsilon_{yy} = 0$, $\varepsilon_{zz} = 0$. Then

$$\sigma_{yy} = v\sigma_{xx} + v\sigma_{zz}.$$

$$\sigma_{zz} = v\sigma_{xx} + v\sigma_{yy}.$$

By substituting,

$$\sigma_{yy} = \sigma_{zz} = \sigma_{xx}\frac{v(1+v)}{1-v^2}.$$

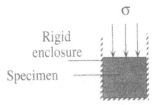

Figure E5.9 Geometry for constrained compression.

So substituting into Hooke's law, the stress–strain ratio for constrained compression, which by definition is the constrained modulus C_{1111}, is

$$\frac{\sigma_{xx}}{\varepsilon_{xx}} = C_{1111} = E\frac{1-v}{(1+v)(1-2v)}.$$

The physical meaning of C_{1111} is the stiffness for tension or compression in the x (or 1) direction, when strain in the y and z directions is constrained to be zero. The reason is that for such a constraint the sum in Eq. 2.8.1 collapses into a single term containing only C_{1111}. The constraint could be applied by a rigid mold, as shown in Fig. E5.9, or if the material is compressed in a thin layer between rigid platens. C_{1111} also governs the propagation of longitudinal waves in an extended medium, since the waves undergo a similar constraint on transverse displacement.

Rubbery materials have Poisson's ratios very close to 1/2, shear moduli on the order of a MPa, and bulk moduli on the order of a GPa. Therefore, the constrained modulus C_{1111} is comparable to the bulk modulus and is much larger than the shear or Young's modulus of rubber.

For a viscoelastic material, apply the dynamic correspondence principle.

$$C^*_{1111} = B^* + \frac{4}{3}G^*.$$

Express this in terms of the loss tangents associated with each modulus.

$$C'_{1111}(1+i\tan\delta_c) = B'(1+i\tan\delta_B) + \frac{4}{3}G'(1+i\tan\delta_G).$$

Although $B' \gg G'$ in the rubbery regime of polymers, we may have $B'' \approx G''$ in the leathery regime as discussed in §7.3.

Problems

5.1 Determine the relationship between the torque history and angular displacement history for torsion of a rod of circular cross section. Does relaxation in Poisson's ratio have any effect?

5.2 Consider the exact three-dimensional elasticity solution [5.2.2] for the displacement field in pure bending of a bar of rectangular cross section.

$$u_x = -\frac{1}{2R}\left\{z^2 + v\left(x^2 - y^2\right)\right\},$$

$$u_y = -\frac{vxy}{R},$$

$$u_z = \frac{xz}{R}.$$

R is the radius of curvature of bending and v is Poisson's ratio.
In what ways will the behavior of a viscoelastic bar differ from that of an elastic one? Consider, in particular, the effect of the viscoelastic Poisson's ratio which is assumed to vary in time.

5.3 Does a stretched rod of *arbitrary* material held at constant extension get fatter or thinner with time? Hint: either make a demonstration based on continuum concepts, or envisage a material microstructure which gives rise to a decreasing Poisson's ratio.

5.4 A slender beam of uniform cross section is extruded horizontally at speed v from a block of rigid material [1.6.5]. The beam droops under its own weight. If the beam is linearly viscoelastic, determine the amount of droop as a function of time.

5.5 A concentrated load P is suddenly applied orthogonal to the surface of a semi-infinite viscoelastic region with a horizontal boundary. Consider the load history as a step function but ignore inertial effects. What is the horizontal motion of points on the surface as a function of time? Suppose the Poisson's ratio is constant. Do points move toward or away from the load? What happens when the load is removed?

5.6 Consider bending of a circular thin plate of thickness 2h and radius a, clamped at the edge. The plate is of a linearly viscoelastic material with relaxation modulus E(t) and Poisson's ratio v(t), neither of which is constant in time. Find the relationship between the deflection history w(t) in response to a uniformly applied force F(t).

5.7 Determine the bulk properties of a viscoelastic material from its shear and tensile properties.

References

5.2.1 Sokolnikoff, I. S., *Mathematical Theory of Elasticity*, Krieger, Malabar, FL, 1983.
5.2.2 Timoshenko, S. P. and Goodier, J. N., *Theory of Elasticity*, McGraw-Hill, NY, 1982.
5.2.3 Fung, Y. C., *Foundations of Solid Mechanics*, Prentice Hall, Englewood Cliffs, NJ, 1968.

5.4.1 Alfrey, T., "Non-homogeneous stresses in viscoelastic media", *Q. Appl. Math.* II(2) 113–119, 1944.

5.4.2 Read, W. T., Stress analysis for compressible viscoelastic materials", *J. Appl. Phys.* 21, 671–674, 1950.

5.4.3 Lee, E. H., "Stress analysis in viscoelastic bodies", *Q. Appl. Math.* 13, 183–190, 1955.

5.6.1 Lakes, R. S. "Foam structures with a negative Poisson's ratio", *Science* 235, 1038–1040, 1987.

5.6.2 Goodier, J. N., "Concentration of stress around spherical and cylindrical inclusions and flaws", *J. Appl. Mech.* 1, 39–44, 1933.

5.6.3 Choi, I. and Horgan, C. O., "Saint Venant end effects for plane deformation of sandwich strips", *Int. J. Solids Struct.* 14, 187–195, 1978.

5.6.4 Lakes, R. S., "Saint Venant end effects for materials with negative Poisson's ratios", *J. Appl. Mech.* 59, 744–746 (1992).

5.7.1 Bohn, E., *The Transform Analysis of Linear Systems*, Addison Wesley, Reading, MA, 1963.

5.7.2 Meeker, T. R. and Meitzler, A. H., "Guided wave propagation in elongated cylinders and plates", in *Physical Acoustics*, Vol. 1A, ed. W. P. Mason, Academic, NY, 1964, pp. 111–167.

5.7.3 Main, I., *Vibrations and Waves in Physics*, Cambridge University Press, England, 1978.

5.7.4 Chen, C. P. and Lakes, R. S., "Apparatus for determining the properties of materials over ten decades of frequency and time", *J. Rheol.* 33(8), 1231–1249, 1989.

5.7.5 Kolsky, H., *Stress Waves in Solids*, Clarendon Press, Oxford, 1953.

5.8.1 Graham, G. A. C., "The correspondence principle of linear viscoelasticity theory for mixed boundary value problems involving time-dependent boundary conditions", *Q. Appl. Math.* 26, 167, 1968.

5.8.2 Lin, J. Y. and Westmann, R. A., "Viscoelastic winding mechanics", *J. Appl. Mech.* 56, 821–827, 1989.

5.8.3 Hunter, S. C., "The rolling contact of a rigid cylinder with a viscoelastic half space", *J. Appl. Mech.* 28, 611–617, 1961.

5.8.4 Hunter, S. C., "The Hertz problem for a rigid spherical indentor and a viscoelastic half-space", *J. Mech. Phys. Solids* 8, 219–234, 1960.

5.8.5 Graham, G. A. C., "The contact problem in the linear theory of viscoelasticity", *Int. J. Eng. Sci.* 3, 27–46, 1965.

5.9.1 Goodman, A. M. and Goodall, I. W., "Use of existing steel data in design for creep", in *Creep of Engineering Materials*, ed. C. D. Pomeroy, Mechanical Engineering Publications Ltd., London, 1978.

5.10.1 Lakes, R. S. and Katz, J. L., "Viscoelastic properties of wet cortical bone. III. A non-linear constitutive equation," *J. Biomech.* 12, 689–698, 1979.

5.10.2 Lockett, F. J. and Gurtin, M. E., "Frequency response of nonlinear viscoelastic solids", Brown University Technical Report, NONR 562(10), 562(30), 1964.

E5.5.1 Love, A. E. H., *A Treatise on the Mathematical Theory of Elasticity*, 4th ed., Dover, NY, 1927.

chapter six

Experimental methods

§6.1   Introduction: General requirements

Experimental procedures for viscoelastic materials, as in other experiments in mechanics, make use of a method for applying force or torque; a method for measuring the same; and a method for determining displacement, strain, or angular displacement of a portion of the specimen. If the experimenter intends to explore a wide range of time or frequency, the requirements for performance of the instrumentation can be severe. Elementary creep procedures are discussed first, since they are the simplest. Methods for measurement and for load application are discussed separately, since many investigators choose to assemble their own equipment from components. Many procedures in viscoelastic characterization of materials have aspects in common with other mechanical testing. Therefore, study of known standard methods [6.1.1, 6.1.2] for mechanical characterization of materials is useful. As in other forms of mechanical characterization, it is important that the stress distribution in the specimen be well defined. End conditions in the gripping of specimens are usually not well known. The experimenter often uses elongated specimens for tension, torsion, or bending to appeal to Saint Venant's principle in using idealized stress distributions for the purpose of analysis.

Frequency response is an important consideration for instruments used in viscoelasticity; the transducers [6.1.3] for force generation and measurement, and for measurement of deformation must respond adequately at the frequencies of interest. Resonances in these devices place an upper bound on the frequency at which meaningful measurements can be made. Manufacturers generally quote frequency response in terms of the frequency at which the amplitude ξ drops by 3 dB, the "3 dB point" (ξ (dB) = 20 $\log_{10} \xi$). Since dynamic viscoelastic studies involve a measurement of phase, one should be aware that significant phase shifts in the instrumentation (see §6.7) occur well below the 3 dB point. As for a lower limit on frequency, some instruments such as electromagnetic motion detectors only operate dynamically and have no quasistatic response. In other cases, drift in the electronics will limit the low frequency and long time performance. The materials used in the manufacture of the transducers in the instrumentation are real, not ideal, and therefore exhibit some creep [6.1.3]. This instrumental creep should be incorporated in the instrument specifications and should be taken into account in the interpretation of results. In the event that the instrumentation imposes no limit, the lowest frequencies and longest times accessible are limited by the experimenter's patience.

§6.2 Creep

§6.2.1 Creep: Simple methods

Creep experiments can be performed in a simple manner since a stress which is constant in time can be achieved by deadweight loading. Tensile or compressive creep tests on a rod are the easiest to interpret since the entire specimen is under a stress which is spatially uniform, assuming uniformity at the ends. If there is nonuniformity due to gripping conditions, use of a slender specimen allows the investigator to take advantage of Saint Venant's principle in interpreting results. Uniform stress is advantageous if the investigator has any interest in examining nonlinear response. In bending and torsion creep tests, the stress is not spatially uniform within the specimen. Interpreting such tests is straightforward if the material behaves linearly; in that case the creep compliance can be extracted directly from measurements of force and displacement as developed earlier in Chapter 5. Measurement of displacement in creep can also be performed simply in the domain $t \geq 10$ sec. Among the simplest methods are the micrometer and the microscope. The microscope can be a traveling microscope, mounted on a calibrated stage and focused on a fiduciary mark on the specimen or its grip; or it can be a microscope with a calibrated reticle. Other methods of displacement measurement will be described below. Cantilever bending of a bar is an advantageous configuration in that the end displacement can be made large enough to measure easily while maintaining small strain, by choice of the bar geometry.

If tensile or compressive creep tests are to be done on relatively large specimens of a stiff material, deadweight loading becomes difficult, since large weights are required. A simple hydraulic device has therefore been developed to perform long-term compressive creep tests on concrete [6.2.1, 6.2.2].

In polymers one must precondition specimens by applying load cycles consisting of creep and recovery segments, to achieve reproducible results in creep, as pointed out by Leaderman [6.2.3]. Specimens of biological origin exhibit similar effects.

§6.2.2 Effect of rise time in transient tests

The definition of the creep compliance assumes a step function load history; however, in the physical world the load cannot be applied arbitrarily

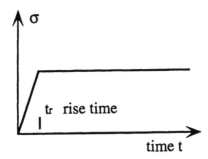

Figure 6.1 Stress input with nonzero rise time for creep.

suddenly. To determine the effect of rise time t, in transient tests, consider a stress history containing a constant load rate segment (Fig. 6.1).

$$\sigma(t) = \begin{cases} 0 \text{ for } t < 0, \\ \dfrac{\sigma_0}{t_r} t \text{ for } 0 < t < t_r, \\ \sigma_0 \text{ for } t > t_r. \end{cases}$$

By substituting in the Boltzmann superposition integral,

$$\varepsilon(t) = \int_0^t J(t-\tau)\frac{d\sigma}{d\tau}\,d\tau = \frac{\sigma_0}{t_r}\int_0^{t_r} J(t-\tau)d\tau. \tag{6.2.1}$$

By the theorem of the mean,

$$\varepsilon(t) = \frac{\sigma_0}{t_r} t_r J(t - t_r\eta) = \sigma_0 J(t - t_r\eta),$$

in which $0 \le \eta \le 1$. If $t \gg t_r$, then $\varepsilon(t) \approx \sigma_0 J(t)$. Since the argument of the creep function differs from t by an unknown but bounded amount, the analysis imposes error limits on the time and the errors are larger for shorter times. In most cases, $t \ge 10t_r$ is considered sufficient so that errors are not excessive [6.2.4]. Analysis similar to the above can also be performed for relaxation. In view of the effect of rise time, one ordinarily begins recording transient data at a time a factor of ten longer than the rise time of the applied load or strain. If the stress history $\sigma(t)$ is known accurately during the initial ramp for $0 < t < t_r$, short time data can be extracted. If the initial ramp is one of constant stress rate or constant strain rate, the analysis for interpretation of the data has been presented in §2.5.

In creep tests by deadweight loading such as described above, the load would be applied in somewhat less than 1 sec, and data collected beginning at 10 sec. A test 3 h in duration (1.08×10^4 sec) covers three decades; 28 h is required for four decades; 11.6 days, for five decades. If mechanical property data are required over many decades of time scale, either great patience in creep tests or shorter time/higher frequency data are needed, or both.

§6.2.3   Creep in anisotropic media

Anisotropic materials have different stiffnesses in different directions. Viscoelasticity in such materials can be studied by conducting creep tests for deformation in different directions in the material. It is possible to perform creep tests in torsion and bending on beams cut in various directions with respect to the material principal axes [6.2.5]. From the results of such tests the elements of the creep compliance tensor S_{ijkl} in the following Boltzmann integral can be inferred:

$$\varepsilon_{ij}(t) = \int_0^t S_{ijkl}(t-\tau)\frac{d\sigma_{kl}}{d\tau}d\tau.$$

§6.2.4   Creep in nonlinear media

Design of experiments intended to explore nonlinear behavior is guided by the kind of constitutive equation considered for the material (see §2.11). Creep according to nonlinear superposition,

$$\varepsilon(t) = \int_0^t J(t-\tau,\sigma(\tau))\frac{d\sigma}{d\tau}d\tau,\tag{6.2.2}$$

may be studied by performing a series of creep experiments at progressively higher stress levels, followed by recovery. The creep compliance depends on stress. Experiments for the study of the more general nonlinearity described by the following Green-Rivlin series involve multiple steps [1.6.10]:

$$\varepsilon(t) = \int_0^t J_1(t-\tau)\frac{d\varepsilon}{d\tau}d\tau + \int_0^t\int_0^t J_2(t-\tau_1,t-\tau_2)\left(\frac{d\sigma}{d\tau_1}\right)\left(\frac{d\sigma}{d\tau_2}\right)d\tau_1 d\tau_2 + \dots .$$

$$\tag{6.2.3}$$

At the simplest level, a single creep test at stress σ_0 on a material obeying a multiple integral representation gives the following [1.6.21]:

$$\varepsilon(t) = J_1(t)\sigma_0 + J_2(t,t)\sigma_0^2 + J_3(t,t,t)\sigma_0^3 + \dots .\tag{6.2.4}$$

A series of creep tests at different stress levels is sufficient to evaluate these kernel functions but only for the case that all time variables are equal.

To evaluate the full effect of nonlinear terms such as $J_2(t - \tau_1, t - \tau_2)$ as a function of the two time variables, stress histories with two steps are applied, for example, creep followed by recovery. A series of creep tests of different *durations* as well as different stress levels may be performed to determine J_2. The relationship between strain and nonlinear kernel functions for a two-step stress history containing a step of magnitude $\Delta\sigma_0$ at time zero and a step of magnitude $\Delta\sigma_1$ at time t_1 is as follows [1.6.10]:

$$\varepsilon(t) = J_1(t)\Delta\sigma_0 + J_2(t,t)(\Delta\sigma_0)^2 + J_3(t,t,t)(\Delta\sigma_0)^3$$

$$+ J_1(t-t_1)(\Delta\sigma_1) + J_2(t-t_1,t-t_1)(\Delta\sigma_1)^2 + J_3(t-t_1,t-t_1,t-t_1)(\Delta\sigma_1)^3$$

$$+ 2J_2(t,t-t_1)(\Delta\sigma_0)(\Delta\sigma_1) + 3J_3(t,t,t-t_1)(\Delta\sigma_0)^2(\Delta\sigma_1)$$

$$+ 3J_3(t,t-t_1,t-t_1)(\Delta\sigma_0)(\Delta\sigma_1)^2 + \dots .$$

$$(6.2.5)$$

A creep and recovery test contains two steps, with $\Delta\sigma_1 = -\Delta\sigma_0$. In a linearly viscoelastic material, only the terms containing J_1 are present, so that recovery follows $J_1(t - t_1)(-\Delta\sigma_0)$ which has the same functional dependence as creep ($J_1(t)\Delta\sigma_0$) only delayed in time and inverted. The terms with J_2 and J_3 show the effect of time–interaction-type nonlinearity; if these are present, recovery does *not* follow the same pattern as creep.

§6.3 Displacement and strain measurement

The micrometers and calibrated microscopes described above are usable for quasistatic measurements such as those performed in slow creep tests. In addition to simplicity, they have the advantage of maintaining their accuracy over very long periods of time. However, they are not suitable for rapid experiments. Other methods are used if rapid response is required, or if measurements are to be made over the surface of a specimen; see, e.g., [6.1.2, 6.3.1, 6.3.2].

Linear variable differential transformers (LVDTs) (Fig. 6.2) consist of several coils of wire surrounding a core of iron or other magnetic material. The core is usually attached to the specimen grip or a connected part of the apparatus. One coil, the primary, is supplied with a sinusoidal electrical signal at several kilohertz. Electrical signals induced in the secondary coils depend on the position of the core. These signals are demodulated, rectified, and filtered to obtain an electrical signal proportional to core displacement. Similar transducers used to detect rotary motion are called rotary variable differential transformers (RVDTs). The upper bound for frequency of the displacement signal may be as high as 1 kHz, depending on the physical

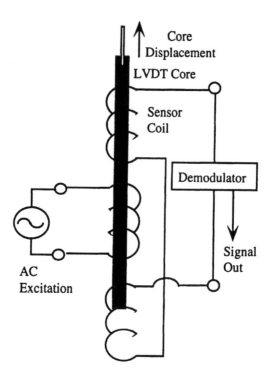

Figure 6.2 Schematic diagram of an LVDT.

size of the LVDT. Studies at low frequencies or long times are limited by drift in the electronics.

Foil strain gages (Fig. 6.3) consist of an electrically conductive film, commonly a metal, upon a flexible substrate. Strain in the gage results in a change in electrical resistance. Strain gages are cemented to the specimen and connected to electronics which convert the resistance change to a voltage proportional to strain. Strain gages are too stiff to be used on soft polymers, rubber, or flexible foam; errors due to gage stiffness can even be problematic for glassy polymers. The strain gage itself imposes no upper limit on frequency other than that of the size of the gage in relation to the wavelength of stress waves in the material to which it is cemented. For gages of typical size, this upper limit can be on the order of 1 MHz. However, the electronic amplification system used with strain gages may impose further frequency limits. Strain gages respond to strain from thermal effects as well as strain resulting from stress. For accurate measurements of strain over long periods of time, considerable care must be exercised in the control of temperature and in matching the thermal expansion of the strain gage with that of the material to which it is cemented. Moreover, the cement used to attach the strain gage must be evaluated for possible creep which could introduce further errors. Cyanoacrylate cement intended for strain gage use exhibits lower creep than "superglue" cyanoacrylate cement for general purposes.

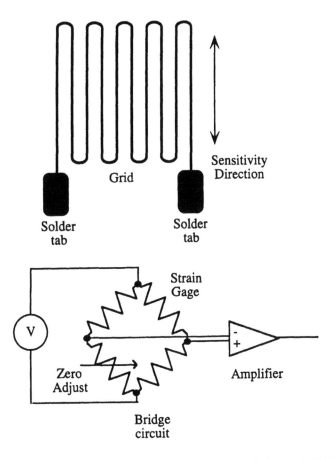

Figure 6.3 Strain gage. (top) Shape of foil strain gage grid. (bottom) Bridge circuit for strain gages.

For tests on rubbery materials or soft biological tissues, foil strain gages are not suitable. One can use a thin rubber tube filled with mercury as a strain gage for tests involving large deformation or compliant materials.

Capacitive transducers (Fig. 6.4) are based on the fact that the capacitance C of a capacitor depends on the spacing between the plates. For a parallel plate capacitor of plate area A and spacing d, $C = \varepsilon_0 A/d$ with ε_0 as the permittivity of free space. This variation of capacitance with plate position can be used to convert linear or angular displacement into an electrical signal. In a typical application, the capacitive transducer is excited by a high frequency sinusoidal electrical signal and is connected to electronics which determine the capacitance, usually by determining the current which passes through the transducer. The signal is then rectified and filtered; the frequency response is therefore limited [6.3.3].

Extensometers (Fig. 6.5) are devices clipped or otherwise attached to a specimen in tension to determine the relative displacement between the two

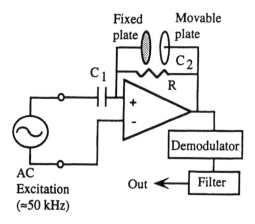

Figure 6.4 Capacitive displacement transducer and electronics.

attachment points. They may be based on strain gages or LVDTs. Since they are attached off the specimen axis, response at high frequency is not to be expected; an upper bound from 30 to 100 Hz is as much as can be expected. Moreover, extensometers impose spring loads of 14 to 100 g (0.14 to 0.98 N) and so are suitable only for relatively stiff specimens.

Optical methods can be extremely sensitive and offer the advantage that little or no mass needs to be attached to the specimen. To measure axial motion, one mirror of a Michelson interferometer (Fig. 6.6) can be attached to the specimen [6.3.4]. One fringe corresponds to half a wavelength of light. The light can be converted to an electrical signal via a light-sensitive diode. Fractional fringes can be readily measured and corresponding displacement can be inferred via a digital counter or computer. The wavelength of light from the commonly used helium neon laser is 632.8 nm. Axial motion can also be measured using fiber optic sensors in which white light from one bundle of fibers is directed to the specimen. Reflected light from the specimen is then detected and the resulting signal is amplified electronically. This reflected light depends sensitively on displacement. Deformation of compliant materials such as rubber or soft biological tissue may be monitored with the aid of fiduciary marks on the specimen. For example, ink spots may be applied with a microscopic pipette, and observed via a calibrated microscope or a video system.

As for angular displacements, they are readily measured either by interferometric or "optical lever" methods [6.3.5]. An interferometer incorporating one or more right-angle prisms as reflective elements offers great sensitivity to angular displacement [6.3.6]. In the optical lever a beam of light is reflected from a mirror attached to the specimen. For creep tests a light beam may be reflected to a distant target and its displacement measured with a caliper. For dynamic tests it is expedient to convert the light to an electrical signal. One way of doing this involves projecting the image of a grating on another grating. The resulting light intensity signal is then

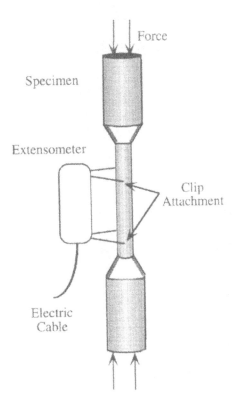

Figure 6.5 Extensometer attached to a specimen. The electric signal out is proportional to the displacement between the clip points.

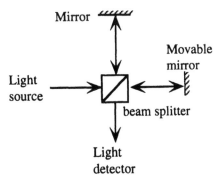

Figure 6.6 Michelson interferometer configuration for determining linear displacement. Moving mirror is attached to a specimen surface (not shown).

converted to an electric signal by a photodiode. Light beam displacement can also be converted to an electrical signal by a split photodiode connected to a subtractor preamplifier. Semiconductor sensors sensitive to the position of an incident light beam are commercially available, but they have limited frequency response.

Motion can also be evaluated by electromagnetic methods: a coil or loop of wire is fixed to the specimen and immersed in a magnetic field. The voltage induced in the coil is proportional to the rate of change of magnetic flux according to Faraday's law. The induced voltage is expressed as the integral of electric field vector **E** around a closed loop which can represent the coil. The magnetic flux is the surface integral of magnetic field vector **B** over a surface spanned by the loop,

$$\int \mathbf{E} \cdot \mathbf{d\ell} = -\frac{d}{dt} \int \mathbf{B} \cdot \mathbf{ds}. \tag{6.3.1}$$

Depending on the specific configuration, this electrical signal may result from axial velocity or from angular velocity. There is no static response in this method; it is intrinsically dynamic and is suitable for resonant vibration or wave procedures.

§6.4 Force measurement

Force and torque transducers, known as load cells and torque cells, respectively, involve measuring the displacement or strain of a deformable substrate, typically steel. Representative arrangements of strain gages for load cells and torque cells are shown in Fig. 6.7. Alternative forms of load cell include the proving ring which is compressed along a diameter, and configurations involving bending of plates. Torque cells are based on solid or hollow shafts, or cruciform arrangements fitted with strain gages. As with other transducers [6.1.3], load cells have characteristics of linearity, linear range, sensitivity, and frequency (or time) response. They typically contain a bar or plate of metal, usually steel, upon which strain gages are cemented. The detected strain signal on the bar is proportional to the force upon it, provided the metal and the strain gages are loaded below their proportional limit, one factor which limits the linear range or "capacity". Overload capability is limited by yield in the metal parts of the transducer. Since steel itself is not perfectly elastic but exhibits a small viscoelastic response, testing machines which use load cells of this type are not suitable for the study of materials with extremely low loss. Torque cells are similar to load cells in their use of strain gages; the metal portion and the strain gage orientation are configured differently so that the detected strain signal is a response principally to torque and not to any superposed axial load (Fig. 6.7). The compliance of the load cell is important as discussed in §6.10.2. The load cell should be much stiffer than any specimen tested. It is also important that the load cell be sensitive only to the kind of load to be measured and not to other loads. In biaxial tests both tension and torsion may be applied. The torque channel should be sensitive only to torque, not to axial load. This characteristic is referred to as absence of cross talk.

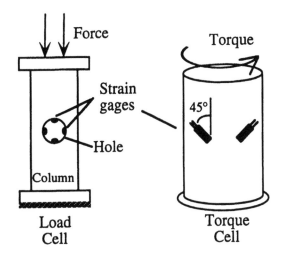

Figure 6.7 Load cell and torque cell structures.

Load cells based on piezoelectric crystals are also available; they offer superior stiffness for dynamic studies. However, there is no response at zero frequency, and there are phase errors at low frequency, though the frequency range can extend below 0.001 Hz.

§6.5 Load application

Methods of load application for creep include deadweights as described above, electromagnetic methods, and hydraulic methods. In the electromagnetic approach, controlled electric current may be passed through a small coil of wire (which generates a magnetic dipole moment) immersed in an external magnetic field. This coil experiences a torque. Alternatively, a permanent magnet (which generates a magnetic dipole moment) is placed within a coil through which a controlled electric current passes. The magnet experiences a torque. The small coil or magnet can exert a force or torque on the specimen to which it is attached, depending on the specific geometry. The torque τ on an object of magnetic moment μ in a magnetic field vector **B** is [6.5.1]:

$$\tau = \mu \times \mathbf{B}. \tag{6.5.1}$$

The force **F** is given by

$$\mathbf{F} = \text{grad}\{\mu \cdot \mathbf{B}\}. \tag{6.5.2}$$

These devices are equivalent to electric motors in which the motion is stalled by the specimen.

In hydraulic systems, an electric pump drives a fluid through a hydraulic pipe system to a movable piston which applies force to the specimen. The fluid is controlled by a valve which in turn is controlled by an electrical device.

Load application in relaxation experiments can be achieved purely mechanically by means of an eccentric cam actuated by a handle. This system provides a displacement which is constant in time after it is applied. The load is free to decrease as the specimen relaxes; it is measured independently.

§6.6   Environmental control

Viscoelastic properties depend on temperature; therefore, some form of temperature control is usually used in viscoelastic measurements. If the method of time–temperature superposition is to be used, then capability for conducting tests at several constant temperatures is required. Alternatively, the experimenter may measure dynamic properties at constant frequency, and may perform a temperature scan.

As for the precision of temperature control, polymers are very sensitive to temperature, particularly in the transition region. It is therefore desirable to control temperature to within 0.1°C over the length of the specimen and for the duration of the experiment. To achieve such control, a stream of air can be passed through a temperature control device, and directed to the specimen through a system of baffles. In tension–compression tests, a length change can arise from creep or from thermal expansion. Excellent temperature control is required in such studies to obtain good creep or relaxation results. Since thermal expansion of an isotropic material has no shear or torsional component, torsion tests are less demanding of accurate temperature control than are tension–compression tests.

Many polymers, as well as materials (such as wood) of biological origin, are sensitive to hydration as well as to temperature. Humidity control is important for such materials; in the study of tissue such as bone, the specimen should be kept fully hydrated under physiological saline solution. The creep behavior of concrete is also dependent on hydration.

Metals which exhibit magnetic behavior are sensitive to external magnetic fields. If electromagnetic methods are used to apply force or torque to the specimen, care must be taken that magnetic fields in the vicinity of the specimen are controlled.

§6.7   Subresonant dynamic methods

§6.7.1   Phase determination

§6.7.1.1   Phase measurement

In subresonant dynamic methods a sinusoidal load is applied at frequencies well below the lowest resonance of the specimen including attached inertia, if any. The test frequency must also be well below any resonances in the instrument. Under these conditions, the phase angle ϕ between force and displacement (or between torque and angular displacement) is the same as the loss angle δ, as derived in §3.5. Direct measurement of phase is necessary in these methods.

Phase measurement can be performed by determining the time delay between the sinusoids on an oscilloscope (for high frequencies) or on a chart recorder (for low frequencies). This approach is suitable for materials with a loss tangent greater than about 0.1; there is a limit in the resolution of the phase. A better method of phase determination is by evaluation of the elliptical stress–strain curve, which is called a Lissajous figure (Fig. 3.2). If the phase is sufficiently large, it may be easily determined from the width of the elliptical curve. In materials with low loss the ellipse is nearly a straight line. The middle of the ellipse can be magnified on an oscilloscope, computer display, or recorder to achieve improved phase resolution, provided that the ratio of signal to noise is sufficiently high. It is possible by such magnification to measure tan δ less than 0.01 in the subresonant domain.

Phase can be measured at high resolution by detecting the zero-crossing of the load and displacement signals and measuring the time delay with a high-speed timer [6.7.1]. A time accuracy of 1 μsec is claimed, which corresponds to a phase uncertainty $\delta\phi \approx 6 \times 10^{-5}$ at 10 Hz, and even better at lower frequencies. Use of such a method requires signals which are low in noise, drift, and DC offset. To that end it is necessary to minimize electrical noise and noise from mechanical sources such as building vibration.

An interesting approach to achieving better resolution involves subtracting a signal of zero phase to fatten the ellipse [6.7.2] and facilitate measurement (Appendix 6). This approach is also limited by the ratio of signal to noise.

A two-phase lock in amplifier can be used to achieve good phase resolution ($\delta\phi \approx 10^{-3}$ to 10^{-4} or better); however, not all models offer such resolution. Lock in amplifiers are capable of rejecting noise. It is a painstaking process to use most lock in amplifiers below 10 Hz; however, some instruments are capable of measurements down to 0.001 Hz. Good phase resolution can also be achieved by Fourier transformation of the signals for stress and strain [6.7.3, 1.6.1]. This procedure makes use of the fact that the phase difference between two sinusoids is equal to the phase difference between their Fourier transforms, provided that the data sample contains a sufficient number of cycles [6.7.3]. A digital fast Fourier transform with 16-bit resolution gives a phase resolution of $\delta\phi \approx 2\pi/2^{16} = 9.59 \times 10^{-5}$ radians (rad) provided the signal occupies the full scale.

As for dynamic measurements at low frequency, a mathematical sinusoid has no beginning or end; by contrast, an experimentally applied sinusoid must have a starting point. This raises an issue of concern in interpreting dynamic experiments, particularly those at low frequency. An analytical solution of the problem has been presented, based on Laplace transform methods [1.6.1, 6.7.4]. Results of a numerical analysis, in which the Boltzmann integral of the stress history was evaluated using a computer, are shown in Fig. 6.8. The response to the experimental "sinusoid" beginning at time zero converges rapidly, within one or two cycles, to a sinusoid, even in

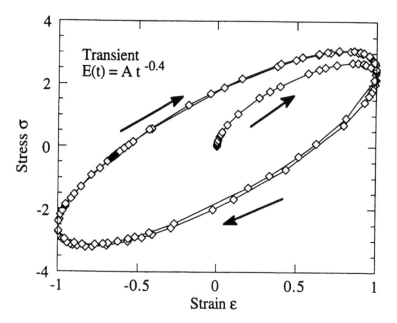

Figure 6.8 Transient behavior in response to a load history $\sigma = 0$ for $t < 0$, $\sigma = A$ sin $(2\pi vt)$ for $t \geq 0$, calculated for a material with power law relaxation.

a material with a high tan δ. Low-frequency dynamic tests are limited by the patience of the experimenter and by drift in electronics.

§6.7.1.2 *Phase errors*

Phase measurements are influenced by phase shifts in any electronics used to process the signals. Consider a low-pass electric filter consisting of a resistor R and capacitor C in series. The output voltage is taken across the capacitor. Such a filter may be used deliberately to reduce noise superposed on an electric signal, or the low-pass characteristic may be inherent in the amplifier or signal conditioning electronics used. The input voltage is v_i and the current $\mathcal{I}$ is

$$\mathcal{I} = \frac{v_i}{R + \dfrac{1}{i\omega C}}.$$

The output voltage is $v_o = \mathcal{I}Z = \mathcal{I}/i\omega C$, with Z as the electrical impedance of the capacitor and $i = \sqrt{-1}$, so the gain with the cutoff angular frequency $\omega_0 = (RC)^{-1}$ is

$$G = \frac{v_o}{v_i} = \frac{1}{1 + i\omega RC} = \frac{1}{1 + i\dfrac{\omega}{\omega_0}}.$$

Here RC is the time constant. The absolute value of the gain is

$$|G| = \frac{1}{\sqrt{1 + \left\{\dfrac{\omega}{\omega_0}\right\}^2}}.$$

For $\omega = \omega_0$, $|G| = 1/\sqrt{2}$. The phase ϕ is given by

$$\tan\phi = \left\{\frac{\omega}{\omega_0}\right\}.$$

If $\omega = 0.1\,\omega_0$, then $|G| = 0.995$ and $\tan\phi = 0.1$. If $\omega = 0.01\,\omega_0$, then $|G| = 0.99995$ and $\tan\phi = 0.01$. If this filter is used as part of a viscoelasticity test apparatus, then phase shifts occur at frequencies significantly below the cutoff frequency. If not properly taken into account, these phase shifts can cause significant error in the inference of $\tan\delta$.

§6.7.1.3 Consistency checks

In experimental studies on viscoelastic materials, the Kramers–Kronig relations are useful in performing consistency checks on stiffness and damping data. The Kramers–Kronig relations are also useful when only one dynamic viscoelastic function is available, and the other one is desired.

§6.7.2 Nonlinear materials

The stress–strain curve under dynamic loading is elliptical only if the material is linearly viscoelastic. If nonlinearity is present, the response will no longer be sinusoidal, and the stress–strain curve (hysteresis loop, Fig. 3.8) is not elliptical. Energy dissipation per cycle can then be determined from the area within the hysteresis loop, and expressed as specific damping capacity (§3.4) rather than a loss tangent.

Experiments performed in torsion or bending present some complications for nonlinear materials since the strain distribution is not uniform. This nonuniformity must be taken into account to properly interpret the results of dynamic damping experiments [6.7.5, 6.7.6].

§6.7.3 Rebound test

Materials may be rapidly screened for their effective $\tan\delta$ by measuring the height H_1 of rebound of a steel ball dropped from height H_0 on a block of material. Alternatively, the ball is made of the viscoelastic material to be tested, and the substrate upon which it is dropped is made stiff and massive. For small damping the approximate relation, as follows (Eq. 10.10.5), is sufficient:

$$\tan \delta \cong \frac{1}{\pi} \ln\left(\frac{H_0}{H_1}\right).$$

(6.7.1)

The effective frequency for this is proportional to the inverse of the impact time t_{imp}. For higher damping, Eq. 10.10.4 is the most accurate. This interpretation is based on assuming that the dissipation of mechanical energy is entirely due to viscoelasticity rather than yield, cracking, or generation of waves in the block.

§6.8   Resonance methods

§6.8.1   General principles

Resonance methods are eminently suitable for low-loss materials. By contrast, small $\tan \delta$ values are difficult to measure in the subresonant domain. The resonant behavior of materials with small $\tan \delta$ 10^{-3} and 10^{-6} differs dramatically as we have seen in §3.5. The difference is used as the basis for experimental methods.

Resonances in tension–compression, torsion, or bending can be examined to characterize materials. The relation, developed in §3.5,

$$\tan \delta \approx \frac{1}{\sqrt{3}} \frac{\Delta \omega}{\omega_0},$$

for the half-width $\Delta \omega$ of the structural compliance peak, is used to interpret the results. One may also infer damping from the free-decay time $t_{1/e}$ of vibration of period T,

$$\tan \delta \approx \frac{1}{\pi} \frac{T}{t_{1/e}}.$$

These relations are valid for small $\tan \delta$. As for the specimen stiffness, it is inferred from the resonance frequency ν_0 in conjunction with the specimen geometry. Observe that in the resonance scheme there is no need to measure the force or torque; but if it is not measured, it must have constant amplitude over the frequency range associated with the resonance peak. A drawback of this approach is that only one frequency is accessible in lumped configurations with a large inertia attached to the specimen, and only a few nearby frequencies are accessible in the case of a specimen without attached inertia. If the attached inertia is comparable to the specimen inertia, the higher harmonics are reduced in amplitude (Fig. 6.9), and all the resonance modes occur at lower frequencies. If the attached inertia is large compared with that of the specimen, only the lowest mode is observable.

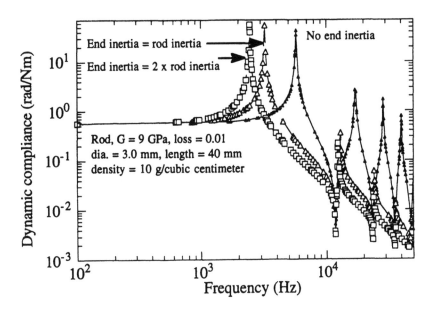

Figure 6.9 Calculated effect of end inertia on the resonance frequencies of a cylindrical rod, fixed at one end and free at the other, in torsion. The rod is of a metal alloy, with G = 9 GPa and tan δ = 0.01; and it is 40 mm long and 3 mm in diameter, with density 10 g/cm³.

The specimen may be fixed at one end and free at the other end (fixed–free configuration), or it may be free at both ends (free–free configuration). In the fixed–free approach, some of the vibration energy may be transmitted into the "fixed" support, leading to measured values of tan δ which are greater than actual values. Error due to such energy loss in fixed–free systems may be quantified by considering the transmission coefficient [6.8.1] for power which is

$$T_p = 1 - |R|^2 = \frac{4}{\left[\sqrt{\dfrac{Z_2}{Z_1}} + \sqrt{\dfrac{Z_1}{Z_2}}\right]^2}.$$ (6.8.1)

Z is the mechanical impedance and the subscript indicates the material. R is the reflection coefficient.

$$R = \frac{Z_1 - Z_2}{Z_1 + Z_2}.$$

By considering the transmission coefficient for power T_p as a parasitic specific damping or energy ratio Ψ_p under the assumption that all the power transmitted into the support rod is lost, the parasitic loss tangent is

$$\tan \delta_p = \frac{1}{2\pi} \Psi_p = \frac{1}{2\pi} T_p. \tag{6.8.2}$$

For longitudinal waves, Z is the ratio of driving force to particle velocity [6.8.2]; for a rod of radius r, density ρ, and complex Young's modulus E^*, Z depends on the cross-sectional area.

$$Z = \pi r^2 \sqrt{\rho E^*}. \tag{6.8.3}$$

The impedance for torsion is the ratio of torque to angular velocity; it depends on the polar moment of inertia.

$$Z = \frac{1}{2} \pi r^4 \sqrt{\rho G^*}. \tag{6.8.4}$$

This is in contrast to the case of plane shear waves in which there is no mismatch in area or moment of inertia for which

$$Z = \sqrt{\rho G^*}$$

represents the characteristic impedance.

The difference in diameter between the specimen and the support as well as the difference in stiffness gives rise to an impedance mismatch which reduces parasitic loss error resulting from transmission of waves into the support. In torsion the impedance goes as the fourth power of diameter. Therefore, the desired impedance mismatch can be large in torsion. (See Examples 6.3 and 6.4.)

Even so, for materials of very low loss, the free–free approach is preferable. In free–free resonance methods, the specimen is supported at one or more nodes (at which the vibration amplitude is zero) by a compliant support such as a thread. There is still some background damping due to losses in the support (which is of nonzero size), but that error is usually lower than in the fixed–free approach. Specific methods are discussed below.

§6.8.2   *Particular resonance methods*

A torsion specimen with a large attached inertia can be made to resonate below 1 Hz. Some experimenters advocate changing the attached inertia to obtain different frequencies, but this is cumbersome and it is difficult to obtain a wide range of frequency. A short specimen without any attached inertia may resonate at a frequency exceeding 100 kHz. Resonance methods are commonly used in conjunction with a scan of temperature; results consist of a plot of stiffness and loss at a nearly constant frequency vs. temperature. A full characterization of the material, however, entails study over a range of frequency.

Free–free resonance methods are suitable for materials of low loss. In the example shown in Fig. 6.10, the driving force is generated by an inter-action between eddy currents induced by the coils and the field of the permanent magnet. This approach is suitable for specimens which are elec-trically conducting. The specimen is shown suspended at two nodal points by fine threads or wires, and vibrates in its second or higher mode. It is also possible to drive the resonance via an electromagnetic shaker attached to one support wire, and to detect the vibration with an electromagnetic pickup device supporting the other wire. In this method, due to Forster, the wires must be displaced slightly from the nodal points to drive and detect the vibration. However, if the support wires are too far from the nodes, their parasitic contribution to tan δ becomes excessive [6.8.3]; the error has been evaluated theoretically.

If the specimen is not electrically conductive, a small magnet or a ferro-magnetic armature [1.6.23] may be cemented to each end, as shown in Fig. 6.11. The specimen is shown suspended at the central nodal point by fine threads or wires, and vibrates in its first mode or a higher mode. The purpose of the polarizing magnet in the coil assembly is to provide a bias magnetization in the armature. As a result, the interaction between the oscillating magnetic field due to current in the coil and the bias field gives an oscillating force on the specimen at the driving frequency. This oscillating force can excite longitudinal, flexural, or torsional vibration, depending on the geometrical arrangement as shown in Fig. 6.11. These magnetic approaches permit larger excitation amplitudes than the eddy current method. Coupling between specimen and driver and sensor coils leads to background loss, since some electrical energy is dissipated in the armature and magnetic pole pieces. This background loss can be minimized by using magnetic materials of low hysteresis. Both sensor and driver coils are usually encased in magnetic shielding of high permeability metal, to reduce pickup of noise and of electrical signals from the driver or heater windings. In all methods which use electromagnetic induction into coils of wire, there is no static response since the induced voltage depends on the specimen velocity.

The resonance approach can be adapted to high-loss materials for which a sharp resonance peak cannot be obtained. The procedure involves prepar-ing a specimen of stiff, low-loss material and applying a layer of the high-loss material. The properties of the high-loss layer can be extracted from resonance experiments on the composite sandwich [1.6.18]; see §9.3 for anal-ysis of viscoelastic laminates.

Microsamples of bone and other stiff biological materials have been studied [6.8.4] with a fixed–free electromagnetic torsion resonance device. The device provides drive torque via a miniature coil of fine wire immersed

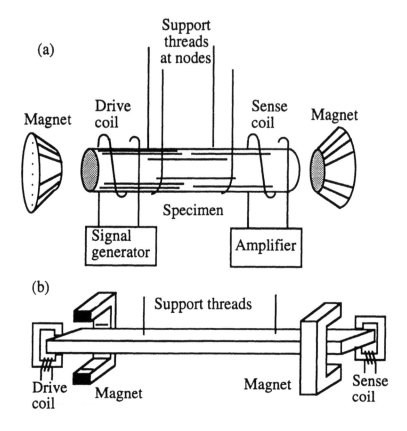

Figure 6.10   A configuration for resonance of an electrically conducting bar, free at both ends. Excitation and detection are via eddy currents. (a) Setup for longitudinal vibration. (b) Setup for flexural vibration.

in the field of a large permanent magnet. Angular displacement is measured from the motion of a laser beam reflected from a mirror attached to the specimen. Microresonance studies in soft biological tissues have been conducted by embedding small beads of magnetic material in the tissue and oscillating them with a magnetic field [6.8.5].

The composite piezoelectric oscillator [6.8.6–6.8.8] is an interesting approach to resonance. This approach is called the piezoelectric ultrasonic composite oscillator technique (PUCOT). The device consists of two piezoelectric crystals and a specimen cemented together (Fig. 6.12). The fundamental resonant frequency can be from 20 to 120 kHz depending on the length of the crystals. Quartz is preferred for the crystals since it has a very low tan δ. One crystal is driven electrically to induce vibration and the other crystal is monitored for electrical signals induced by strain. The wires enter at the crystal midpoints, which are nodes for free–free vibration. Since the wires are of nonzero size, they allow some vibration to leak away and so represent a source of parasitic damping. Therefore, the experimenter makes

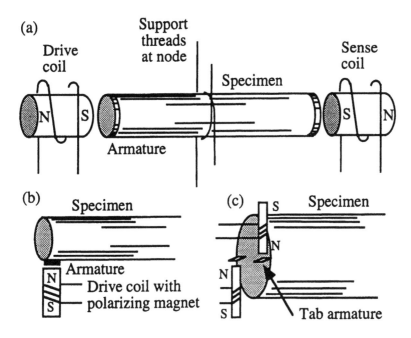

Figure 6.11 A configuration for resonance of a bar, free at both ends. Excitation and detection via magnetic interaction between external coils and a ferromagnetic armature cemented to the specimen. (a) Setup for longitudinal vibration. (b) Setup for flexural vibration. (c) Setup for torsional vibration.

a measurement of damping of the supported quartz crystals alone to evaluate the parasitic damping. Tension–compression or torsion can be achieved depending on the type and orientation of crystal. Viscoelastic properties of the specimen are inferred from electrical measurements on the sensor crystal. A correction for parasitic damping is incorporated in the calculation. The method is intrinsically dynamic, and gives no response in the subresonant or static regimes. Variations of this approach have incorporated low-temperature capability [6.8.9]. A low-temperature variant of the piezoelectric composite oscillator uses a stalk which is slender (0.5 mm in diameter compared with 4 mm for the quartz crystal), and compliant, made of copper–beryllium alloy, to support the oscillator and provide thermal contact [6.8.9]. Background loss with this system is $\tan \delta_{parasitic} < 2 \times 10^{-6}$ at 2 K. Capability for high temperatures exceeding those tolerable by crystalline quartz can be achieved by attaching the specimen to the piezoelectric crystals by a long stalk of fused quartz, and placing the specimen in a furnace. The composite piezoelectric oscillator can also be used in a free-decay mode (see §3.6) by modulating the drive waveform by a gate signal.

Free decay of resonant vibration (see §3.6) is also used to determine viscoelastic properties at lower frequencies. A device consisting of two large pendulums supported by ball bearings has been used to characterize

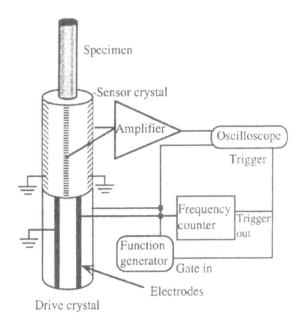

Figure 6.12   Composite torsional piezoelectric oscillator.

polymethyl methacrylate (PMMA) at about 1.5 Hz over a range of temperatures [6.8.10]. Motion of the pendulums induces torsion in the specimen.

Resonance experiments have been applied to polymers [6.8.11], metals,
fibers [1.6.18], and single crystals. Modern refinements include the addition
of more sophisticated electronics to aid in data acquisition and analysis
[6.8.12]. Study of fibers by resonance methods [1.6.18] is done in an evacuated
chamber [6.8.13] for thin fibers in which parasitic damping due to air viscosity is excessive. Vibration may be excited electrostatically by applying an
oscillatory voltage across the plates of a capacitor, one plate of which is fixed
to the specimen. Vibration amplitude is determined from electrical measurements on the capacitor.

Resonance methods can be difficult to use above 100 kHz because a
multiplicity of vibrational modes can be excited. Some analysis schemes have
been presented [6.8.14] to interpret experiments in the frequency range
between 100 kHz and 1 MHz, above which ultrasonic wave methods become
applicable.

§6.8.3   *Special methods for low-loss materials*

For materials of low-loss (tan $\delta < 10^{-4}$), free-decay methods are easier to use
than the method of resonance half width. The reason is that the resonance
peak is so sharp for low-loss materials that it can be tedious to scan through
it. Special experimental precautions are necessary in studying low-loss

materials [6.8.15, 6.8.16]. It is necessary to eliminate all spurious sources of damping other than that in the bulk matter of the specimen. Errors can arise due to radiation of sound energy into the air; these errors cause the apparent loss to be greater than the true material loss. To eliminate such errors, it may be necessary to perform the experiments in an evacuated chamber. For $\tan \delta \leq 10^{-4}$, precautions of this type are usually necessary. It is possible to calculate the damping due to air friction. In particular [6.8.16],

$$\tan \delta_{\text{air}} = \frac{2P}{\pi \rho c} \left[\frac{C_P}{C_V} \frac{\mu}{RT} \right]^{1/2}, \tag{6.8.5}$$

with P as the gas (air) pressure; μ is the mean molecular weight of the gas, ρ is the density of bar material, c is the sound velocity in the bar, R is the gas constant, and C_P and C_V are the heat capacities of the gas at constant pressure and volume, respectively. The formula is valid if the mean free path of the gas molecules is much less than the wavelength of the sound waves in the gas. That assumption would be valid for oscillations at 10 kHz and for gas pressures greater than 0.001 Torr. For a better vacuum, a different equation applies. Bending procedures are more vulnerable to error due to air damping than torsion procedures, since the translational motion associated with bending causes more motion of air. Care in specimen preparation is also required. In particular, defects such as microcracks, spalls, absorbed molecules, and dislocations concentrated in a surface layer can result in elevated apparent damping in low-loss materials.

The resonating specimen must be supported in some way, and that support can introduce spurious losses. In fixed–free cantilever bending vibration, care is required that any grip for the specimen is not only tight but also massive [6.8.15]. Errors in fixed–free vibration are discussed above. In free–free vibration, specimens are commonly supported at a vibration node, where displacement is minimal, to reduce these errors. Use of a fine thread for support rather than a solid nodal mounting serves to minimize loss due to the support. If the thread is greased to reduce transmission of shear stress, the spurious loss is reduced further [6.8.15]. Moreover, if the plane of bending vibration is horizontal rather than vertical, parasitic loss is reduced since the support threads undergo rotation rather than extension. Capacitive devices are commonly used to excite and measure vibration in low-loss materials. Although little power is available to excite vibration, little is needed. These transducers offer the advantage of contributing minimal spurious damping. Even so, a careful experimenter will calculate the spurious damping due to electrical losses in the transducer. The above experimental methods have been used to measure losses as small as $\tan \delta < 10^{-9}$ in single crystal sapphire and silicon at low temperature [6.8.16].

§6.8.4 *Resonant ultrasound spectroscopy*

Resonant ultrasound spectroscopy involves the measurement of natural frequencies for a number of a sample's normal modes of vibration [6.8.17]. Elastic moduli are inferred via an inverse calculation and a knowledge of the sample shape and mass; mechanical damping is inferred from the width of the resonance curves. If the sample is a slender rod and the vibration mode is known from the excitation conditions, calculation of material properties is straightforward as described above. The mode structure of a vibrating sphere, cube, or parallelepiped is complex. With the advent of microcomputers capable of rapid computation, it has become feasible to numerically calculate moduli and damping values from the complex mode structure of single crystal or isotropic specimens. In a typical embodiment of the method, a rectangular parallelepiped or cubic sample is supported by transducers at opposite corners, as shown in Fig. 6.13. Corners are used since they always move during vibration (they are never nodes) and they provide elastically weak coupling to the transducers. A piezoelectric polymer film such as polyvinylidene fluoride (PVDF) is suitable as a transducer. One transducer excites the vibration, and the other converts the resonant motion to an electrical signal. The specimen is supported by contact pressure; it does not require cementing or the use of coupling agents. The method is capable of determining, from one specimen, all the complex modulus tensor elements C^{*}_{ijkl} of an anisotropic material, but is not capable of static or low-frequency studies. As an example, resonant ultrasound spectroscopy, used in the study of $Ni_{80}P_{20}$ has disclosed tan δ in the range 10^{-4} to 2×10^{-3} at 320 kHz over a range of temperature [6.8.18].

§6.9 *The problem of achieving a wide range of time or frequency*

§6.9.1 *Rationale*

As discussed in Chapters 4 and 10, it is desirable to know the properties of materials over as wide a range of time or frequency as possible. Materials may be expected to support loads for periods from hours to many years, and they may be used under conditions in which acoustic behavior is important. The acoustic frequencies which can be perceived by a young human ear are 20 Hz to 20 kHz, corresponding to time scales of milliseconds or shorter. If the material is subject to ultrasonic testing, frequencies in the megahertz range will be of interest. Determination of material properties over a wide range is problematic, since each experimental method typically covers no more than three or four decades, and the lumped resonance methods permit characterization at only one frequency. Specific instrumentation intended for a wide range of time and frequency is discussed in some detail in §6.10.6.

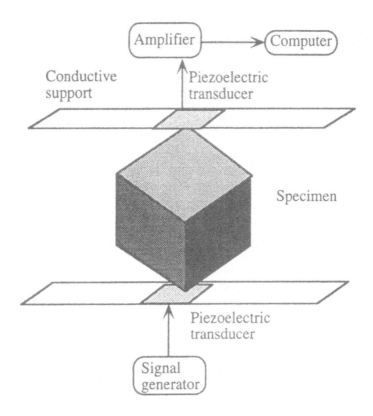

Figure 6.13 Simple configuration for resonant ultrasound spectroscopy.

§6.9.2 *Multiple instruments: long creep*

A wide range of time or frequency may be achieved by performing experiments with a variety of devices, each one covering a portion of the range. This approach is cumbersome and may require preparation of several types of specimen; nevertheless, some experimenters have done such tests. Koppelmann [6.8.11] described several flexural and torsional pendulum devices and wave propagation devices for the study of polymers. The upper limit of time studied in a creep test is limited only by the patience (or lifetime) of the experimenter. Some creep tests have extended to 1 year, or even 20 years, as presented in §7.6. If, in addition, the load is applied rapidly and the response at short times is recorded electronically, a wide range can be achieved. Letherisch [6.9.1], by this approach, obtains a ten-decade range of several milliseconds to 1 year. Extension of the range on the short time/high-frequency end is achieved by the use of multiple instruments or in a single instrument by minimizing inertia and by back-calculation through specimen resonances, as discussed below in §6.10.6.

§6.9.3   Time–temperature superposition

For certain types of material, it is possible to infer behavior over a wide range of equivalent time or frequency from experiments conducted at different temperatures. For a class of materials called *thermorheologically simple*, a change in temperature is equivalent to a shift of the behavior on the time or frequency axis (time–temperature superposition), as presented in §2.7. To experimentally determine if a material is thermorheologically simple, one may perform a set of creep, relaxation, or dynamic tests at different temperatures; and plot the results as illustrated in Fig. 2.8. If the various curves can be made to overlap by horizontal shifts on the log time axis, the material is considered thermorheologically simple. The curve constructed by time–temperature shifts is called a *master curve*, as illustrated in Fig. 2.9. Amorphous polymers are most amenable to this approach.

Materials in which the relaxation is dominated by a thermally activated process, obeying the following Arrhenius equation, are thermorheologically simple as demonstrated in Example 6.9.

$$\nu = \nu_0 \exp{-\frac{U}{RT}}, \qquad (6.9.1)$$

with U as the activation energy, ν as frequency for a feature such as a peak, and R = 1.986 cal/mol K as the gas constant.

If the curves do not overlap, the material is thermorheologically complex. Plazek [6.9.2–6.9.4] has pointed out that given data within a fairly narrow experimental window (three decades or less) the test for thermorheological simplicity can only be definitive in its failure. That is, such an experiment is capable of demonstrating thermorheological complexity but it cannot demonstrate simplicity. Plazek showed that nearly perfect superposition was obtained for data for polystyrene over a restricted range of 3.4 decades, but that data over 6 decades could not be superposed.

If there are multiple relaxation mechanisms, each with its own dependence on temperature, the material will not be thermorheologically simple. If the material exhibits a phase change in the temperature range considered, or if it undergoes a chemical change such as decomposition, oxidation, or light-induced cross-linking, then it will not be thermorheologically simple. These issues could be problematic in inferring behavior at very long times. There is a further caveat associated with high frequencies or short times. It is possible to generate master curves for frequencies as high as 10^{28} Hz as discussed in §7.3. For comparison, a frequency ν of 10^{12} Hz corresponds, for a typical wave speed c of 3 km/sec (via $\lambda\nu = c$), to a wavelength λ of 3 nm, or just a few atoms. Physically realizable sound waves in real materials cannot be achieved at frequencies of about 10^{13} Hz and higher, therefore, since the corresponding wavelength would be smaller than one atom, which is not possible since sound entails a disturbance of the atoms. Before that point is reached, new relaxation processes such as scattering of waves become operative.

§6.10   Test instruments for viscoelasticity

§6.10.1 Servohydraulic test machines

Electronically controlled hydraulic systems are used to produce load in commercially available testing machines (e.g., Instron Co., Canton, MA; MTS Systems Co., Eden Prairie, MN); large forces are readily generated. Such machines are available for tension–compression or tension–compression and torsion tests. These machines are commonly used to conduct stress–strain testing to fracture of structural materials. Servohydraulic machines are equipped with transducers, usually LVDTs, to measure the appropriate displacements; and are fitted to accept signals from extensometers and strain gages. Signals associated with force, displacement, or strain may be used as a feedback input to a servocontroller which causes the variable in question to follow the control signal. Experiments in creep, relaxation, or dynamic loading may be conducted, depending on the control signal chosen. The upper bound on frequency depends both on the particular transducers used, as described above, and on the hydraulic system. If a large load amplitude is required, the frequency range is reduced. Ordinarily it is difficult to exceed 10 Hz. Special servohydraulic instruments have a claimed upper bound of 200 Hz. Creep and relaxation tests are certainly possible, but since these test machines are expensive, many experimenters are reluctant to use them for lengthy creep tests. Servohydraulic test machines are versatile; however, many experimenters choose to build their own instruments for detailed viscoelasticity studies, particularly if a wide range of time or frequency is required.

§6.10.2 A relaxation instrument

A relaxation instrument intended for the study of glassy polymers over three decades of time used an eccentric cam system actuated by a handle for load application [6.10.1], and an axial load cell of the strain gage type for force measurement. Applied deformation was measured with a dial micrometer. In relaxation experiments it is necessary that the specimen strain be held constant. To evaluate the quality of the experiment it is necessary to consider the compliance of the load cell. In this example, the axial load cell had a 1.3 kN capacity at full scale and a deflection at full-scale load of 0.01 mm. Since the experiment is quasistatic, the force in the specimen is equal to the force in the load cell. A typical 76-mm-long sample of stiff polymer experienced a strain of 1% at a load of 440 N. Here the specimen end displacement is 0.01×76 mm = 760 µm, while the load cell deflection is 0.01 mm (440 N/1.3 kN) = 3.4 µm, or only 0.0045 of the total. The rigidity of the entire load frame is as important as that of the load cell; moreover, for tension experiments thermal expansion of the load frame can introduce error. For that reason, the load frame was made robust and was enclosed in a temperature-controlled chamber independent of the temperature control system for the specimen.

The specimen was gripped at the ends with jaws of a flared collet design and serrated to enhance grip capability. The instrument also provided an independent loading system and a torque cell of the strain gage type for torsional relaxation. Torsional and axial degrees of freedom were decoupled by an elaborate system of bearings. Signals from the force and torque transducers were processed with high-frequency carrier amplifiers. The stiffness of the entire instrument was checked by measuring relaxation in steel which is known to exhibit negligible relaxation in comparison with polymers.

§6.10.3 Driven torsion pendulum devices

A biaxial driven torsion pendulum [6.10.2, 6.10.3] is shown in Fig. 6.14. It permits viscoelastic measurements in torsion, in the presence of a superposed axial static load. Biaxial experiments are of interest in the study of polymers since the tensile stress alters the free volume which alters the relaxation kinetics. The torque was generated electromagnetically: a large electromagnet provided a field for a magnesium rotor frame containing many turns of fine wire through which an amplified electric current was passed. The lower part of the specimen was mounted to a reaction torque sensor of the strain gage variety. Angular displacement was measured using an LVDT core supported by a wire wrapped around a rotor disc concentric with the torsion axis. Signals from the angular displacement and torque transducers were processed with high-frequency carrier amplifiers. Transient experiments were conducted using an electronic servo-controller. The rise time for relaxation tests was 60 to 100 msec, so data were taken beginning at 1 sec. The amplifiers exhibited zero drift of about 1% per day and sensitivity drift of 1.5% per day which limited transient experiments to less than one day. Dynamic experiments employed a sinusoidal input signal to the torque motor and a lock-in amplifier to measure phase. Phase measurements at lower frequencies down to about 0.001 Hz were done using an oscilloscope to examine elliptical Lissajous figures; the analysis is shown in §3.2. The system permitted a resolution in tan δ of 0.001 from 0.5 to 100 Hz, the lower frequency limit being due to the lock-in amplifier and the upper limit being due to resonances in the LVDT assembly. Resonance of the specimen combined with added inertia occurred at 30 to 60 Hz. It was possible to take measurements through this resonance since the torque at the bottom of the specimen was measured, not the driving torque on the inertia member. Resonance due to elasticity of the torque sensor combined with the inertia of the specimen, driver, and other parts occurred at about 700 Hz. Resonance in the LVDT system occurred at about 100 Hz, which dictated the upper limit of frequency. The overall range of time and frequency accessible with this instrument was eight decades.

An automated instrument was developed for dynamic testing from 10^{-5} to 1 Hz, and for transient studies [6.10.4]. The device was a torsion pendulum containing a magnet and mirror fixed to a long stalk attached to the top of the specimen, which is contained within a temperature-controlled furnace.

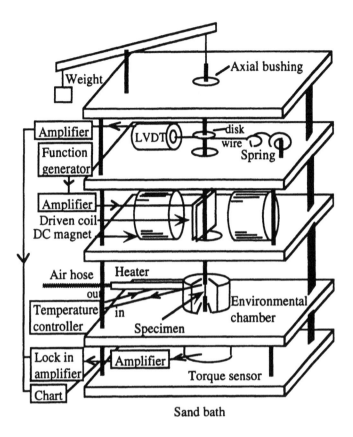

Figure 6.14   Biaxial driven torsion pendulum. (Adapted from Myers et al. [6.10.2].)

Current in a Helmholtz coil was used to apply the torque. Representative results for a polymer and an oxide glass were reported [6.10.4].

§6.10.4 Other dynamic instruments

A device for dynamic torsional measurements on polymers over a five-decade range from 0.01 Hz to 1 kHz used an electromagnetic drive system in which a coil carrying an electric current was immersed in a static magnetic field [6.10.5]. Angular displacement was measured via a capacitive sensor. The specimen was attached to a lumped inertia element and the dynamic material properties were extracted from the measured structural compliance and phase via the lumped system equations such as those developed in §3.5. Resonance was typically at about 1 kHz; measurements could be made at, near, or well below resonance. The specimen was restrained by pivots supported on leaf springs to allow axial expansion with temperature changes. Spurious phase shifts were introduced by the capacitive sensor and these were corrected by subtracting the "background" associated with a low-loss metal specimen.

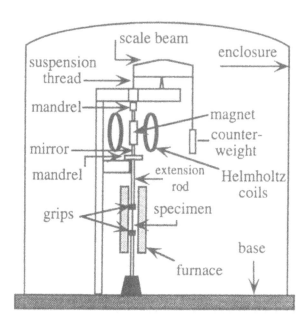

Figure 6.15   Driven torsion pendulum. (Adapted from Woirgard et al. [6.7.1].)

Subresonant instruments include that of Woirgard et al. [6.7.1], which involves phase measurement at high resolution, and that of D'Anna and Benoit [6.10.6], which allows measurement at cryogenic temperatures. The frequency range is limited to subaudio values. The approach of Woirgard et al. [6.7.1] provides a six-decade range of frequency from 10^{-5} to 10 Hz at strain amplitudes in the range 10^{-6} to 10^{-5}. It is an inverted torsion pendulum of the Kê type. Torque was generated by the action of electric current in a Helmholtz coil on a permanent magnet fixed to the specimen via an extension rod. The specimen was surrounded by furnace windings well below the driving coil. The assembly was stabilized via a weight acting through a rocker shown as a scale beam in Fig. 6.15. Angular displacement was measured by reflecting a beam of light from a mirror fixed to the specimen on a differential light-sensitive diode. Excellent phase resolution was attained by measuring the time delay between the zero-crossing of the load and displacement signals with a high-speed timer. A time accuracy of 1 μsec was claimed, which corresponds to a phase uncertainty $\delta\phi \approx 6 \times 10^{-5}$ at 10 Hz, and even better at lower frequencies. Control of electronic and mechanical noise is essential in this approach; building vibration amplitude was at most 0.1 μm. Noise of mechanical origin was reduced by the suspension system which suppresses transverse vibrations. Transverse vibrations can be problematic since they occur at considerably lower frequency than torsional vibrations and so are more easily excited by movement of the supporting base. The upper frequency limit was about 10 Hz since the pendulum natural frequency was 100 to 500 Hz. Tuning through the resonance was not done

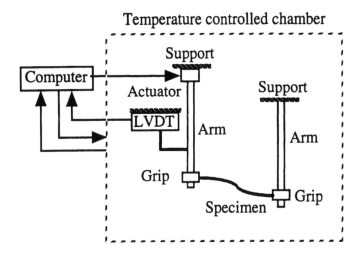

Figure 6.16 Schematic diagram of a commercial dynamic mechanical analyzer.

in view of the complex arrangement of stalks which would give a resonance structure difficult to interpret.

§6.10.5 Commercial instrumentation

Instrumentation specifically intended for viscoelastic characterization of materials is commercially available. The "Dynamic Mechanical Analyzer" [6.10.7] applies flexural load to a bar-shaped specimen via an electromagnetic driver coupled to the specimen by sample arms and clamps (Fig. 6.16). Deflection is measured by LVDTs. Temperature and load history are controlled by a computer. Creep and stress relaxation experiments can be performed, as well as low-frequency dynamic measurements. The mass of the sample arms and gripping fixtures limits the frequency range to less than about 10 Hz. However, for thermorheologically simple materials, an extended time or frequency range can be obtained by performing experiments at different temperatures under computer control.

§6.10.6 Instruments for a wide range of time and frequency

§6.10.6.1 Rationale

Measurements over a wide range of frequency are preferable to the more common approach of varying the temperature, for the following reasons [6.7.1]. (i) Temperature-related methods are restricted to activated processes. Viscoelastic spectra due to thermoelasticity or magnetic flux diffusion cannot be obtained by temperature scans. (ii) The underlying viscoelastic theory is established for isothermal conditions and is not directly applicable if temperature is varied continuously. (iii) Temperature variations can introduce important structural changes in the specimen during testing.

Most available experimental methods for characterizing viscoelastic materials are applicable over restricted portions of the time and frequency domains, usually no more than three or four decades. In the case of thermorheologically simple materials, such a limitation can be circumvented by performing a series of tests at different temperatures and computing a "master curve" of effective viscoelastic behavior. When a master curve is to be verified or in thermorheologically complex materials, direct measurements over many decades of time–frequency are required. Even for materials which are thought to be thermorheologically simple, a wide range of time and frequency is of use. Plazek [6.9.2–6.9.4] suggested that experimental data within a fairly narrow experimental window (three decades or less) are capable of demonstrating thermorheological complexity but cannot demonstrate simplicity. Plazek showed that nearly perfect superposition was obtained for data for polystyrene over a restricted range of 3.4 decades, but that data over 6 decades could not be superposed.

§6.10.6.2 *Various instruments*

In most reports of measurements over a wide range, many different apparatus are used, each of which covering two or three decades of the time–frequency domain [1.6.18, 6.10.8]. However, several authors reported instruments covering wider ranges, as follows. The multiple lumped resonator concept of Birnboim has been applied by Schrag and Johnson [6.10.9] to viscoelastic measurements on *fluids*. In this approach, torsional oscillations are set up in a system of cylindrical inertia members joined by rods. Angular displacement is measured by reflecting light from a mirror on the oscillator and modulating the light intensity via Ronchi gratings. This is a resonant method which provides data at discrete frequencies from 100 to 8300 Hz: two decades. Quasistatic experiments cannot be done with this apparatus; however, modifications of the Birnboim apparatus permit low-frequency studies down to 0.01 Hz [6.10.10]. Recent developments in signal processing permit good precision down to 10^{-6} Hz [6.10.11]. An extended frequency range exceeding seven decades has been obtained for fluids [6.10.12]. The fluid specimen is sandwiched between discs of piezoelectric ceramic, and mechanical properties inferred from electrical impedance measurements. The discs deform radially inducing shear deformation in the small sample. Since the specimen and discs are so small, the lowest apparatus resonance is about 100 kHz, permitting subresonant measurements to 50 kHz.

As for solids, Letherisch [6.9.1] obtained ten decades in creep by extending the tests to 1 year and by using fast transducers to capture the initial transient. Moreover, six decades were obtained via an automated creep apparatus described by Plazek [6.10.13]; it uses electromagnetic drive and optical measurement of rotation. It covers a range of time from about half a second to several days. The approach of Woirgard et al. [6.7.1] provides six decades of frequency from 10^{-5} to 10 Hz via subresonant phase measurement. About 6.5 decades in the frequency domain has been attained using two apparatuses [6.10.14], one from 0.0002 to 30 Hz, and a second from 20 Hz to 1 kHz.

A biaxial driven torsion pendulum [6.10.2] after modification [6.10.3] provided eight decades up to 100 Hz.

§6.10.6.3 Instrument for eleven decades

An experimental apparatus (Fig. 6.17) and analysis scheme was developed for determining the viscoelastic properties of a solid material isothermally, with a single apparatus, over more than 11 decades of time and frequency [6.10.15, 6.10.16], from 100 kHz in resonant harmonics to creep over several days. For an experimenter with the patience to extend creep tests to 1 year, the range extends to 13 decades. Torque is applied to the specimen electromagnetically and its deformation is determined by electronic measurement of displacement of a laser beam. Resonances are eliminated from the torque and angle measuring devices by this approach. The effect of resonances remaining in the specimen itself are corrected by a numerical analysis scheme based on an analytical solution which is applicable to homogeneous cylindrical specimens of any degree of loss. Stiffness and damping are calculated by numerical solution of an exact relationship for the torsional rigidity (ratio of torque M^* to angular displacement Φ) of a viscoelastic cylinder of radius R length L, and density ρ with an attached mass of mass moment of inertia I_{at} at one end and fixed at the other end [6.10.14]:

$$\frac{M^*}{\Phi} = \left[\frac{1}{2}\rho\pi R^4\right]\left[\omega^2 L\right]\frac{\cot\Omega^*}{\Omega^*} - I_{at}\omega^2, \qquad (6.10.1)$$

where

$$\Omega^* = \sqrt{\frac{\rho\omega^2 L^2}{KG^*}},$$

and K is a geometrical constant (equal to 1 for a cylindrical specimen with circular cross section).

For low-loss materials the method of resonance half widths is applicable.

$$\tan\delta \approx \frac{1}{\sqrt{3}}\frac{\Delta\omega}{\omega_0}, \qquad (6.10.2)$$

in which $\Delta\omega$ is the full width of the resonance curve (of structural compliance vs. angular frequency) at half maximum and ω_0 is a resonant angular frequency. If the magnet at the end has a sufficiently small inertia, higher order harmonics can be studied using the same method to extend the frequency range.

The apparatus is capable of creep, constant load rate, subresonant dynamic, and resonant dynamic experiments in bending and torsion. The range of equivalent frequency for torsion is from less than 10^{-6} Hz (limited

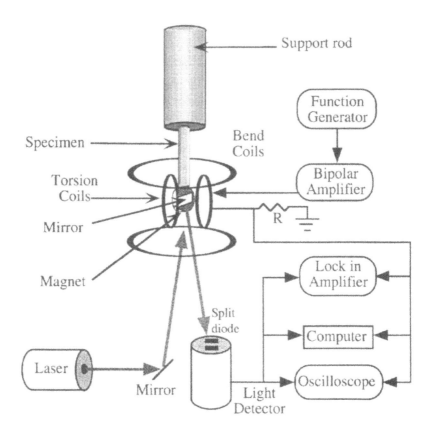

Figure 6.17 Instrument for 11 decades of time and frequency, capable of transient creep tests and dynamic subresonant and resonant tests to 100 kHz.

by the patience of the experimenter in conducting slow oscillatory or long-term creep tests) to about 10^5 Hz (limited by the ratio of signal to noise in the study of resonances of high order). The capability of the recent version of the apparatus was illustrated with measurements on indium–tin alloys (Fig. 7.11) over more than 11 decades up to 100 kHz [6.10.17]. Refinements to this instrument include the use of a microcomputer to control subresonant dynamic experiments and display expanded Lissajous figures, and improved isolation from mechanical and thermal perturbations from the environment. Recently a high-resolution digital lock-in amplifier was used to achieve further improvements in phase resolution.

The apparatus is a modified version of a micromechanics apparatus used earlier for study of microsamples of foams and composites in bending and torsion. The specimen is driven by an electromagnetic torque acting on a permanent magnet at one end, and the angular displacement is measured optically. The rationale for the torsion geometry is that the dynamic boundary value problem for the specimen, for which all dimensions are finite [1.6.1,

6.10.14], has an exact solution. The wide range of effective frequency is obtained as follows [6.10.15, 6.10.16]:

1. A fixed–free specimen geometry is used in combination with near-zero drift and zero friction methods for torque generation and angular displacement measurement. The lower bound on frequency (or equivalently, the upper bound on creep time) is dictated by the experimenter's patience, not the instrument.
2. The fundamental resonance frequency of the specimen is made as high as possible by minimizing the inertia of the torque generator (a high-intensity neodymium iron boron magnet fixed to the specimen end) and of the angular displacement measuring system (a mirror 3-mm square fixed to the specimen end); and by using the shortest possible specimen.
3. Resonances in the apparatus itself are eliminated or are moved to frequencies well above the range of interest.
4. The torsion or bending configuration allows a large impedance mismatch between the specimen and support rod, to minimize parasitic damping and to reduce any coupling to instrument resonances. See Examples 6.3 and 6.4.
5. The geometry of specimen and attachment is made as simple as possible. The governing equation for the specimen resonance is therefore sufficiently simple that it can be uniquely inverted numerically. The inversion method extracts the viscoelastic properties from the measured dynamic compliance and phase in the vicinity of specimen resonance, even for high-loss materials.

Representative experimental results over more than 11 decades are presented in §7.4, Fig. 7.10. Creep results are shown on a time scale linked to the main frequency scale by the relation $t = 1/2\pi v$. When both creep and dynamic results are available, they may also be plotted on a common frequency scale using approximate or exact conversion relations as presented in Chapter 4.

§6.11   Wave methods

Wave methods are suitable for studies at high frequencies above those attainable by resonance methods. Surveys of wave methods are given in References [5.7.5, 6.11.1]. The stiffness is inferred from the wave speed c and the density ρ. The speed c_T of shear or transverse waves in an unbounded isotropic elastic medium is given in References [6.11.2, 6.11.3].

$$c_T = \sqrt{\frac{G}{\rho}}, \qquad\qquad (6.11.1)$$

and the speed c_L of longitudinal or dilatational waves is

$$c_L = \sqrt{\frac{\left(B + \frac{4}{3}G\right)}{\rho}}, \qquad (6.11.2)$$

in which G is the shear modulus and B is the bulk modulus of the material. Observe that

$$B + \frac{4}{3}G = C_{1111},$$

which is the elastic modulus tensor element for constrained axial compression. In viscoelastic solids, the moduli are complex numbers, G^* and B^*. Interpretation of the modulus elements is discussed in Example 5.9. The storage modulus is inferred from the wave speed and the density, and the loss tangent is inferred from the attenuation α as presented in §3.7. Specifically, $\alpha \approx (\omega/2c)\tan \delta$, and the exact version is

$$\alpha = \frac{\omega}{c}\tan\frac{\delta}{2}. \qquad (6.11.3)$$

Ultrasonic methods are suitable for frequencies from 0.5 to 20 MHz and beyond. Special techniques permit study to more than 100 GHz.

The attenuation α is measured in nepers per unit length or in decibels (dB) per unit length. One neper is a decrease in amplitude of a factor of $1/e$. This corresponds to $\alpha z = 1$ in Eq. 3.7.11. The word "neper" is a degradation of the name of the inventor of the natural logarithm, Napier [6.11.4].

Attenuation of ultrasonic waves can be measured by comparing the signal amplitude transmitted through two samples of different length [6.11.5], as shown in Fig. 6.18. The electrical signal is a group of sinusoidal oscillations known as a tone burst. The attenuator is used to isolate the signal generator from the reflection of the waveform at the transducer. The velocity c is determined from the difference Δt in transit times of a particular zero-crossing in the signal, and the known lengths ℓ_1 and ℓ_2 of the specimens,

$$c = \frac{(\ell_1 - \ell_2)}{\Delta t}.$$

The attenuation α, in units of nepers per unit length is determined from the magnitudes of the signals: A_1 through a specimen of length ℓ_1, and A_2 through a specimen of length ℓ_2.

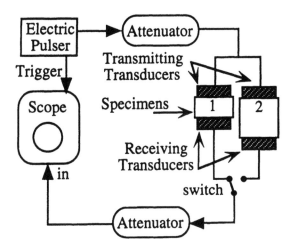

Figure 6.18 Measurement of ultrasonic attenuation.

$$\alpha = \frac{\ln\left(A_1/A_2\right)}{\left(\ell_1 - \ell_2\right)}.$$

One may also use a single transducer on one face of a specimen of parallelepiped shape to emit a waveform which reflects off the back surface of the specimen [6.11.4, 6.11.6]. The transducer is excited with bursts of electrical oscillations at its resonant frequency. The bursts must be short enough that successive echoes can be distinguished, but they must have at least several cycles so that the frequency is well defined. The waveform is received by the same transducer. A series of echoes is observed, and the attenuation is inferred from the decrease in amplitude of the echoes. The experimenter must take care that spurious sources of energy loss do not influence the results. In this echo method, quartz is favored for the transducer material. Quartz has a relatively weak electromechanical coupling; therefore, little energy from the sound wave is converted back into electricity and lost at the transducer interface. The quartz plate is bonded to the specimen or, if only longitudinal waves are to be used, coupled to it with a thin layer of oil or grease. Alternatively, the quartz transducer may be bonded to a slab-shaped specimen with salol (phenyl salicylate) which melts at 42°C. Moreover, the transducer should not be impedance-matched to the driving and detection circuitry since any energy absorbed by the circuit will increase the apparent attenuation. Commercial transducers for nondestructive testing (NDT) are not appropriate in this method because they exhibit strong coupling. Therefore, the transducer extracts considerable energy from the sound wave at each echo. Moreover, these transducers are heavily damped to achieve broadband response.

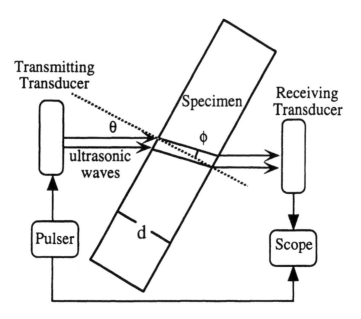

Figure 6.19   Ultrasonic method for determining complex bulk and shear moduli. The transducers and specimen are immersed in a fluid such as water.

In attenuation measurements, a long buffer rod may be interposed between the transducer and the specimen. The rationale for this approach is to eliminate parasitic energy loss from sound waves entering the transducer. A broadband NDT-type transducer can be used in the buffer rod approach. Attenuation is inferred from the magnitude of echoes called A, B, and C in order of their time delay following the driving electric pulse. Echo A is a reflection from the buffer–specimen interface. Echo B is from the specimen–air interface. Echo C has reflected once from the buffer–specimen interface and twice from the specimen–air interface. The reflection coefficient R for the rod–specimen interface must be known. It may be calculated from the echoes as follows.

First, normalize the echo amplitudes, retaining their sign.

$$\underline{A} = \frac{A}{B},$$

$$\underline{C} = \frac{C}{B}.$$

The reflection coefficient is

$$R = \sqrt{\frac{\underline{AC}}{\underline{AC} - 1}}.$$

Finally, the attenuation is

$$\alpha = \frac{1}{2}\frac{1}{\ell}\ln\left\{-\frac{R}{C}\right\}.$$

Waves can also be transmitted through long rods or plates [6.11.7, 6.11.8]. As for rods, bending waves and longitudinal waves are dispersive (the wave speed depends on frequency) even for an elastic material. For longitudinal waves in a rod, the wave speed c_L is

$$c_L = \sqrt{\frac{E}{\rho}}, \qquad (6.11.4)$$

provided that the wavelength is much larger than the rod diameter. For shorter wavelengths smaller than the rod diameter, the Poisson effect cannot readily occur, giving rise to dispersion. Torsional waves of lowest order are nondispersive in an elastic medium, and their velocity is the shear wave velocity given above; in a viscoelastic medium the dispersion results from viscoelastic effects only. Again, in viscoelastic materials the elastic modulus is a complex quantity, so that the displacement is

$$u(x,t) = u_0 \exp i\omega\left\{\frac{x}{c}-t\right\}\exp\left\{-x\frac{\omega}{2c}\tan\delta\right\} = u_0 \exp i\omega\left\{\frac{x}{c}-t\right\}\exp\{-\alpha x\},$$

$$(6.11.5)$$

and the loss tangent is inferred from the attenuation α; Eq. 6.11.3.

Ultrasonic methods for determining the complex bulk and shear moduli were presented by Waterman [6.11.9] as shown in Fig. 6.19. The method involves transmitting longitudinal ultrasonic waves through a slab of specimen material. The slab is immersed in a fluid which serves as the medium for wave propagation. The slab is then rotated through known angles. Oblique incidence converts part of the incident longitudinal wave to a transverse wave, so that the speed and attenuation of both waves can be measured. A variant of this approach [6.11.10] involving comparison of amplitudes A_1 and A_2 transmitted through otherwise identical specimens of different thicknesses L_1 and L_2 allows simpler interpretation since the loss due to surface wave reflection does not need to be considered since it is the same for both specimens. When the specimen is held perpendicular to the ultrasound beam, only longitudinal waves are generated in the specimen. The attenuation α_L of longitudinal waves, in decibels per centimeter, is given by

$$\alpha_L = \alpha_{liquid} + 20(L_2 - L_1)^{-1}\log(A_1/A_2). \qquad (6.11.6)$$

The wave speed is calculated by allowing for the time delay in the liquid. If the ultrasound beam is incident at an oblique angle ψ, shear waves are generated in a beam displaced from the longitudinal beam. If the angle exceeds a critical angle, the longitudinal wave is totally reflected in the specimen and only a shear wave is propagated. For oblique incidence, allowance is made for refraction of the ultrasonic wave, to calculate the speed c_T and attenuation α_T (in decibels per centimeter) of shear (transverse) waves as follows:

$$c_T = c_{\text{liquid}}\left[\left(\cos\psi - c_{\text{liquid}}\frac{\Delta t}{\Delta L}\right)^2 + \sin^2\psi\right]^{-1/2},$$

$$\alpha_T = 20(L_2 - L_1)^{-1}\left\{1 - \left[c_T\sin\psi/c_{\text{liquid}}\right]^2\right\}^{1/2}\log(A_1/A_2),$$

with Δt as the time delay and ΔL as the specimen thickness or difference in thickness.

Kolsky [5.7.5] has reviewed some wave methods and adduced articles by Nolle [6.11.11, 6.11.12] who studied rubbery materials from 0.1 Hz to 120 kHz using five different experimental methods. From 0.1 to 25 Hz, the specimen provided a restoring force for a beam rocking on a knife edge. From 10 to 500 Hz a vibrating reed-type resonance method was used. In this method the frequency can be changed only by changing the size or shape of the specimen. Longitudinal wave methods in thin strips were used for higher frequencies from 1 to 40 kHz. For 12 to 120 kHz, a compound magnetostrictive resonator was used, in which a rubber–metal sandwich was examined. Buna rubber, over a range of frequency and temperature, the α transition in modulus, and the corresponding α peak in loss vs. temperature, was demonstrated to be sharpest at low frequency and progressively less sharp at higher (ultrasonic) frequency. Kolsky [6.11.13] also demonstrated the excitation of pulsed waveforms via an explosive charge on the end of a rod, and the interpretation of results for nonsinusoidal waveforms.

Ultrasonic waves are usually generated and detected by piezoelectric transducers; these are commonly used for ultrasonic studies and are available to generate longitudinal or shear waves. The thickness of the piezoelectric element governs its natural frequency. Piezoelectric transducers are available for frequencies between 0.5 and 20 MHz, the frequency range most commonly used for nondestructive evaluation of machine parts and for diagnostic ultrasonic diagnosis of disorders in the human body. Electromagnetic methods are also used, particularly at lower frequencies.

Piezoelectric transducers may be prepared as thin layers for ultrasonic frequencies as high as 1 GHz. Ultrasonic attenuation measurements at frequencies as high as 440 GHz have been performed as follows [6.11.14]. A pulsed laser generates a short light pulse of picosecond duration. This light

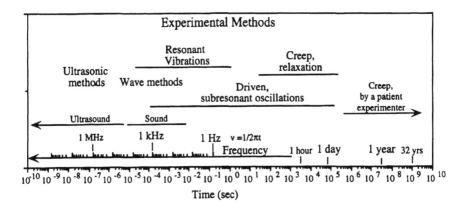

Figure 6.20 Summary of experimental methods in the time and frequency domains.

is directed at a solid surface where it is absorbed, raising the temperature of the surface a few degrees. The sudden temperature change induces a pulse of thermal stress, which propagates into the solid as an ultrasonic wave. This wave is detected as it is reflected back to the surface by an optical method which uses a probe pulse of laser light to determine the strain-induced change in reflectivity due to the ultrasonic wave. The method has been used to characterize the properties of thin films and small-scale structures with features as small as 200 nm.

§6.12 Summary

A brief review of some experimental methods has been presented. Although experimental methods are available over 20 decades of time and frequency (Fig. 6.20), most individual methods permit study of three decades or less for a given specimen. Study of an extended range requires multiple apparati, use of time–temperature superposition, or special techniques.

Examples

Example 6.1

A durometer is a device used to measure the stiffness of rubbery materials. The durometer has a flat surface which the user presses to the rubber specimen. A spring-loaded protruding probe causes an indentation in the rubber. The observer reads the value of the indentation from a dial gage, which reads from 0 to 100, linked to the probe. A spring within the durometer provides the indenting force exerted by the probe. If the rubber is viscoelastic, the dial reading changes with time after the durometer is pressed to the rubber specimen.

Calculate the relationship between the time-dependent dial reading and the creep function.

Hint: First calculate the relationship between the dial gage reading and the stiffness of an elastic material considered as a spring of unknown stiffness. *Hint:* The sum of the deflections of the specimen and the internal spring as an experiment is conducted may be approximated as a step function in time. The internal spring deflection is observable as the dial reading. The specimen deflection is not directly observed.

Solution

For the durometer, the sum of the spring displacement X_1 and the specimen displacement X_2 is a step function in time: $X_1 + X_2 = X_{tot}\mathcal{H}(t)$. The force in the spring equals the force in the specimen since inertial effects are neglected in the quasistatic regime: $F_1 = F_2$. So, with k_1 as the stiffness of the spring and k_2 as the stiffness of the specimen,

$$k_1 X_1(t) = \int_0^t k_2(t-\tau)\frac{dX_2}{d\tau}\,d\tau. \qquad (E6.1.1)$$

Take the Laplace transform

$$k_1 X_1(s) = sk_2(s)X_2(s) = sk_2(s)\big(X_2(s)+X_1(s)\big) - sk_2(s)X_1(s). \qquad (E6.1.2)$$

Incorporate

$$X_1(t) + X_2(t) = X_{tot}\,\mathcal{H}(t), \qquad (E6.1.3)$$

and take the Laplace transform

$$k_1 X_1(s) = sk_2(s)X_{tot}\frac{1}{s} - sk_2(s)X_1(s). \qquad (E6.1.4)$$

Define a new (dimensionless) x_1 as $(X_1)/(X_{tot})$. Then

$$k_1 x_1(s) = k_2(s)\big(1 - sx_1(s)\big), \qquad (E6.1.5)$$

so

$$x_1(s)\big(k_1 + sk_2(s)\big) = k_2(s), \qquad (E6.1.6)$$

so

$$x_1(s) = \frac{k_2(s)}{k_1 + sk_2(s)}. \qquad (\text{E6.1.7})$$

Rewrite this in a form amenable to power series expansion and recognize from Eq. 2.4.2 that $J(s)k_2(s) = 1/s^2$, with J as the specimen compliance and k_2 as its stiffness.

$$x_1(s) = \frac{\frac{1}{s}}{1 + \frac{k_1}{sk_2(s)}} = \frac{1}{s}\{1 - sk_1 J(s) + s^2 k_1^2 J(s)J(s) - \ldots\}. \qquad (\text{E6.1.8})$$

By transforming back to the time domain,

$$x_1(t) = 1 - k_1 J(t) + k_1^2 \int_0^t J(t - \tau)\frac{dJ}{d\tau}d\tau - \ldots . \qquad (\text{E6.1.9})$$

The result is an exact series representation thus far. If $k_2 \gg k_1$ for all time, then the series can be approximated by retaining only the first two terms. Here the spring is much more compliant than the specimen material, so most of the deformation occurs in the spring and little in the specimen so that the durometer reading approaches 100%. In that case, the normalized spring displacement $x_1(t)$ follows the time dependence of the creep function.

Example 6.2

Suppose experimental data are available for both creep and dynamic properties of a material. Suppose the creep results are converted to dynamic form with the relations given in Chapter 4, and that reasonable agreement is observed in the region of overlap. Suppose moreover that the real and imaginary parts of the dynamic compliance are found to obey the Kramers–Kronig relations. Can one conclude that the experimental results are valid?

Answer

The experimental results are not necessarily valid. If phase angles of electrical origin occur in the transducers or the preamplifier circuits, they will also obey the interconversion formulae (since parasitic phase angles are also due to causal processes) but will generate errors in the results. Independent calibration of the phase behavior of the instrumentation is required.

Example 6.3

A specimen of diameter d_1, length L_1, shear modulus G_1, and damping tan δ_1 is supported by a rod of diameter d_2, length L_2, shear modulus G_2, and damping tan δ_2 in a fixed–free torsional configuration. How much error is introduced in the measured stiffness and damping by the fact the support rod is not infinitely rigid? Assume the frequency is well below any specimen or structural natural frequency.

Solution

The structural compliance is defined as the ratio of the angle θ to the torque τ of specimen rod and the support rod is the sum of the compliances. Consider the elastic case first, in terms of the compliances $J_1 = 1/G_1$ and $J_2 = 1/G_2$. At low frequency the static relations are appropriate. The structural compliance is as follows:

$$\frac{\theta}{\tau} = \frac{32}{\pi}\left\{J_1\frac{L_1}{d_1^4} + J_2\frac{L_2}{d_2^4}\right\}.$$

Apply the dynamic correspondence principle.

$$\frac{\theta^*}{\tau} = \frac{32}{\pi}\left\{J_1'(1+i\tan\delta_1)\frac{L_1}{d_1^4} + J_2'(1+i\tan\delta_2)\frac{L_2}{d_2^4}\right\}.$$

The fractional error ξ_s in stiffness is the real part of the second term divided by the real part of the first term.

$$\xi_s = \frac{G_1'd_1^4L_2}{G_2'd_2^4L_1}.$$

The observed phase angle φ between torque and angular displacement is as follows:

$$\tan\varphi = \frac{\mathrm{Im}\left\{\dfrac{\theta}{\tau}\right\}}{\mathrm{Re}\left\{\dfrac{\theta}{\tau}\right\}} = \frac{J_1'\dfrac{L_1}{d_1^4}\tan\delta_1 + J_2'\dfrac{L_2}{d_2^4}\tan\delta_2}{J_1'\dfrac{L_1}{d_1^4} + J_2'\dfrac{L_2}{d_2^4}},$$

$$\tan\varphi = \frac{\tan\delta_1 + \dfrac{G_1'\,d_1^4L_2}{G_2'\,d_2^4L_1}\tan\delta_2}{1 + \dfrac{G_1'\,d_1^4L_2}{G_2'\,d_2^4L_1}}.$$

Since ξ_s is small for a specimen sufficiently thinner than the support rod,

$$\tan \varphi \approx \left[\tan \delta_1 + \frac{G_1'}{G_2'} \frac{d_1^4 L_2}{d_2^4 L_1} \tan \delta_2 \right] \left\{ 1 - \frac{G_1' d_1^4 L_2}{G_2' d_2^4 L_1} \right\}.$$

The quantity in the { } brackets represents a multiplicative error in the loss. The term containing $\tan \delta_2$ in the [ ] brackets is a parasitic damping, $\tan \delta_p$. The fractional error ξ_d associated with this parasitic damping is as follows:

$$\xi_d = \frac{\tan \delta_p}{\tan \delta_1} = \frac{G_1'}{G_2'} \frac{d_1^4 L_2}{d_2^4 L_1} \frac{\tan \delta_2}{\tan \delta_1}.$$

Use of a stiff material such as steel or tungsten for the support rod entails $\tan \delta_2 \approx 10^{-3}$. In view of the limited resolution for specimen damping in the subresonant regime, and the ratio of specimen diameter, the parasitic damping error is usually negligible.

For example, if $d_1 = 3.18$ mm, $d_2 = 12.7$ mm, $L_1 = 30$ mm, $L_2 = 120$ mm, $G_1 = 15.4$ GPa (a metal alloy specimen), $G_2 = 154$ GPa (a tungsten support rod), then $\xi_s = 1.56 \times 10^{-3}$, which is a small fractional error in stiffness. Moreover, the parasitic damping, at

$$\tan \delta_p = 1.56 \times 10^{-6},$$

is too small to easily be resolved in the subresonant domain. If $\tan \delta_1 = 10^{-3}$ and $\tan \delta_2 = 10^{-3}$, then $\xi_d = 1.56 \times 10^{-3}$ which is a small fractional error in damping. In the subresonant regime (no inertial effects or waves), phase uncertainty due to electrical and mechanical noise is more of a limitation. These errors in modulus and phase are much smaller in importance in the study of polymers, which are more compliant and higher in damping than the alloy considered above.

By contrast, in the higher frequency regime, wave transmission into the support rod can be problematic, particularly in view of the fact that at resonance one can resolve small values of $\tan \delta$ by the method of resonance half width or the method of free decay of vibration. An illustration is given in the following example.

Example 6.4

Consider parasitic damping at high frequency due to transmission of waves into the support rod in a fixed–free torsion device. Calculate the parasitic damping using the following data:

Case 1: The support rod is steel, with G_2 = 78 GPa, ρ_2 = 7.9 g/cm³, r_2 = 6.2 mm, and material 1; and a specimen is aluminum alloy with G_1 = 27 GPa, ρ_1 = 2.7 g/cm³, and r_1 = 0.8 mm.

Case 2: The support rod is tungsten, with G_2 = 154 GPa, d_2 = 12.7 mm, L_2 = 120 mm, ρ_2 = 19.3 g/cm³; and a specimen is an alloy with G_1 = 15.4 GPa, d_1 = 3.18 mm, L_1 = 30 mm, ρ_1 = 7 g/cm³.

Solution

For case 1, the transmission coefficient for power is T_p = 4.7 × 10⁻³ based on Eq. 6.8.1. By considering this as a parasitic specific damping or energy ratio Ψ_p under the assumption that all the power transmitted into the support rod is lost, the parasitic loss tangent is

$$\tan\delta_p = \frac{1}{2\pi}\Psi_p = 7.5\times10^{-4}. \qquad \text{(case 1)}$$

This is comparable to damping seen in pure aluminum and is larger than damping seen in many aluminum alloys. The parasitic loss can be reduced by using tungsten in the support rod, since it is stiffer (E = 400 GPa) and denser (ρ = 19.3 g/cm³) than steel.

For case 2, we have the same alloy specimen considered in the prior example and the same tungsten support rod.

$$\tan\delta_p = 4.8\times10^{-4}. \qquad \text{(case 2)}$$

Here in the resonant modality the parasitic damping is much larger than the value found above under quasistatic, subresonant conditions. Moreover, at a resonance, the experimenter can resolve much smaller values of $\tan\delta$ by the method of resonant half width or by the method of free decay of vibration. Therefore, the fixed–free configuration can be problematic in the study of low-loss stiff materials.

One may use a larger diameter support rod to increase the impedance mismatch and so reduce the parasitic loss in the fixed–free configuration. Even so, a free–free configuration is recommended for low-loss materials, though such methods are not amenable to static measurements. The fixed–free approach is adequate for high-loss metals which are usually less stiff than aluminum. For such metals the error would be correspondingly small, and a negligible percentage of the total damping. Polymers have even lower stiffness and higher damping, so the parasitic loss in a fixed–free configuration is negligible if the specimen is slender.

Example 6.5

Discuss the trade-off between the upper frequency attainable and the phase resolution of the torsion apparatus of Woirgard et al. [6.7.1] shown in Fig. 6.15. Refer to parts of the apparatus shown in the diagram.

Answer

Phase resolution is improved by the suppression of noise due to mechanical and electrical causes. Mechanical noise is reduced by constraints which suppress parasitic bending vibration. Support of the specimen at both ends reduces parasitic vibration but also adds inertia which reduces the resonant frequencies. Resonant frequencies are also lowered by the fact that the specimen is isolated from the driving magnet by a long stalk, which allows the specimen to be heated without heating the magnet. The stalks and grips contribute a considerable inertia, as well as a complex resonance structure. Therefore, measurements are conducted in the subresonant regime, well below the lowest resonance frequency. Restriction of the frequency range has a side effect of reducing electrical noise, hence improving phase resolution.

Example 6.6

Ultrasonic attenuation is presented as 10 dB/μsec at 10 MHz, and the velocity is 6 km/sec. What is the attenuation in nepers per millimeter, and what is tan δ?

Solution

To convert units divide the attenuation by the velocity, so

$$\alpha = 10 \frac{dB}{\mu \sec} \frac{1}{6 \frac{km}{\sec}} = 1.67 \cdot 10^3 \frac{dB}{m}.$$

But from §3.7, α (dB/cm) = 8.68 α (neper/cm).

$$\alpha = 192 \frac{neper}{m}.$$

Also,

$$\alpha = \frac{\omega}{c} \tan \frac{\delta}{2}.$$

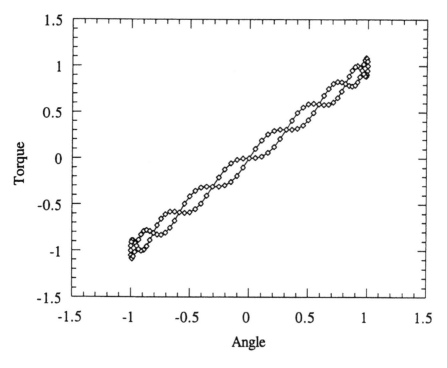

Figure E6.1   An observed Lissajous figure of unusual shape.

With $\omega = 2\,\pi \cdot 10^6\,\text{Hz}$, and $c = 6$ km/sec,

$$\tan \delta = 0.037.$$

Example 6.7

During dynamic subresonant torsional testing of a material, the Lissajous figure shown in Fig. E6.1 suddenly appears following a small change in driving frequency. What might be the cause?

Answer

Since the figure is not elliptical, some form of nonlinearity is present. Specifically, a high order Fourier harmonic of the driving frequency is present. One possibility is that a resonance has been excited by a weak Fourier harmonic of the driving signal, which is in the subresonant domain. If the material is of low loss, the magnification, proportional to $(\tan \delta)^{-1}$ of any input, is large. In that case even a small nonlinearity in any part of the system can excite the resonance. If such a pattern only appears at frequencies an integer fraction of a natural frequency of the specimen, then the resonant phenomenon is identified with the specimen itself. The nonlinearity could be in the specimen, the transducers, the signal generator, or the electronics.

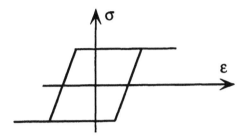

Figure E6.2 A stress–strain plot inferred from observations of torque and angular displacement.

Example 6.8

During dynamic torsional testing of a material, the Lissajous figure shown in Fig. E6.2 is observed. What might be the cause?

Answer

This is clearly a nonlinear response of the saturation type. There are several possible causes.

1. Transducers can saturate if driven beyond their range of linear behavior. Examine the specifications or calibration curve for the load transducer to see if this is the case.
2. If the specimen is deformed sufficiently, it may yield. Calculate the strain level and consider the type of material.
3. Some kinds of material, particularly those with magnetoelastic coupling, can exhibit nonlinear hysteresis behavior.
4. If the specimen is gripped in the apparatus, a stick–slip phenomenon in the grip can give the impression of material nonlinearity. Tighten the grips and observe whether the nonlinear behavior persists.

Example 6.9

How does one obtain the activation energy for thermally activated processes experimentally?

Answer

Do experiments at different temperatures. The shift factor for a thermally activated process can be obtained as follows [1.6.21]. Consider that at temperatures T_1 and T_2 the peak loss occurs at frequencies v_1 and v_2, respectively. Then, suppose an Arrhenius equation (§2.7) applies, which arises when the rate-limiting step of the relaxation consists of movement of an atomic or molecular process over an energy barrier.

$$v = v_0 \exp{-\frac{U}{RT}}, \tag{E6.9.1}$$

with U as the activation energy and $R = 1.986$ cal/mole K as the gas constant used if U is to be expressed in calories per mole. When expressed in energy per unit temperature one speaks of the Boltzmann constant $k = 1.38 \times 10^{-16}$ erg/K. The Boltzmann constant may also be written in units of electron volts (eV), defined as the energy acquired by an electron in traversing a potential difference of 1 V.

$$k = 8.64 \times 10^{-5} \, eV/K.$$

By considering different frequencies and taking a ratio,

$$\frac{v_1}{v_2} = \frac{\exp\left(-U/RT_1\right)}{\exp\left(-U/RT_2\right)}, \tag{E6.9.2}$$

$$\frac{\ln v_1}{\ln v_2} = \ln a_T = \frac{U}{R}\left\{\frac{1}{T_2} - \frac{1}{T_1}\right\}. \tag{E6.9.3}$$

So the activation energy U is given by

$$U = R \frac{\ln v_1}{\ln v_2}\left\{\frac{1}{T_2} - \frac{1}{T_1}\right\}^{-1}. \tag{E6.9.4}$$

One may also perform experiments at more than two temperature values. Suppose the temperature at which a loss peak occurs is called T_p; one may plot ln v vs. $1/T_p$, and compute a slope. If a straight line is observed, the slope is interpreted as U/R.

Example 6.10

What is the appearance of a Debye peak of a thermally activated material when plotted as a function of temperature?

Solution

Consider the Debye peak

$$\tan \delta = A \frac{\omega\tau}{1+\omega^2\tau^2}, \tag{E6.10.1}$$

with the time constant

$$\tau = \tau_0 \exp \frac{U}{RT},$$ (E6.10.2)

in which τ_0 is a characteristic time. Here the experimentalist maintains the angular frequency ω constant in Eq. E6.10.1 and varies the temperature T. The time constant may also be written $\tau^{-1} = \nu_0 \exp(-U/RT)$ with ν_0 as a characteristic frequency. Due to the exponential dependence on temperature, the time constant may be varied over a wide range by changing the temperature. This aspect facilitates the experimental characterization of this class of materials. By multiplying both sides of Eq. E6.10.2 by ω and taking the natural log,

$$\ln \omega\tau = \ln \omega\tau_0 + \frac{U}{R}\frac{1}{T}.$$ (E6.10.3)

At the maximum of a Debye peak, $\omega\tau = 1$, so for the peak at temperature T_{peak},

$$\ln \omega\tau_0 + \frac{U}{R}\frac{1}{T_{peak}} = 0.$$

The full width at half maximum of that peak on an inverse temperature scale is [1.6.23]

$$\Delta(T^{-1}) = 1.144(2.303\,R/U) = 2.635\,R/U.$$ (E6.10.4)

The width can be used to infer the activation energy provided that the peak is in fact a Debye peak.

Example 6.11

How might the viscoelastic Poisson's ratio of an isotropic material be determined experimentally from shear and axial tests?

Answer

In Example 5.6, the correspondence principle was used to obtain the following, in the time domain and in the frequency domain, respectively:

$$v(t) = \frac{1}{2}\int_0^t E(t-\tau)\frac{dJ_G(\tau)}{d\tau}d\tau - 1.$$ (E6.11.1)

$$v^* = \frac{1}{2}(E' + iE'')(J'_G - iJ''_G) - 1. \qquad\qquad \text{(E6.11.2)}$$

Experimental data do not cover the full range of times from zero to infinity. Therefore, a portion of the range of integration in Eq. E6.11.1 is inaccessible. If the material under consideration were reasonably well understood, the investigator could attempt to write bounds on the strength of relaxation in the short time region, and calculate the effect on the integral. A better choice would be to work in the frequency domain, following Eq. E6.11.2. The calculation is done pointwise, for each frequency, and is simpler and less demanding of input data.

Example 6.12

Determine the tan δ from the graph in Fig. 6.8. Compare with the value expected from the transient data from which this plot was calculated.

Solution

By referring to Fig. 3.2, use $\sin\delta = \dfrac{A}{B}$ and measure the corresponding dimensions on the elliptical figure. By making measurements with a millimeter scale or with a micrometer, $\delta = \sin^{-1}\dfrac{38.5\ \text{mm}}{56\ \text{mm}} = 0.63$, so tan $\delta = 0.74$. The relaxation function was given as a power law, $E(t) = At^n$, with n = 0.4. The inferred phase is $\delta = n\dfrac{\pi}{2} = 0.628$, so that tan $\delta = 0.73$, a satisfactory agreement in view of the nature of the measurement.

Example 6.13

Calculate the bulk properties of an isotropic viscoelastic material from the shear and tensile properties. What is the effect of a 5% error in the measured stiffness? What is the effect of a 5% error in the measured loss tangent? Discuss the results.

Solution

The relation between bulk modulus B, shear modulus G, and Young's modulus E for an isotropic elastic material is as follows:

$$B = \frac{GE}{3(3G-E)}. \qquad\qquad \text{(E6.13.1)}$$

By passing to the compliance formulation, with $J_E = E^{-1}$ and $J_G = G^{-1}$, to simplify inversion of the Laplace transform

$$\kappa = B^{-1} = \frac{3(3G - E)}{GE} = 3(3J_E - J_G). \qquad (E6.13.2)$$

By applying the correspondence principle,

$$s\kappa(s) = 3(3sJ_E(s) - sJ_G(s)). \qquad (E6.13.3)$$

Inverting to obtain the time domain behavior gives

$$\kappa(t) = 3(3J_E(t) - J_G(t)). \qquad \textbf{(E6.13.4)}$$

As for the bulk *modulus* in the time domain, since $B\kappa = 1$, with the correspondence principle,

$$B(s)\kappa(s) = s^{-2}. \qquad (E6.13.5)$$

By transforming back, the following time domain convolution relation between bulk creep compliance and bulk relaxation modulus is obtained.

$$\int_0^t \kappa(t - \tau)B(\tau)d\tau = t. \qquad (E6.13.6)$$

In the frequency domain, applying the dynamic correspondence principle gives

$$\kappa^*(\omega) = 3(3J_E^*(\omega) - J_G^*(\omega)). \qquad \textbf{(E6.13.7)}$$

Moreover

$$\kappa^* = \{B^*\}^{-1}. \qquad (E6.13.8)$$

Consider now errors in the input data for the frequency domain case. Suppose Poisson's ratio v is 0.3: so with $E = 2G(1 + v)$, for an elastic material, if $G = 1$ GPa, then $E = 2.6$ GPa, $J_G = 1$ GPa^{-1}, and $J_E = 0.3846$ GPa^{-1}. So the bulk compliance is $\kappa = 0.4615$ GPa^{-1}. Suppose there is a –5% error in J_G; then there is a 32% error in κ. Suppose there is a –5% error in J_E; then there is a

–37% error in κ. For a viscoelastic material, we have $\kappa(1 - i \tan \delta_\kappa) = 3(3J_E(1 - i \tan \delta_E) - J_G (1 - i \tan \delta_G))$. Suppose $\tan \delta_E = \tan \delta_G = 0.01$. Then, for a –5% error in $\tan \delta_G$, there is a 32% error in $\tan \delta_\kappa$. Errors in the input data are magnified since the bulk compliance is written in terms of a difference of terms of similar magnitude. This situation becomes even more severe if Poisson's ratio of the material is larger. Consequently, input data of considerable precision and accuracy in both magnitude and phase are required to infer bulk properties from shear and axial data.

Problems

6.1 A vibrating bar of a continuous material has an infinite number of resonance frequencies, yet it is usually possible to use only the lowest few in experiments. Why?

6.2 (a) Determine the creep properties of a foam earplug (see §10.2) using available materials and methods. If earplug foam is unavailable, obtain an alternate flexible foam. Design the experiment so that data are obtained over as many decades of time scale as possible, and so that the error bars are not excessively large. If you have already done a crude creep experiment in Problem 2.4, refine your technique here. It is not necessary to have sophisticated laboratory electronics to do creep tests. Be creative. Also write down what you do and make diagrams of your apparatus. Prepare error estimates for your creep results. Is your specimen sensitive to changes in relative humidity?

(b) Determine the creep properties of a foam earplug at several temperatures. If your apparatus is reasonably compact, it can be placed in a refrigerator. Does time–temperature superposition apply to the material? If so, prepare a master curve.

6.3 Consider phase measurement in relation to the problem of phase resolution. How does the signal subtraction method described in Appendix 6 differ in effectiveness from a measurement based on the width of an amplified Lissajous figure?

6.4 Grip a thin rod of plastic or other inert material between your teeth and tweak it into "cantilever" bending. Describe your observations. Discuss implications regarding the effect of grip conditions on measurements of free decay of vibration in cantilever bending.

6.5 A durometer is a device used to measure the hardness of materials. Durometers used for compliant materials such as rubbers contain a spring-loaded probe connected to a dial gage. The experimenter presses the device on the material to be tested and reads the indentation from the dial. If a durometer is available, observe the dial reading as a function of time following the application of the instrument to a viscoelastic rubber. Write down the data. Prepare a computation outline. Can these results be interpreted as creep?

References

6.1.1 Dally, J. W. and Riley, W. F., *Experimental Stress Analysis*, 2nd ed., McGraw-Hill, NY, 1978.

6.1.2 Whitney, J. M., Daniel, I. M., and Pipes, R. B., *Experimental Mechanics of Fiber Reinforced Composite Materials*, SASE/Prentice Hall, Englewood Cliffs, NJ, 1982.

6.1.3 Norton, H. N., *Handbook of Transducers*, Prentice Hall, Englewood Cliffs, NJ, 1989.

6.2.1 L'Hermite, R., "What do we know about the plastic deformation and creep of concrete?", *Bull. RILEM* 1, 22–51, 1959.

6.2.2 Bazant, Z., *Mathematical Modeling of Creep and Shrinkage of Concrete*, J. Wiley, NY, 1988, p. 30.

6.2.3 Leaderman, H., in *Rheology*, Vol. 2, ed. F. R. Eirich, Academic, NY, 1958.

6.2.4 Turner, S., "Creep in glassy polymers", in *The Physics of Glassy Polymers*, ed. R. H. Howard, J. Wiley, NY, 1973.

6.2.5 Johnson, A. F., "Bending and torsion of anisotropic beams", *Int. J. Solids Struct.* 9, 527–551, 1973.

6.3.1 Kobayashi, A. S., ed., *Manual of Engineering Stress Analysis*, 3rd ed., Prentice Hall, Englewood Cliffs, NJ, 1982.

6.3.2 Holman, J. P., *Experimental Methods for Engineers*, 6th ed., McGraw-Hill, NY, 1994.

6.3.3 Doebelin, E. O., *Measurement Systems: Application and Design*, McGraw-Hill, NY, 1975.

6.3.4 Steel, W. H., *Interferometry*, 2nd ed., Cambridge University Press, London, 1986.

6.3.5 Jones, R. V., "Some developments and applications of the optical lever", *J. Sci. Instr.* 38, 37–45, 1961.

6.3.6 Shi, P. and Stijns, E., "New optical method for measuring small angle rotations", *Appl. Optics* 27, 4342–4344, 1988.

6.5.1 Jackson, J. D., *Classical Electrodynamics*, J. Wiley, NY, 1962.

6.7.1 Woirgard, J., Sarrazin, Y., and Chaumet, H., "Apparatus for the measurement of internal friction as a function of frequency between 10^{-5} and 10 Hz", *Rev. Sci. Instr.* 48, 1322–1325, 1977.

6.7.2 Brennan, B. J. , "Linear viscoelastic behaviour in rocks", in *Anelasticity in the Earth*, ed. F. D. Stacey, M. S. Paterson, and A. Nicholas, American Geophysical Union, Washington, DC, 1981.

6.7.3 Kinra, V. K. and Wren, G. G., "Axial damping in metal–matrix composites. I. A new technique for measuring phase difference to 10^{-4} radians", *Exp. Mech.* 32, 163–171, 1992.

6.7.4 Gottenberg, W. G. and Christensen, R. M., "Prediction of the transient response of a linear viscoelastic solid", *J. Appl. Mech.* 33, 449, 1966.

6.7.5 Lazan, B., "Effect of damping constants and stress distribution on the resonance response of members", *J. Appl. Mech.* 75, 201–209, 1953.

6.7.6 Graesser, E. J. and Wong, C. R., "Analysis of strain dependent damping in materials via modeling of material point hysteresis", DTRC-SME-91-34, David Taylor Research Center, U.S. Navy, 1991.

6.8.1 Cremer, L. and Heckl, M. *Structure Borne Sound*, 2nd ed., Springer Verlag, Berlin, 1986.

6.8.2   McSkimin, H. J. "Ultrasonic methods for measuring the mechanical properties of liquids and solids", in *Physical Acoustics*, Vol. 1A, ed. E. P. Mason, Academic, NY, 1964, pp. 271–334.

6.8.3   Wachtman, J. H., Jr. and Tefft, W. E., "Effect of suspension position on apparent values of internal friction determined by Forster's method", *Rev. Sci. Instr.* 29, 517–520, 1958.

6.8.4   Frasca, P., Harper, R. A., and Katz, J. L., "Micromechanical oscillators and techniques for determining the dynamic moduli of microsamples of human cortical bone at microstrains", *J. Biomech. Eng.* 103, 146–150, 1981.

6.8.5   Amblard, F., Yurke, B., Pargellis, A., and Leibler, S., "A magnetic manipulator for studying local rheology and micromechanical properties of biological systems", *Rev. Sci. Instr.* 67, 818–827, 1996.

6.8.6   Quimby, S. L., "Experimental determination of viscosity of vibrating solids", *Phys. Rev.* 25, 558–573, 1925.

6.8.7   Marx, J., "Use of the piezoelectric gauge for internal friction measurements", *Rev. Sci. Instr.* 22, 503–509, 1951.

6.8.8   Robinson, W. H., Carpenter, S. H., and Tallon, J. L., "Piezoelectric method of determining torsional mechanical damping between 40 and 120 kHz", *J. Appl. Phys.* 45, 1975–1981, 1974.

6.8.9   Cahill, D. G. and Van Cleve, J. E., "Torsional oscillator for internal friction data at 100 kHz", *Rev. Sci. Instr.* 60, 2706–2710, 1989.

6.8.10 Iwayanagi, S. and Hideshima, T., "Low frequency coupled oscillator and its application to high polymer study", *J. Phys. Soc. Jpn.* 8, 365–368, 1953.

6.8.11 Koppelmann, V. J., "Über das dynamische elastische Verhalten hochpolymerer Stoffe", *Kolloid Z.* 144, 12–41, 1955. (In German).

6.8.12 Garrett, S. L., "Resonant determination of elastic moduli", *J. Acoust. Soc. Am.* 88, 210–221, 1990.

6.8.13 Simpson, H. M. and Fortner, B. E., "The dynamic modulus and internal friction of a fiber vibrating in the torsional mode", *Am. J. Phys.* 55, 44–46, 1987.

6.8.14 Tverdokhlebov, A., "Resonant cylinder for internal friction measurement", *J. Acoust. Soc. Am.* 80, 217–224, 1986.

6.8.15 Bishop, J. E. and Kinra, V. K., "Some improvements in the flexural damping measurement technique", in *M³D: Mechanics and Mechanisms of Material Damping*, ed. V. K. Kinra, and A. Wolfenden, ASTM, Philadelphia, PA, 1992, ASTM STP 1169.

6.8.16 Braginskii, V. B., Mitrofanov, V. P., and Panov, V. I., *Systems with Small Dissipation*, University of Chicago Press, IL, 1985.

6.8.17 Maynard, J., "Resonant ultrasound spectroscopy", *Phys. Today* 49, 26–31, 1996.

6.8.18 Kuokkala, V. T. and Schwarz, R. B., "The use of magnetostrictive film transducers in the measurement of elastic moduli and ultrasonic attenuation of solids", *Rev. Sci. Instrum.* 63, 3136–3142, 1992.

6.9.1   Letherisch, W., "The rheological properties of dielectric polymers", *Br. J. Appl. Phys.* 1, 294–301, 1950.

6.9.2   Plazek, D. J., "Temperature dependence of the viscoelastic behavior of polystyrene", *J. Phys. Chem.* 69, 3480–3487, 1965.

6.9.3   Plazek, D. J., "Oh, thermorheological simplicity, wherefore art thou?", *J. Rheol.* 40, 987–1014, 1996.

6.9.4   Plazek, D. J., "Magnetic bearing torsional creep apparatus", *J. Polym. Sci. A* 2(6), 621–638, 1968.

6.10.1   Sternstein, S. S. and Ho, T. C., "Biaxial stress relaxation in glassy polymers", *J. Appl. Phys.* 43, 4370–4383, 1972.

6.10.2   Myers, F. A., Cama, F. C., and Sternstein, S. S., "Mechanically enhanced aging of glassy polymers", *Ann. NY Acad. Sci*, 279, 94–99, 1976.

6.10.3   Lakes, R. S., Katz, J. L., and Sternstein, S. S., "Viscoelastic properties of wet cortical bone I. Torsional and biaxial studies", *J. Biomech.* 12, 657–678, 1979.

6.10.4   Etienne, S., Cavaille, J. Y., Perez, J., and Salvia, M., "Automatic system for micromechanical properties analysis", *J. Phys. Colloq.* C5, suppl. 10, 42, C5-1129–C5-1134, 1981.

6.10.5   Wetton, R. E. and Allen, G., "The dynamic mechanical properties of some polyethers", *Polymer* 7, 331–365, 1966.

6.10.6   D'Anna, G. and Benoit, W., "Apparatus for dynamic and static measurements of mechanical properties of solids and of flux lattice in type II superconductors at low frequency (10^{-5}–10 Hz) and temperature (4.7–500 K), *Rev. Sci. Instr.* 61, 3821–3826, 1990.

6.10.7   TA Instruments, Inc., 109 Lukens Drive, New Castle, DE 19720 (instrument was formerly made by DuPont).

6.10.8   Koppellmann, V. J. "Uber die Bestimmung des dynamischen Elasticitats-moduls und des dynamischen Schubmoduls im Frequenzbereich 10^{-5} bis 10^{-1} Hz", *Rheol. Acta* 1, 20–28, 1958.

6.10.9   Schrag, J. L. and Johnson, R. M. "Application of the Birnboim multiple lumped resonator principle to viscoelastic measurements of dilute macromolecular solutions", *Rev. Sci. Instr.* 42, 224–232, 1971.

6.10.10   Schrag, J. L. and Ferry, J. D., "Mechanical techniques for studying viscoelastic relaxation processes in polymer solutions", *Faraday Symp. Chem. Soc.* 6, 182–193, 1972.

6.10.11   Winther, G., Parsons, D. M., and Schrag, J. L., "A high-speed, high precision data acquisition and processing system for experiments producing steady state periodic signals", *J. Polym. Sci. B: Polym. Phys.* 32, 659–670, 1994.

6.10.12   Christensen, T. and Olsen, N. B., "A rheometer for the measurement of a high shear modulus covering more than seven decades of frequency below 50 kHz", *Rev. Sci. Instr.* 66, 5019–5031, 1995.

6.10.13   Plazek, D. J., "Magnetic bearing torsional creep apparatus", *J. Polym. Sci. A* 2(6), 621–638, 1968.

6.10.14   Gottenberg, W. G. and Christensen, R. M., "An experiment for determination of the material property in shear for a linear isotropic viscoelastic solid", *Int. J. Eng. Sci.* 2, 45–56, 1964.

6.10.15   Chen, C. P. and Lakes, R. S., "Apparatus for determining the properties of materials over ten decades of frequency and time", *J. Rheol.* 33(8), 1231–1249, 1989.

6.10.16   Brodt, M., Cook, L. S., and Lakes, R. S., "Apparatus for determining the properties of materials over ten decades of frequency and time: refinements", *Rev. Sci. Instr.* 66, 5292–5297, 1995.

6.10.17   Lakes, R. S. and Quackenbush, J., "Viscoelastic behaviour in indium tin alloys over a wide range of frequency and time", *Philos. Mag. Lett.* 74, 227–232, 1996.

6.11.1 McSkimin, H. J., "Ultrasonic methods for measuring the mechanical properties of liquids and solids", in *Physical Acoustics*, Vol. 1A, ed. E. P. Mason, Academic, NY, 1964, pp. 271–334.

6.11.2 Thurston, R. N., "Wave propagation in fluids and normal solids", in *Physical Acoustics*, Vol. 1A, ed. E. P. Mason, Academic, NY, 1964, pp. 1–110.

6.11.3 Gorelik, G. S., *Oscillations and Waves*, Fizmatgiz, Moscow, 1959.

6.11.4 Papadakis, E., "The measurement of ultrasonic attenuation", in *Physical Acoustics*, ed. R. N. Thurston, and A. D. Pierce, Academic, NY, 1990, pp. 107–155.

6.11.5 Truel, R., Elbaum, C., and Chick, B., *Ultrasonic Methods in Solid State Physics*, Academic, NY, 1966.

6.11.6 Chung, D. H., Silversmith, D. J., and Chick, B. B., "A modified ultrasonic pulse echo overlap method for determining sound velocities and attenuation of solids", *Rev. Sci. Instr.* 40, 718–720, 1969.

6.11.7 Meeker, T. R. and Meitzler, A. H., "Guided wave propagation in elongated cylinders and plates", in *Physical Acoustics*, Vol. 1A, ed. E. P. Mason, Academic, NY, 1964, pp. 111–167.

6.11.8 McSkimin, H. J., "Notes and references for the measurement of elastic moduli by means of ultrasonic waves", *J. Acoust. Soc. Am.* 33, 606–615, 1961.

6.11.9 Waterman, H. A., "Determination of the complex moduli of viscoelastic materials with the ultrasonic pulse method (Part I)", *Kolloid Z. Z. Polym.* 192, 1–16, 1963.

6.11.10 Hartmann, B. and Jarzynski, J., "Immersion apparatus for ultrasonic measurements in polymers", *J. Acoust. Soc. Am.* 56, 1469–1477, 1974.

6.11.11 Nolle, A. W., "Methods for measuring the dynamic mechanical properties of rubber-like materials", *J. Appl. Phys.* 19, 753–774, 1948.

6.11.12 Nolle, A. W., "Dynamic mechanical properties of rubberlike materials", *J. Polym. Sci.* 5, 1–54, 1949.

6.11.13 Kolsky, J., "The propagation of stress pulses in viscoelastic rods", *Philos. Mag.* 1, 693–711, 1957.

6.11.14 Lin, H., Maris, H. J., Freund, L. B., Lee, K. Y., Luhn, H., and Kern, D. P., "Study of vibrational modes of gold nanostructures by picosecond ultrasonics", *J. Appl. Phys.* 73, 37–45, 1993.

chapter seven

Viscoelastic properties of materials

§7.1   Introduction

The purpose of this chapter is to present the viscoelastic behavior of representative real materials so that the reader can gain a sense of orders of magnitude of the effects. Engineers and scientists who deal with elastic behavior of materials are aware of the moduli of various common materials. Similarly, knowledge of the viscoelastic properties of particular materials is essential to rationally apply them.

§7.2   Overview: some common materials

The loss tangents of some well-known materials at various temperatures and frequencies are presented in Table 7.1. Most of the data are at room temperature, denoted rt in Table 7.1 except as noted, and at audio or subaudio frequencies. Further comparison of materials [7.2.1–7.2.11] is given in the stiffness–loss map in Fig. 7.1. Since viscoelastic properties depend on frequency, each material at a given temperature is associated with a *curve*, not a point, in a stiffness–loss map. If only a narrow range of time or frequency is available, the curve may appear to degenerate into a point, particularly if the damping is small. Properties of many materials have also been compiled by Ashby [7.2.12] including charts of damping vs. stiffness, as well as stiffness and strength vs. density.

The diagonal line in Fig. 7.1 represents $E'' = E' \tan \delta = 0.6$ GPa. Most materials occupy the region to the left of that line. The product $E' \tan \delta$ represents a figure of merit for damping layers as discussed in §10.3.

Table 7.1 Loss Tangent of Common Materials at Various Temperatures T and Frequencies ν

Material	T	ν	tan δ	Ref.
Sapphire	4.2K	30 kHz	2.5×10^{-10}	[7.2.1]
Sapphire	rt	30 kHz	5×10^{-9}	[7.2.1]
Silicon	rt	20 kHz	3×10^{-8}	[7.2.1]
Quartz	rt	1 MHz	$\approx 10^{-7}$	[7.2.2]
Aluminum	rt	20 kHz	$<10^{-5}$	[7.2.3, 7.2.4]
Cu–31%Zn	rt	6 kHz	9×10^{-5}	[7.2.3, 7.2.4]
Steel	rt	1 Hz	0.0005	[1.6.22]
Aluminum	rt	1 Hz	0.001	[1.6.22]
Fe–0.62%V	33°C	0.95 Hz	0.0016	[7.2.3, 7.2.5]
Basalt	rt	0.001–0.5 Hz	0.0017	[7.2.6]
Granite	rt	0.001–0.5 Hz	0.0031	[7.2.6]
Glass	rt	1 Hz	0.0043	[1.6.22]
Wood	rt	1.5–8 Hz	0.0083	[7.2.7]
Wood	rt	≈ 1 Hz	0.02	[7.10.6]
Bone	37°C	1–100 Hz	0.01	[7.2.8]
Lead	rt	1–15 kHz	0.029	[7.2.7]
PMMA	rt	1 Hz	0.1	[7.2.9, 7.2.10]
PMMA	135°C	0.5 Hz	1.7	[7.2.11]

§7.3   Polymers

§7.3.1   Shear and extension in amorphous polymers

For illustrative purposes we consider only a few of the many available polymers, particularly polymethyl methacrylate (PMMA) for which many different methods of characterization have been applied. The properties of many others are described in Reference [1.6.18]. The general features of the viscoelastic response of a polymer are shown in Fig. 7.2. Polymers consist of long-chain molecules. A *linear polymer* is one in which the polymer molecule is a long chain with no branch points or side appendages. In a *cross-linked polymer* the chains are linked at various points. In *amorphous* polymers there is no long-range order in the molecular arrangement. Amorphous polymers have a glass transition temperature but no melting temperature. They are stiff below that temperature and rubbery or viscous above it. Glassy polymers such as PMMA and polystyrene are brittle; a second component can be added to improve toughness. Many amorphous polymers are considered to be thermorheologically simple. PMMA is a representative amorphous linear polymer. At room temperature, $|G^*| \approx 1$ GPa and tan δ ≈ 0.1 at 1 Hz [7.2.9, 7.2.10]. A plot of viscoelastic properties at constant frequency vs. temperature [7.2.11] is shown in Fig. 7.3. PMMA exhibits a large peak [7.3.1] (called an α peak) in the loss tangent at 135°C and a smaller one (a β peak) at 20°C. A plot of properties vs. frequency at constant temperature [7.2.10], shown in Fig. 7.4, shows the β peak in the frequency domain.

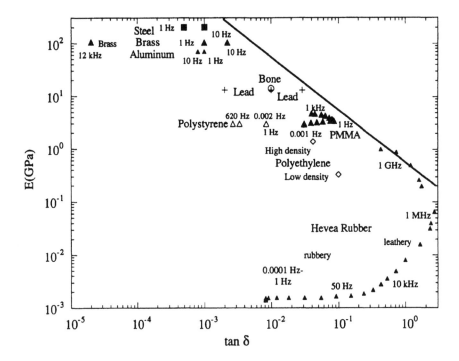

Figure 7.1 Stiffness–loss map for some materials. Temperature is near room temperature. Data points are adapted from the following sources: polymethyl methacrylate (PMMA), β peak, various frequencies, [7.2.9]; lead, 1 to 15 kHz [7.2.7]; single-crystal lead, 64 kHz, [7.4.7]; bone, 1 to 100 Hz [7.2.8]; steel, 1 Hz [1.6.22]; aluminum, 1 Hz [1.6.22]; Hevea rubber [1.6.18, p. 50]; and polystyrene, 0.001 Hz to 1 kHz [1.6.18, p. 468].

A master curve derived from relaxation tests at various temperatures (Fig. 7.5) [7.3.2] is shown in Fig. 7.6. At ambient temperatures and short times, PMMA is comparatively stiff; loss as well as dispersion is relatively small: it exhibits the *glassy* state of an amorphous polymer. At higher temperatures or longer times, the stiffness decreases rapidly with increasing temperature or time: the *transition* or *leathery* region. In the transition region the loss tangent can attain values exceeding one. At yet higher temperatures and longer times, the stiffness is low as are the dispersion and the loss: the *rubbery* region. Polymers for which the rubbery region occurs at ambient temperatures are called rubbers; such polymers exhibit glassy behavior at sufficiently low temperatures or high frequencies. The effect of any cross-links and of molecular weight [7.3.3] generally manifests itself at long times or low frequencies.

Although amorphous polymers are ordinarily viewed as thermorheologically simple, deviations from time–temperature superposition are known [7.3.1]. For example, in a study of properties of PMMA vs. frequency and temperature, the α peak hardly shifted as the frequency was changed from

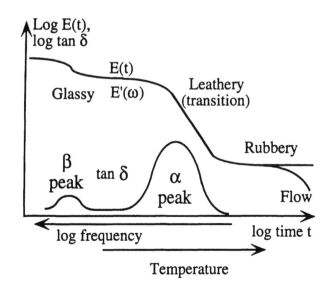

Figure 7.2   General form of viscoelastic behavior of a polymer.

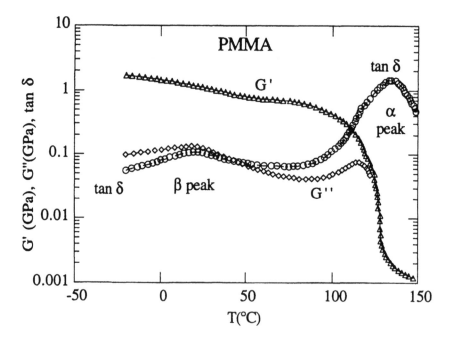

Figure 7.3   Dynamic properties of PMMA at 1 Hz vs. temperature. (Adapted from Iwayanagi and Hideshima [7.2.11].)

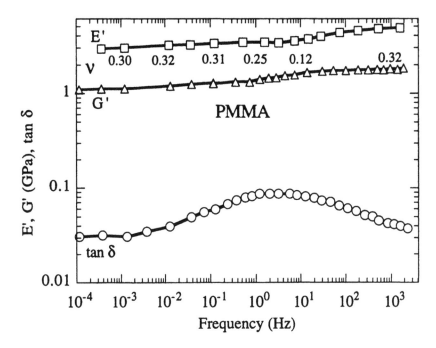

Figure 7.4 Dynamic properties (Young's modulus E', shear modulus G', and loss tangent tan δ for both torsion and tension) of PMMA at room temperature vs. frequency, in the glassy region. (Adapted from Koppelmann [7.2.9].) ν is the Poisson's ratio calculated from the axial and shear properties.

7 to 60 Hz, but the β peak shifted markedly to higher temperature, so much as to coalesce at 60 Hz with the α peak [7.3.4]. This behavior is indicative of different activation energies for the peaks: on the order of 80 to 200 kcal/mol for the α peak and 29 kcal/mol for the β peak [7.3.1]. In this vein, viscoelastic and dielectric relaxation phenomena are to be distinguished from other frequency-dependent phenomena, such as the infrared spectra of polymers. Infrared spectra result from resonances of chemical bonds, and the frequencies at which these resonances occur are quite insensitive to temperature changes [7.3.5].

Rubbery materials (elastomers) can differ significantly in their viscoelastic behavior. Properties of several types of rubber used for vibration damping and shock isolation have been presented [7.3.6]. Natural rubber has a low loss tangent of about 0.03 at room temperature, up to about 100 Hz increasing to 0.2 at 10 kHz; neoprene rubber has a loss tangent of 0.1 to 0.15 at room temperature, up to about 100 Hz. Shear moduli on the order of 1 MPa are representative in the plateau region.

High-loss rubbers (viscoelastic elastomers) can be prepared by incorporating plasticizing agents in the rubber or by a nonstoichiometric composition.

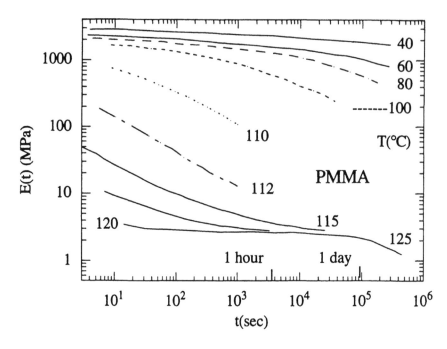

Figure 7.5   Relaxation properties of PMMA at various temperatures. (Adapted from McLoughlin and Tobolsky [7.3.2].)

It is possible to achieve a loss tangent exceeding 1.0 (Fig. 7.7) over one or more decades of frequency. Further results for rubbery materials given in Reference [7.3.8] disclose loss peaks for natural rubber at 10 GHz at room temperature; 10 kHz for polyisobutylene, and near zero to 10 kHz for plasticized polynorbornene. Since the loss peak of natural rubber is at such a high frequency (and the glass transition temperature is correspondingly low (–73°C)), the loss tangent of natural rubber is low at room temperature and at low frequency. In many elastomers, the peak in the loss vs. temperature is sharp. The loss peak of elastomers can be broadened by mixing incompatible components to generate localized fluctuations in concentration on a scale of 10 to 50 Å, and a range of relaxing environments. Some specific polymeric compositions for high damping have been described in the patent literature. An energy absorbing polyurethane [7.3.9, 7.3.10] can be made by reacting a mixture of linear and branched polyols, a polyisocyanate, and optionally a blowing agent (to make foam). The isocyanate index should be from 65 to 85. Such viscoelastic elastomers have been used in shoe insoles for the absorption of impacts [7.3.11]. Chlorosulfonated polyethylene [7.3.12], widely used commercially [7.3.13], allows the glass transition temperature to be varied from –35°C to near room temperature by control of the chlorine content. These polymers are often prepared with carbon black as a filler, which stiffens the material in the rubbery regime and reduces the peak damping. Further discussion of the control of damping in polymers is presented in §8.6.

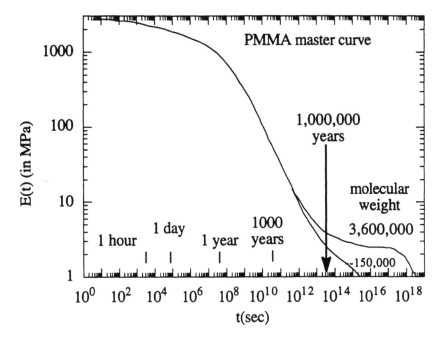

Figure 7.6 Master curve for PMMA reduced to 40°C. (Adapted from McLoughlin and Tobolsky [7.3.2].)

§7.3.2 Bulk relaxation in amorphous polymers

Polymers exhibit relaxation in their bulk properties as well as in their shear properties [7.3.14]. Experimental studies of shear and bulk response have been conducted using ultrasonic wave procedures. The total change in shear stiffness G' or extensional stiffness E' through the α transition between glassy and rubbery consistency can exceed a factor of 1000. By contrast, the corresponding change in the bulk stiffness B' is about a factor of two. Master curves for shear and bulk behavior of polyisobutylene disclose loss modulus peaks of similar magnitude in shear (G" ≈ 0.44 GPa) and volumetric deformation (B" ≈ 0.72 GPa) at the same reduced frequency (about 100 MHz at 25°C). The loss peak in shear is broader than the one for bulk deformation. This difference is attributed to an extra physical mechanism involving entanglement of molecules; in shear the molecules can exhibit large relative motions not possible in volumetric deformation. Similar studies on polystyrene and polymethyl methacrylate [7.3.15] disclose a difference in the temperature or reduced frequency associated with the peaks for G" and B".

The difference in the magnitude of relaxation in shear modulus and bulk modulus through the α transition gives rise to a corresponding change in Poisson's ratio which goes from about 1/3 to values approaching 1/2 as time increases or frequency decreases through the transition. By contrast in the β peak, observe that Poisson's ratio of PMMA, as inferred

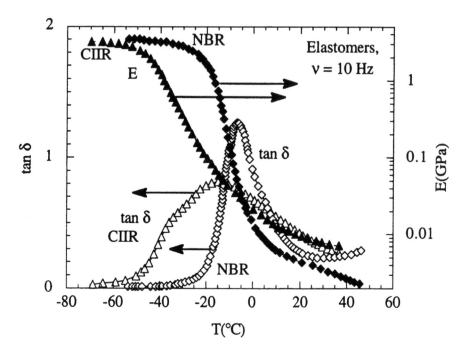

Figure 7.7 Behavior of viscoelastic elastomers used as damping layers. Viscoelastic elastomers, cured to a Shore A hardness of 55. Nitrile-epichlorohydrin rubber (ASTM designation NBR-ECO) and a chlorobutyl rubber (CIIR). (Adapted from Capps and Beumel [7.3.7].)

from the shear and uniaxial tensile properties [7.2.10], is not monotonic in frequency (Fig. 7.4). Therefore, Poisson's ratio is not monotonic in the time domain through the β transition. As for comparisons between tension and torsion in relaxation, PMMA relaxes more slowly in tension than in torsion [7.3.16]. In biaxial tests, torsional relaxation rates decrease with superposed tensile strain, but tensile relaxation is unaffected by a super-posed torsion strain.

Bulk properties also are important in determining the longitudinal wave speed and attenuation, since, as discussed in §5.7 and Example 5.9, the constrained modulus which governs longitudinal waves is $C_{1111} = B + \frac{4}{3}G$.

For example, at ultrasonic frequencies, PMMA exhibits a longitudinal wave speed of $c_L = 2.74$ km/sec and an attenuation of $\alpha = 0.018$ neper/mm at 1 MHz and at 21.1°C [7.3.17]. The stiffness, extracted via Eqs. 5.7.32 and 5.7.33 from the density $\rho = 1.184$ g/cm³ and the wave speed c_L, is $C_{1111} = c_L^2 \rho = 8.89$ GPa. From the wave attenuation one may infer tan $(\delta/2) = \alpha v / 2\pi v = 7.85 \times 10^{-3}$, so tan $\delta = 1.57 \times 10^{-2}$, which is considerably smaller than values observed for shear and uniaxial tension properties at acoustic frequencies and below. The constrained modulus C_{1111} can have a

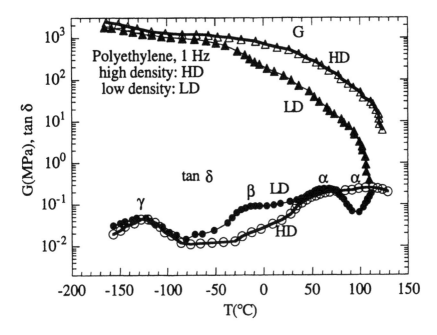

Figure 7.8 Polyethylene shear modulus and tan δ as a function of temperature. (Adapted from Flocke [7.3.19].)

very different loss tangent than Young's modulus or shear modulus because it depends on the bulk modulus as seen in Example 5.9.

Several epoxy resins exhibited attenuations about 2.6 times larger. Interpretation of C_{111} is discussed in Reference [7.3.18] and in §5.7. Ultrasonic data collected at a variety of temperatures [7.3.17] were used to generate master curves for frequencies as high as 10^{28} Hz. We remark that for a wave speed of 3 km/sec, a frequency of 10^{12} Hz corresponds (via $\lambda\nu = c$) to a wavelength of 3 nm, or just a few atoms. Frequencies of about 10^{13} Hz and higher therefore do not correspond to physically realizable sound waves in real materials, since one cannot have a wavelength smaller than one atom in a sound wave.

§7.3.3 Crystalline polymers

Crystalline polymers exhibit long-range molecular order as a result of regularity and symmetry in the molecular chains. Linear polyethylene (Fig. 7.8) [1.6.18, 1.6.21, 7.3.19], polytetrafluoroethylene, and isotactic polypropylene are crystalline. Many biological polymers such as collagen are also highly ordered. Crystalline polymers are generally not thermorheologically simple, so that a master curve cannot be obtained from tests at different temperatures. The α transition in crystalline polymers tends to be broader than in amorphous polymers. Some results are given in References [1.6.18, 7.3.1, 7.3.20].

§7.3.4 Aging: other relaxations

Polymers exhibit an aging behavior in their viscoelastic properties as a function of time following polymerization [7.3.21, 7.3.22]. The effect of aging time t_a is to slow the characteristic retardation times τ_c by a multiplicative factor:

$$\tau_c = \tau_0 t_a^\mu,\qquad (2.9.2)$$

in which μ is the Struik shift factor and τ_0 is a retardation time. In such cases, the effect of aging is represented by a shift of the creep curves along the log time axis. The rate of aging in glassy polymers depends in part on how far below the glass transition temperature T_g the polymer is used.

Similarities in the *shape* of the relaxation behavior of metals and polymers have been observed [7.3.23], though polymers ordinarily exhibit much more relaxation than metals.

The thermal expansion of polymers is time dependent, as anticipated in the discussion of constitutive equations in §2.10. Following a temperature change, the strain may continue slowly to change in the same direction or may slowly change in the opposite direction [7.3.24]. In rubbery materials the time-dependent expansion depends on the degree of prestretch [7.3.25].

§7.4 Metals

§7.4.1 Linear regime

At small stresses and at low temperatures far below the melting point most metals behave in a nearly elastic manner. Viscoelasticity manifests itself in the form of a small but nonzero value of the loss tangent as shown in Table 7.1. The corresponding creep or relaxation is small; considerable precision is needed to detect it. Viscoelastic effects in metals used for structural purposes are usually much smaller than those in polymers. Metals often exhibit a series of peaks in the loss tangent at various frequencies [1.6.22, 7.2.3]. A variety of causal mechanisms are responsible for the damping as discussed in Chapter 8. Results for many metals are compiled in Reference [7.2.3].

Single crystals of metals such as copper, zinc, and lead have been examined by a number of investigators [7.4.1–7.4.8]. Part of the rationale is to eliminate damping due to the boundaries between the crystalline grains found in bulk samples of metal, in an effort to better understand the causes of damping. In single-crystal lead [7.4.7], the loss tangent at 64 kHz depended on heat treatment and changed with time after mounting of the specimen; values from 0.002 to 0.01 were observed at room temperature. Damping increased with temperature.

Metals as they are used in most applications are polycrystalline. The boundaries between crystals or grains in metals give rise to additional damping as discussed in §8.7. Peaks of this type, known as grain boundary peaks

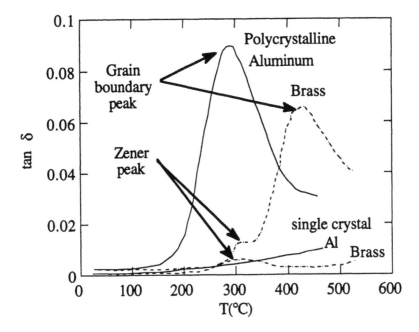

Figure 7.9 Tan δ as a function of temperature for aluminum [7.4.10] and brass [7.4.15], both polycrystalline and single-crystal, at low frequency, adapted from Kê.

(Fig. 7.9), have been observed in a variety of metals, including aluminum [7.2.3, 7.4.9, 7.4.10]. Grain boundary peaks occur at relatively high temperatures, a significant fraction (referred to absolute zero, degrees Kelvin) of the melting temperature. The ratio of sample absolute temperature to melting temperature is called the homologous temperature. The effect of the grain boundaries differs if the grain and specimen sizes are comparable (macrocrystalline materials); the grain boundaries are called "bamboo boundaries" in this case [7.4.11–7.4.13]. In macrocrystalline materials, the loss peak height is proportional to the number of grains; by contrast, if the grains are small, the loss peak height is independent of grain size.

Boron fibers, which are used in high-performance composites, exhibit a torsional loss tangent of about 0.001 at room temperature and a frequency of 5 Hz; the shear modulus is 100 to 160 GPa [7.4.14].

As for alloys, grain boundary effects for brass are as follows. In polycrystalline α-brass (70% copper, 30% zinc), there is a large peak [7.4.15] of about 0.066 in the loss tangent at 430°C, at 0.5 Hz; this peak is absent in single-crystal brass; under the same conditions, the loss is less than 0.0025. Polycrystalline brass at room temperature exhibits a small loss tangent of about 0.001. Alloys can also exhibit damping due to stress-induced ordering of the different atoms constituting the alloy. In α-brass, a peak in the loss tangent [7.4.15] due to this cause occurs at 300°C and at 0.5 Hz: tan δ is 0.013 in polycrystalline brass and 0.006 in single-crystal brass.

Some aluminum alloys exhibit very low damping. Aluminum alloy 2090 exhibits $Q \approx 3.1 \times 10^5$ (tan $\delta \approx 3.2 \times 10^{-6}$) in torsion at 1 kHz at room temperature (300 K) [7.4.16] and 5×10^6 (tan $\delta = 2 \times 10^{-7}$) at 0.5 K. Alloy 2090 is a strong aluminum lithium alloy which has been developed for aircraft applications. Aluminum alloy 6061, commonly used for pipelines and structural applications, exhibits $Q \approx 2.8 \times 10^5$ (tan $\delta \approx 3.6 \times 10^{-6}$) in torsion at room temperature. Aluminum alloy 5056, commonly used for rivets and window screen, exhibits $Q \approx 2.2 \times 10^5$ (tan $\delta \approx 4.5 \times 10^{-6}$) in torsion at room temperature. Young's moduli for these alloys are E = 78 GPa for 2090, 69 GPa for 6061, and 72 GPa for 5056.

Aluminides [7.4.17] consisting of 51 to 58 atomic % iron and the balance aluminum exhibit Young's moduli from 260 GPa at 300 K to 200 GPa at 988 K at kHz frequencies; loss tangents are from 3×10^{-4} to 10^{-2} at high temperature. For a uranium–0.75w%Ti alloy [7.4.18] at 40 to 80 kHz, and at 25°C, E = 193 GPa, G = 81 GPa, and ν = 0.17; moduli decrease with temperature and Poisson's ratio increases to about 0.5 at 650°C.

Resonance dispersion has been reported in polycrystalline metals such as lead and aluminum [7.4.19] in the range 100 Hz to 5 kHz. The sharp resonance-type peaks observed in the loss were considered to be due to microinertial effects in the metal itself rather than due to structural resonance. Crystalline polymers have also been reported to exhibit such microresonance effects [1.6.18].

§7.4.2 Nonlinear regime

§7.4.2.1 Dynamic

The dynamic mechanical loss in metals often depends on strain amplitude. In the presence of nonlinearity the response to a sinusoidal input is no longer sinusoidal. Nevertheless, since the losses are usually small in metals, the usual terminology associated with linear materials is still used by most writers. One source of nonlinearity is that loss due to dislocation motion increases with strain amplitude, even at small strains. In single-crystal zinc [7.4.8], the damping increased with strain along one path and then decreased with strain along a different path. Moreover, the damping increased following application of a static stress. Loss peaks due to grain boundary motion also are strain dependent [7.4.20]. Magnetoelastic coupling also gives rise to nonlinear behavior. In 0.42% carbon steel [7.4.21], tan δ was less than 10^{-3} for a peak strain less than 10^{-4}, rising to 8.4×10^{-3} for a peak strain of 10^{-3}.

Hysteresis in general refers to a lag between cause and effect; however, in the context of metals, some authors such as Smithells [7.2.3] in the metals field use the following specific definition. Metals may exhibit *hysteresis damping* in which the stress–strain curve for oscillatory loading has cusps rather than being elliptical as in linear behavior. Values for hysteresis damping are normally quoted for a specific, high, stress amplitude level, e.g., 34.5 MPa shear stress [7.2.3]. Hysteresis damping is usually independent of frequency, but is strain dependent and achieves a relative maximum for a particular strain amplitude. Metals exhibiting considerable hysteresis damping include cast irons, nickel titanium alloys, ferromagnetic alloys, and manganese copper alloys. Smithells also defines *anelastic damping* as damping which is independent of strain amplitude and which exhibits a peak with respect to frequency. Recall that anelastic materials are viscoelastic materials which exhibit full recovery following creep, a definition rather more general than that of Smithells.

Damping in metals depends on whether they have been subjected to plastic deformation prior to the damping test. Pure iron [7.4.22] exhibits an effective torsional tan δ below about 0.003 for peak strains below 10^{-4}. A prior plastic deformation strain of 10^{-3} gives rise to increased amplitude-dependent damping, up to tan δ = 0.03 for a peak strain of 6×10^{-5}. Plastic deformation at higher temperatures gives rise to increased damping after that deformation. The effect is attributed to dislocations.

§7.4.2.2 Creep

Metals exhibit significant creep at high temperatures and high stresses. The behavior is nonlinear at the stress levels associated with structural members. In an effort to simplify matters so that realistic problems become tractable, many workers have adopted an equation of state approach [1.6.16] in which the creep strain depends only on the conditions of temperature T and stress σ at the present time t.

$$\varepsilon(t) = \sum_{i=1}^{n} f_i(T) g_i(\sigma) h_i(t). \tag{7.4.1}$$

In this approach, memory of past events is not incorporated, in contrast to the Boltzmann superposition integral of linear viscoelasticity which we have thus far emphasized. A specific equation of state commonly used [1.6.16] is the Bailey–Norton law intended to model primary and secondary creep.

$$\varepsilon(t) = A\sigma^m t^n, \tag{7.4.2}$$

with A, m, and n dependent on temperature and m > 1; n < 1. Other functions [7.4.23] of stress include $g_i(\sigma) = A \sinh (\sigma/\sigma_0)$ or $g_i(\sigma) = A \exp (\sigma/\sigma_0)$. Polynomial functions of time are also used: $h_i(t) = t^n + Ct + Dt^p$ with n as a fraction, often about 1/4 and p an integer often considered to be 3. Creep in metals becomes substantial at temperatures above about 0.3 T_m in which T_m is the absolute melting temperature (degrees Kelvin); above 0.5 T_m, the material is at a high homologous temperature, and creep can be rapid [7.4.23]. The specific dependence on temperature is usually assumed to be

$$f_i(T) = \exp{-\frac{U}{RT}}, \tag{7.4.3}$$

in which U is the activation energy, R is the gas constant, and T is the absolute temperature. The state variable approach is only an approximation to the true behavior, but it is sufficient for many purposes.

As for specific metals, most structural metals such as steel and aluminum do not exhibit much creep at room temperature and at low stress. Metals with a low melting point, such as solders, creep rapidly at ambient temperature [7.4.24]. Common lead–tin solders have eutectic compositions or 60%Sn–40%Pb. These solders exhibit primary, secondary, and tertiary creep at room temperature. The secondary creep is thermally activated according to Eq. 7.4.3.

§7.4.3   *High damping metals and alloys*

We consider here some special materials which exhibit relatively high effective loss tangents combined with high stiffness. Such materials are used to absorb vibration. In many of these metals, damping is inherently nonlinear even for strain below 10^{-4}. Most writers do not attempt a full nonlinear analysis of these materials. In view of the nonlinearity, some authors refer to the specific damping capacity $\Psi = \Delta W/W$, the energy dissipated per cycle divided by the maximum energy stored per cycle as a measure of damping. Recall (§3.4) that for linear materials, $\tan \delta = \Psi/2\pi$.

A variety of high damping metals are available [7.4.25–7.4.28]. Materials in the nonlinear regime have been compared by tabulating data obtained at an oscillatory stress amplitude one tenth of the tensile yield stress. Effective loss tangents under these conditions ranged from 0.5×10^{-3} for 1100-F aluminum alloys, to about 0.1 for magnesium alloys [7.4.25]. Early work on high damping alloys dealt with ferromagnetic materials [7.4.29] in which the damping occurs via coupling between stress and motion of magnetic domain walls.

A class of manganese copper alloys is of interest in applications which require high damping. These alloys are given the trade name "Sonoston" and are used for ship propellers used in naval applications. Loss tangents as large as 0.014 occur at a stress amplitude of 4000 psi (28 MPa) [7.4.25]. Mn–Cu alloys are nonlinear, and the loss is smaller at small stress amplitude. Hysteresis damping [7.2.3] in the nonlinear domain in these alloys can correspond to an equivalent loss tangent ($\Psi/2\pi$ in terms of the specific damping capacity Ψ) as large as 0.067 for alloys of 60 to 70% Mn at a surface shear stress of 34.5 MPa. Moreover, there is an aging effect in which loss decreases over a period of weeks [7.4.30]. Nevertheless, the damping of Mn–Cu is high for a structural metal (Fig. 7.10).

Damping in all metals tends to increase with stress, and large values of damping occur as the stress approaches the fatigue limit, as noted by Lazan [7.4.27]. It would be risky to exploit this phenomenon to achieve high damping since at such stress levels, damage occurs in the metal.

Pure magnesium exhibits high damping, but it is comparatively weak. Cast commercially available magnesium [7.4.25] exhibited a loss tangent

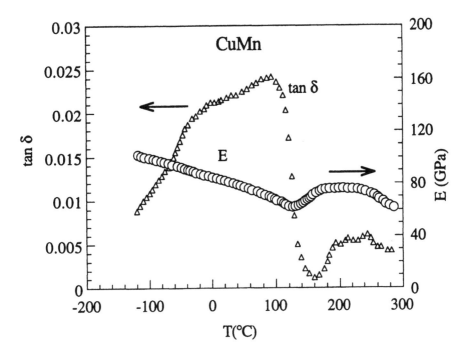

Figure 7.10   Tan δ for a copper–manganese alloy at 1 Hz and a peak strain of 4.3×10^{-4}, aged for 16 h [7.4.30].

(extracted from log decrement measurements in flexural vibration) of 0.15 at a maximum strain of 6×10^{-4}, to 0.03 at a strain of 5×10^{-5}. Alloying elements can be added to improve the strength, but they usually reduce the damping. A cast AZ 81 magnesium alloy [7.4.26] exhibited a loss tangent of 0.016 at a strain of 0.001, down to about 0.0015 at a strain of 1.3×10^{-4}. Special alloys of magnesium have been prepared to achieve high damping. Magnesium–zirconium alloys with about 0.5% zirconium exhibit damping comparable to that of pure magnesium, but with higher strength [7.4.31]. Forged Mg 4.5% Ce exhibits an effective loss tangent (extracted approximately from reported specific damping capacity) of 0.07 in torsion near 10 Hz at a stress of 14 MPa. The effect is nonlinear so the effective loss tangent is only 0.011 at a stress of 0.86 MPa [7.4.32]. The crystals were favorably aligned for high damping in shear in the forged samples. Alloys of magnesium with nickel, copper, aluminum, and tin have been studied in flexural vibration [7.4.33]. The highest loss tangent observed at small strain in these alloys was less than 0.002, but damping increased with amplitude.

Nickel–titanium alloys exhibit a large hysteresis damping at high stresses. Nitinol, which consists of 55% Ni and 45% Ti, has an equivalent loss tangent of 0.04 at a surface shear stress of 69 MPa [7.2.3]. This alloy is also called "memory metal" in view of its ability to return to a prior configuration after it is heated or cooled.

A variety of magnetic alloys exhibit high damping, and some of these have been made commercially. Pure polycrystalline nickel exhibits a Young's modulus of about 200 GPa and a peak loss tangent of 0.046 at 100 kHz [7.4.34]. If a magnetic field is applied, the loss tangent decreases. A Co–35%Ni alloy at 1 Hz in torsion exhibited a loss of $\delta = 0.075$ at 10^{-4} shear strain γ, rising to a peak of 0.18 at $\gamma = 2.3 \times 10^{-4}$. An alloy of Co–20%Fe exhibited $\delta = 0.048$ at $\gamma = 10^{-4}$ and a peak $\delta = 0.11$ at $\gamma = 3.5 \times 10^{-4}$. More recently, an alloy commercially known as Vacrosil, containing 12%Cr, 3%Al, and 85%Fe, exhibited a loss tangent [7.4.35] (defined as Q^{-1}) of 0.006 at 2 kHz and $\varepsilon = 10^{-6}$ over a wide range of temperatures from –200°C to more than 300°C. A strain-dependent peak in the loss, up to 0.1, was observed in this material at 20 Hz at $\varepsilon = 7 \times 10^{-5}$. Other alloys, Fe10Cr and Fe12Cr, exhibited similar peaks of lower amplitude; magnetoelastic alloys based on FeCo exhibited qualitatively similar behavior, tan δ to 0.075 at $\varepsilon = 1.3 \times 10^{-4}$ and about 1 Hz for Fe–25%Co [7.4.36]. FeMo [7.4.37] and FeMoCr [7.4.38] alloys are in the same category: tan δ up to 0.059 in Fe–6%Mo. A superposed static tensile stress reduces damping in these alloys [7.4.39]. In these alloys the loss depends on the ambient magnetic field strength, heat treatment, and strain amplitude of vibration.

In certain compositions of CdMg (33 atomic % Mg), high damping has been observed at room temperature [7.4.40, 7.4.41]. The peak loss is tan $\delta = 0.14$ at 1 Hz. The peak in tan δ was of magnitude 0.14 at 20°C and 0.75 Hz; it was verified to be a Debye peak in the frequency domain [7.4.42]. The peak for CdMg was about 20°C wide at constant frequency and the activation energy U was inferred to be 19,000 cal/mol [7.4.41]. The loss appears to be due to a stress induced ordering of the different atoms. However, this has not to the writer's knowledge been exploited practically. In 81%Cu–19%Pd alloy, tan $\delta \approx 0.1$ to 0.24 at 2 Hz was claimed near and below room temperature [7.4.43]. Following annealing, the loss was reduced to about 0.003 at room temperature and even less at lower temperature. The loss was attributed to a Hasiguti effect in which interaction between dislocations and point defects gives rise to loss in cold worked face centered cubic (fcc) metals. In alloys of titanium and vanadium (Ti–20%V) a peak in tan δ of 0.022 at 10 kHz and at 81%Cu–19%Pd 130K is observed [7.4.44]. At room temperature, tan $\delta \approx 0.001$ and $E \approx 97$ GPa. Alloys with more vanadium exhibit less loss.

Metals of low melting point are at a high homologous temperature $T_H = T/T_m$ at ambient room temperature $T = T_r$; consequently, they are expected to exhibit substantial viscoelastic response. For elements, such as Cd, In, Pb, and Sn, and alloys of low melting point, a high homologous temperature occurs at room temperature. Lead is popularly viewed as a high-loss metal, but it exhibits a relatively small peak tan $\delta = 0.015$ in bending [7.4.45] and tan $\delta = 0.005$ to 0.016 in the audio range in torsion [7.4.46]. Cadmium exhibits a substantial loss tangent of 0.03 to 0.04 over much of the audio range of frequencies. Indium is the softest metal which is stable in air [7.4.47]. The primary reason for its softness at room temperature is its low melting point $T_m = 429$ K, so the homologous temperature is high: $T_H = T_r/T_m = 0.69$.

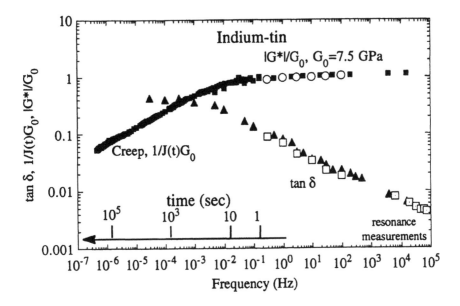

Figure 7.11   Viscoelastic behavior of an alloy of low melting point, a eutectic indium–tin alloy at room temperature. (Adapted from Lakes and Quackenbush [7.14.2].) Results are for a peak strain less than 10^{-5}, measured directly with no appeal to time–temperature superposition.

Young's modulus is $E = 12.6$ GPa at $T = 300$ K and 18.4 GPa at 80 K. Indium has a yield point of only about 3 MPa, so that the strain at yield is 2.4×10^{-4}. Indium exhibits substantial creep, even at low temperature [7.4.48]. Lead is commonly thought to be a high damping metal; however, compared with other metals of low melting point such as indium, tin, and cadmium, it is not the highest in damping. Indium also exhibits high loss, particularly at low frequency [7.4.46, 7.4.49]. It was studied in connection with attempts to make composites with high loss [7.4.49]. In these composites, indium was cast into the interstices of copper foams of conventional and negative Poisson's ratio behavior. Indium–aluminum alloys have been made, with the aim of achieving high loss [7.4.50]. The Al–In alloys were found to contain embedded particles of indium. Tan δ was on the order of 0.1 at 0.5 Hz and above 140°C, and there was a loss peak observed at the melting temperature ($\approx$160°C) of the indium inclusions. Tan δ was not as large at room temperature. Low melting point alloys, including PbSn and InSn, are used in soldering. Such alloys can exhibit substantial viscoelasticity (Fig. 7.11). The effective performance of solders in electronic devices is related to their viscoelastic behavior (creep); flow or cracking of a solder joint can lead to failure of the device containing that joint.

     Aluminum–zinc alloys have been studied for high damping applications [7.4.51, 7.4.52]. These alloys have from 3.5 to 28% aluminum and small amounts of copper and magnesium. At 20°C, damping is independent of

Table 7.2   Composition by Weight of a Typical Creep-Resistant Alloy.

Element	Content (%)	Element	Content (%)
Ni	59	Mo	0.25
Co	10	C	0.15
W	10	Si	0.1
Cr	9	Mn	0.1
Al	5.5	Cu	0.05
Ta	2.5	Zr	0.05
Ti	1.5	B	0.015
Hf	1.5	S	<0.008
Fe	0.25	Pb	<0.0005

Adapted from Ashby and Jones [7.4.56].

strain from 10^{-6} to 10^{-4}, but increases with temperature up to about 270°C [7.4.51]. The highest damping at 20°C and at 10 Hz is 0.009 for SPZ alloy (78% Zn, 22% Al) and 0.005 for ZA27 (25–28% Al) alloy.

Gray cast iron, as described by Millett et al. [7.4.53] exhibits a loss tangent of about 0.011 over a range of frequencies, and it contains platelet-shaped inclusions of graphite with a loss tangent of 0.015. The graphite gives rise to most of the loss; the underlying mechanism is stated to be dislocation loop motion (see Chapter 8).

§7.4.4   Creep-resistant alloys

Metals used at high temperature, specifically at an absolute temperature which is greater than about 0.6 of the melting point, tend to creep noticeably. Aluminum alloys [7.4.54] are available for use up to about 200°C; titanium alloys, to 600°C; and stainless steels, to 850°C. In many applications, creep is to be minimized. To that end, a variety of creep-resistant alloys has been developed [1.6.27]. The creep resistance of aluminum [7.4.55] is improved by incorporating particles of alumina (Al_2O_3) or of aluminum carbide. Although this alloy creeps less at high stress than pure aluminum does, it has a higher tan δ at subaudio frequencies. Nickel-based superalloys are used at higher temperatures than are aluminum alloys. A representative superalloy has the following composition, by weight, as shown in Table 7.2. Such an alloy [7.4.56] as used in turbine blades is expected to withstand a stress of about 250 MPa, and to creep no more than 0.1% over 30 h at 850°C. Since these materials melt at $T_m = 1280$ °C, the homologous temperature is high: $T/T_m = 0.72$. By contrast, most ordinary materials creep a great deal at such temperatures. Materials that may be used at temperatures up to 1000°C include nickel-based superalloys; refractory metals such as W–Re alloys, which have the disadvantage of being dense; ceramics based on TiC, ZrC, and Al_2O_3, which are brittle; and ceramic composites, which are also brittle.

§7.5 Rocks and ceramics

Rocks ordinarily exhibit relatively small loss tangents. For basalt, tan δ = 0.0017 and for granite tan δ = 0.0031 for small strain (10^{-6}) and at frequencies between 0.001 and 1 Hz [7.2.6]. For comparison, frequencies of seismological interest are from about 3×10^{-4} to 1 Hz corresponding to periods from about 1 h to 1 sec. Other studies of rock are reviewed in Reference [7.2.7]: granite exhibited tan δ = 0.005 from 140 Hz to 1.6 kHz, and limestone exhibited tan δ ≈ 0.02 from 50 to 120 Hz. A small amount of water in porous rock can substantially increase the loss; for dry sandstone, tan δ = 0.05 independent of frequency in the audio range, but with 0.4% water, tan δ increases with frequency up to 0.35 at 1.7 kHz.

Rocks at the elevated temperatures prevailing in the interior of the earth can exhibit relatively large loss tangents in the absence of hydration. Naturally deformed peridotite has a loss tangent exceeding 0.05 at 5 Hz and small strain above 900°C [7.5.1]. Loss tangents for *in situ* rock have been inferred from seismic data [7.5.2], tan δ = 0.0023 for waves of period 3000 sec, to tan δ = 0.0085 for waves of period 100 sec. Rock has also been studied at ultrasonic frequencies [7.5.3]. Westerly granite exhibits tan δ rising from 0.01 near 30 kHz to 0.06 at 2.5 MHz. Other rocks such as slate and limestone exhibit similar broad loss peaks in the ultrasonic domain, and the loss is attributed to a Granato–Lücke dislocation process as discussed in §8.8.

The loss tangent of Moon rocks differs from that of Earth rocks; unusually long reverberations were recorded by a seismic station on the Moon from lunar impacts of spacecraft [7.5.4]. Tan δ = 3×10^{-4} was inferred for v ≈ 1 Hz from these observations, compared with values from 3×10^{-3} to 0.1 for most materials of Earth's continental crust. Lunar rocks studied on Earth have tan δ = 0.01 to 0.1 [7.5.5]. The difference is attributed to the outgassing of trace volatiles such as water in the dry lunar environment [7.5.6–7.5.8].

Creep of rock has been studied by mining and civil engineers as well as by geologists [7.5.9]. The concern of the civil engineers is the stability of structures over years to decades, while geologists are concerned with flow properties of rock over millions of years. Rock exhibits primary, secondary, and tertiary creep. The creep is attributed to the initiation and development of microcracks following chemical reactions at the crack tips [7.5.10]. The primary or transient component is often described by a power law or by more involved empirical equations. Creep can be substantial: delayed strain can be three or four times the initial strain over periods of a few hours. Creep in rock is nonlinearly dependent on stress level and on temperature. As for steady-state creep, strain rate also depends on stress level. Empirical relationships derived from metallurgy have been used to model the creep. For example, the value of m in Eq. 7.4.2 for rock is in the range 1 to 5.

Tectonic movement of the earth's plates is associated with steady-state creep at strain rates of 10^{-20} to 10^{-29} per second [7.5.10]. Creep in hot rock can occur by the collective motion of atoms by dislocation glide or climb, as discussed in §8.8; or by the diffusive transport of individual atoms. The constitutive relations associated with these processes differ; the creep rate due to dislocation motion is insensitive to grain size and increases nonlinearly with stress but the creep rate due to diffusion increases dramatically with decreasing grain size and linearly with stress. Creep rate also depends on crystal structure.

In the earth's mantle, creep in hotter regions of the upper mantle is dominated by dislocation movement; creep in the cooler and deeper regions in the mantle is governed by diffusion [7.5.11]. In general, creep by diffusion in polycrystalline materials is associated with low stress levels and small grain size; creep by dislocation movement is associated with larger grain size and higher stress. The mantle is considered to be made of polycrystalline olivine. If the mantle is viewed, as is commonly done, as a stratified linear viscoelastic material, an effective viscosity of 10^{21} Pa · s can be inferred. Moreover, a constitutive equation for steady flow is given as follows:

$$\frac{d\varepsilon}{dt} = A\left(\frac{\sigma}{G}\right)^{n}\left(\frac{b}{d}\right)^{m} \exp\left\{-\frac{U+PV}{RT}\right\}, \qquad (7.5.1)$$

with $G \approx 80$ GPa as the shear modulus; A, n, and m as a constant; $b \approx 0.5$ nm as the length of the Burgers vector; d as the grain size; U as the activation energy; V as the activation volume; and R as the universal gas constant.

Synthetic ceramics, due to their high melting point, tend to exhibit relatively little damping and creep at room temperature; however, they can exhibit significant viscoelastic effects at elevated temperature [7.5.12, 7.5.13]. Alumina [7.5.13] at 200 kHz exhibited $E = 436$ GPa and $\tan \delta = 9 \times 10^{-4}$ at room temperature (298 K), but $E = 246$ GPa and $\tan \delta = 1.05 \times 10^{-2}$ at 1473 K. Silicon carbide at 120 kHz exhibited $E = 450$ GPa and $\tan \delta = 4 \times 10^{-4}$ at room temperature (293 K), but $E = 414$ GPa and $\tan \delta = 1.25 \times 10^{-2}$ at 1322 K.

Ionic crystals exhibit viscoelastic loss which is attributed to dislocation motion. In potassium chloride, internal friction ($\tan \delta$) at 40 kHz increased from 10^{-6} at about 430°C to 3×10^{-4} at 700°C [7.5.14]. Polycrystalline LiF–22%CaF$_2$ eutectic salts (which are used for thermal energy storage) have been examined, again by the piezoelectric oscillator method [7.5.15] but at 80 to 150 kHz. At near room temperature, $E = 116$ GPa, $G = 42.6$ GPa and $\tan \delta = 3.9 \times 10^{-3}$; and $\tan \delta$ increased with temperature. For KCl near room temperature, a broad peak [7.5.16] in $\tan \delta$ was observed, 3×10^{-3} at about 40 MHz; for LiF, there was a similar broad peak, $\tan \delta \approx 10^{-3}$ at about 10 MHz.

§7.6   Concrete

Concrete is an aging material; its properties depend significantly on the time following its preparation [7.6.1]. Concrete is formed via a chemical reaction of water with cement paste. When first prepared, concrete is of a

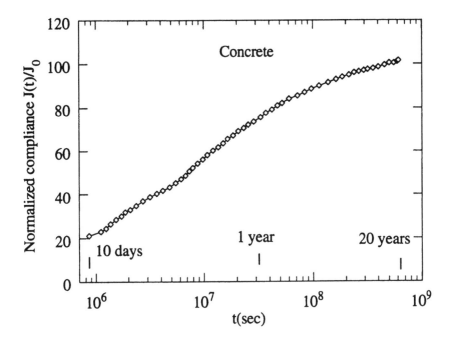

Figure 7.12 Creep of concrete for 20 years. (Adapted from Troxell [7.6.3].)

soft pastelike consistency and it experiences permanent deformation under minimal force. The reaction is about 50% complete after seven days following mixing and continues for more than a year [7.6.2]. The "aggregate" component, sand and gravel, is considered in most models to behave elastically. Concrete progressively shrinks after it is formed, even if no load is applied. The creep in concrete is attributed to stress-induced moisture transfer and to recrystallization in the cement or slip between layers. Microcracking is considered to cause creep at stresses above half of the short-term strength.

Concrete is used in structures which are expected to last a long time; however, the time temperature superposition concept is inapplicable. Creep tests of very long duration are therefore conducted, up to 20 years [7.6.3], as shown in Fig. 7.12.

A variety of empirical relations [7.6.2] have been used to model the creep strain as it depends on time under load $t - \tau$, and on age of loading τ. The effect of age is that creep is slower for older concrete and that effect is modeled by a factor $R = 1/\tau^p$, with p as an empirical material parameter, which would imply (incorrectly) that old concrete does not creep at all. Fits to experimental data give $p \approx 0.1$, but the effect of loading age on creep does not change greatly after about 28 days [7.6.1]. The logarithmic model, shown incorporating the effect of age, fits the early creep reasonably well but overestimates later creep.

$$\varepsilon_{creep} = \frac{1}{\tau^p}\left[c + d\log(1+t-\tau)\right]. \qquad (7.6.1)$$

The power law model is a familiar one and it is successful in modeling early creep, but overestimates later creep even more than the logarithmic model.

$$\varepsilon_{creep} = \frac{1}{\tau^p}(t-\tau)^m. \qquad (7.6.2)$$

A creep function used in design [7.6.1] is as follows, with c and d as empirical constants:

$$\frac{\varepsilon_{creep}(t)}{\varepsilon(0)} = \frac{t^c}{d+t^c}\frac{\varepsilon_{creep}(\infty)}{\varepsilon(0)}. \qquad (7.6.3)$$

Fits to experimental data give c = 0.4 to 0.8, and d = 6 to 30.

An empirical kernel for the constitutive equation for creep aging (Fig. 7.13) of concrete follows [7.6.4, 7.6.5]:

$$J(t,\tau) = \frac{1}{E(\tau)} + e_c^{\infty}\left[1 - (\tau/t)^k e^{-K(t-\tau)}\right], \qquad (7.6.4)$$

with k and K as empirical constants, and E(τ) as the elastic modulus at age τ.

Concrete under large stress can exhibit tertiary creep culminating in failure [7.6.6]. During tertiary creep, concrete accumulates damage, one consequence of which is a progressive increase in Poisson's ratio [7.6.7].

§7.7 Asphalt

Asphalt is usually produced via vacuum distillation of petroleum crude oils [7.7.1]. The resulting residue is used as a binder or cement for mineral aggregates to form mixtures used in paving. Asphalt behaves as a viscoelastic liquid with a shear modulus of about G_0 = 1 GPa in the glassy regime of high frequency and low temperature, as shown in Fig. 7.14. In the terminal regime (10^{-6} to 10^{-7} Hz at 15°C) the loss angle tends to π/2. The following phenomenological model has been proposed [7.7.2] for the normalized complex shear modulus:

$$\frac{G^*}{G_0} = \left\{1 + (i\omega\tau)^{-h} + \xi(i\omega\tau)^{-k}\right\}^{-1}, \qquad (7.7.1)$$

with h = 0.6, k = 0.25, and ξ = 2. The temperature dependence of the mean relaxation times follows an Arrhenius law (from –10 to 30°C) with activation

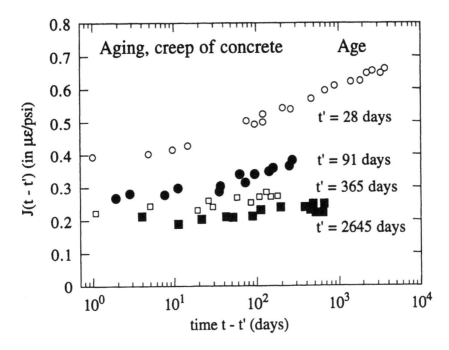

Figure 7.13   Creep as a function of age for concrete. (Adapted from Bazant [7.6.4].)

energies ranging from 125 to 258 kJ/mol. For temperatures above about 45°C, deviations from time–temperature superposition occur and they are more noticeable in the phase angle than in the modulus. Deviations from time–temperature superposition also occur if crystalline fractions are present. Such behavior parallels that seen in polymers.

§7.8   Ice

The behavior of ice is complex [7.8.1]. Stress–strain curves and other properties depend markedly on temperature and on load rate. At ambient conditions the temperature can approach the melting point, where properties depend particularly strongly on temperature.

For ice under steady-state creep conditions [7.8.2], the strain rate is given by

$$\frac{d\varepsilon}{dt} = A\sigma_d^n, \tag{7.8.1}$$

with $n \approx 3$ and $A = 2.3 \times 10^{-25}$ if σ is in units of N/m^2 and $d\varepsilon/dt$ is in sec^{-1}; this is considered valid for stresses greater than 0.1 MPa based on lab studies and 0.01 MPa based on studies of glaciers. Rheology of ice is therefore nonlinear in this stress range. The practical significance of ice rheology is in

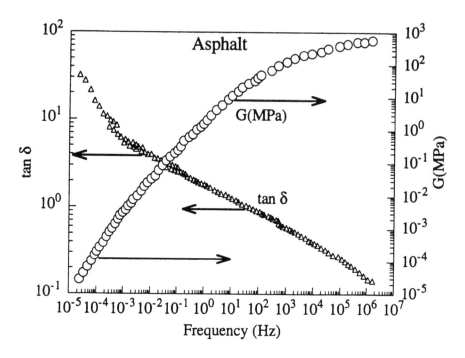

Figure 7.14 Viscoelastic properties of asphalt [7.7.1]. Master curve.

connection with glaciers and with structures built where there is moving ice. If temperature is included in the description [7.8.1],

$$\frac{d\varepsilon}{dt} = A\sigma_d^n \exp-\frac{U}{RT}, \tag{7.8.2}$$

in which U is the activation energy, T is the absolute temperature (in K), and R is the Boltzmann constant.

   As for dynamic studies, the damping for single-crystal ice at the melting point 0°C and at small strain is tan δ ≈ 0.13 at low frequency [7.8.3]. Damping in ice decreases with reductions in temperature [7.8.3, 7.8.4] and increases with strain amplitude above 10^{-4}. Dislocation motion was considered as a cause for the damping. Polycrystalline pure ice at 5 to 8 Hz and 0°C exhibits tan δ ≈ 0.17 and G = 3.3 GPa [7.8.5]; tan δ for natural ice and compressed snow is about three times lower. At –40°C, tan δ for pure ice is 8×10^{-3}, a considerably lower loss than near the melting point. Damping in ice at –1°C has been attributed to water molecule reorientation, peaking at 10 kHz, and dislocation damping peaking at about 10 MHz [7.8.6]. Damping was inferred from ultrasonic measurements of attenuation.

   Saline ice at low frequency exhibits large damping. Specifically, based on elliptical hysteresis loops, tan δ = 0.61 at –30°C [7.8.7] in ice of 0.1‰ salinity.

§7.9   Piezoelectric materials

Piezoelectric materials are those which produce an electric polarization when stressed mechanically, as discussed in §2.10. They are always anisotropic. This coupling between mechanical and electrical degrees of freedom can be responsible for viscoelastic behavior if there is electrical conductivity, as presented in §8.10.

Quartz is a crystalline piezoelectric material used in transducers, in electronic filters, and in frequency standards for radio equipment and timepieces. Its loss tangent is extremely small, 10^{-7} at 1 MHz increasing to 10^{-6} at 15 MHz [7.2.1]. Polyvinylidene fluoride (PVF_2) is a piezoelectric polymer. It exhibits a broad peak in tan δ of 0.09 at about 5 MHz [7.9.1].

Selenium in its crystalline form is a piezoelectric semiconductor [7.9.2] which exhibits dielectric relaxation of about 35% over two decades from 1 to 100 MHz. The shear modulus c_{44} is 18.3 GPa at ultrasonic frequency.

One can obtain amplification rather than attenuation of ultrasonic waves in piezoelectric semiconductors such as cadmium sulfide [7.9.3, 7.9.4]. This corresponds to tan δ < 0, which represents a gain tangent rather than a loss tangent. The energy required for amplification is derived from an externally applied direct current (DC) electric field. The amplification can be as much as 4% per wavelength of the ultrasonic wave; attenuations of comparable magnitude can be achieved depending on the electric field applied; for interpretation in terms of tan δ see §3.8. Consequently, the positivity of the loss tangent as ordinarily expected, depends on the characteristics of the material. It is not logically necessary, and is not always true. So it is not appropriate as a mathematical postulate. The law of conservation of energy is not violated, since in these cases for which tan δ < 0, an external source of energy such as electrical energy is provided. One may speak more generally of "excitable media" as physical, chemical, or biological systems in which energy dissipation is compensated by an energy supply [7.9.5, 7.9.6]. Examples include nerve tissue, heart muscle, the retina of the eye, and the Belousov–Zhabotinsky chemical reaction.

Piezoelectric ceramics exhibit strong piezoelectric effects; they are used in transducers and spark generators for stoves. Shear moduli from 21 to 45 GPa and loss tangents from 0.003 to 0.02 for lead titanate zirconates to 0.09 for lead metaniobate are representative [7.9.7].

§7.10   Biological materials

§7.10.1 Hard tissue: bone

Bone is a complex composite material consisting of mineral (principally hydroxyapatite, $Ca_{10}(PO_4)_6(OH)_2$), protein (principally collagen), water, and polysaccharides. Viscoelastic behavior occurs over a wide range of frequency [7.10.1, 7.10.2] including ultrasonic frequencies as shown. Observe in Fig. 7.15 that the loss tangent of compact bone attains a broad *minimum* over the frequency range associated with most bodily activities.

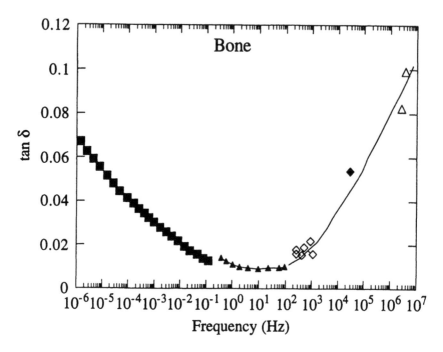

Figure 7.15   Tan δ for human compact bone, adapted from data of Lakes et al.
[7.10.1], for human bone at 37°C (calculated from relaxation, solid squares, ■; directly
measured, ▲); as compiled in Lakes [7.10.2] of Thompson at acoustic frequencies
(diamonds, ◇); of human bone via a piezoelectric ultrasonic oscillator (diamonds, ◆);
and of Adler and Cook at ultrasonic frequency for canine bone at room temperature
(open triangles, △).

Although compact bone is a complex composite material, secondary
creep is a simpler aspect of its viscoelastic behavior, and it appears to be
thermally activated [7.10.3]. Both creep and fatigue in bone give rise to
microscopic damage [7.10.4]. Creep under sufficiently large load, giving rise
to an initial strain in the range 0.003 to 0.007, terminates in fracture [7.10.5].
Bone such as deer antler, which has less mineral than leg bone, endures
higher strains without breaking.

Secondary creep in trabecular bone [7.10.6] follows a power law in stress:
$d\varepsilon/dt = 2.21 \times 10^{33} (\sigma/E_0)^{17.65}$. Creep rupture occurs at a time given by $t = 9.66 \times 10^{-33} (\sigma/E_0)^{-16.18}$. The strength of spongy bone can be substantially
reduced if a stress approximately half the ultimate strength is applied for a
few hours. Such strength reductions might play a role in the etiology of
progressive, age-related collapse of vertebrae.

§7.10.2 Soft tissue: human and animal

Soft tissues all exhibit viscoelastic behavior of various degree [7.10.7]. Muscle
exhibits considerable creep (Fig. 7.16). Moreover, soft tissue can be deformed

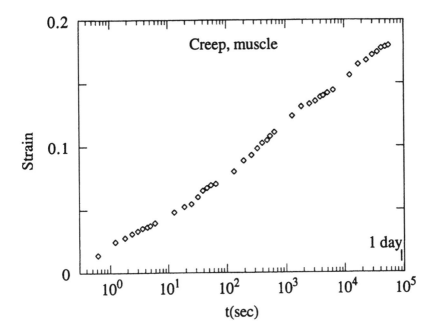

Figure 7.16   Creep for muscle. (Adapted from data of Fung [7.10.7].)

to relatively large strains. During such deformation, soft tissue exhibits non-linear behavior in which stress increases much faster than predicted by Hooke's law.

A nonlinear superposition-type constitutive equation has been proposed for soft tissue [7.10.7]. It is not completely general but it does describe a variety of experimental data. The nonlinearity of the behavior is considered separable from the time-dependent aspect.

$$\sigma(t) = \int_{-\infty}^{t} G(t-\tau) \frac{\partial \sigma^{\text{elastic}}}{\partial \varepsilon} \frac{\partial \varepsilon(\tau)}{\partial \tau} d\tau. \tag{7.10.1}$$

Stress relaxation in muscle [7.10.8] in tension over a limited time scale up to about 6 sec was observed and resolved into two components:

$$E(t) = E_0 \left(a_1 e^{-t/\tau_1} + a_2 e^{-t/\tau_2} \right), \tag{7.10.2}$$

and the time constants were linked to the contraction speed capability of the corresponding living muscle. Cardiac muscle exhibited relaxation over a four-decade range beginning from 0.5 sec [7.10.9]. The longer term portion of the relaxation had a characteristic time which increased as the square of the sample radius. Such behavior was interpreted in light of a poroelastic model (see §8.4) in which the viscoelasticity arises from fluid flow through pores.

Tendons in the leg and ligaments in the arch of the foot are sufficiently compliant that they store significant amounts of energy during walking and running [7.10.10, 7.10.11]. The total energy turnover in each stance phase of a 70-kg man running at 4.5 m · sec⁻¹ is about 100 J. In comparison, 17 J is stored as strain energy in the compliant parts of the arch of the foot, and 35 J is stored in the Achilles tendon. These structures store enough energy to improve the energy efficiency of running in humans [7.10.11]. The load–deformation behavior discloses nonlinearly viscoelastic behavior in that the hysteresis curve is nonlinear and encloses some area. The nonlinear behavior arises from an initial waviness in the collagen fibers in the tendon or ligament; these curves are straightened in the early stages of loading. Energy storage predominates over energy dissipation since the effective loss tangent, considered as $\Psi/2\pi$, extracted from areas in the plots, is only about 0.04.

Most land animals have compliant pads in their paws [7.10.10]. These pads cushion impacts with the ground during running. The cushioning effect arises from compliance more so than damping since the effective loss tangent is only about 0.04.

Tissues in the eye are viscoelastic. The cornea exhibits considerable creep, more than a factor of two over 1 h [7.10.12]. Dynamic studies of the cornea [7.10.13] from 0.1 mHz to 100 Hz disclosed dependence of properties on temperature and hydration. Corneas from nearsighted people were found to be softer (by a factor of 7 at lower frequency and a factor of 4 at higher frequency) than those of emmetropic (normal) people. The lens [7.10.14] of the eye is also viscoelastic. Recovery following release of a load was dependent on time from less than 0.01 sec to more than 180 sec. Recovery did not proceed to zero strain, particularly if a relatively large load had been applied. This phenomenon might play a role in the etiology of myopia in association with much reading or other close work.

Table 7.3 compares properties of several tissues.

Table 7.3 Stiffness and Damping of Some Biological Materials at 10 rad/sec (1.6 Hz)

Material	$\mid G^* \mid$	tan δ	Ref.
Compact bone	4 GPa	0.01	[7.10.1]
Spinal motion segment	45 MPa	0.1	[7.10.16]
Articular cartilage	0.6–1 MPa	0.23	[7.10.17]
Meniscus	100 kPa	0.40	[7.10.18]
Nucleus pulposus	11 kPa	0.45	[7.10.15]
Synovial fluid	0.02 kPa	0.29	[7.10.19]
Vitreous humor	0.002 kPa	0.81–1.43	[7.10.20]

Ultrasonic properties of many kinds of tissues have been determined and compiled [7.10.21]. For example, one study of human muscle at $v = 1$ MHz reports the longitudinal wave velocity c_L as 1566 m/sec and the attenuation α as 0.16 neper/cm. The stiffness based on Eqs. 5.2.1 and 5.7.33 is $C_{1111} = c_L^2 \rho$; so by assuming a density ρ of 1 g/cm³, the stiffness is 2.4 GPa. Compare with the bulk modulus, 2.18 GPa of water alone, and recognize (Example 5.9) that C_{1111} depends on the bulk modulus. For muscle, for that longitudinal mode of deformation, $\tan (\delta/2) = \alpha v / 2\pi c_L = 4 \times 10^{-3}$, so $\tan \delta = 8 \times 10^{-3}$. Liver is similar to muscle in both longitudinal wave velocity and attenuation; however, there is considerable variation in reported results and such variation is typical in biological materials. The shear wave velocity c_T in dog liver at 2.2 MHz is reported as 8.7 m/sec and the attenuation 9200 neper/cm. So the stiffness is $C_{2323} = G = 76$ kPa and $\tan (\delta/2) = 0.58$; $\tan \delta = 1.7$. The loss is sufficiently large that the exact relation for the velocity is required,

$$c_T = \sqrt{\frac{|G^*|}{\rho}} \sec \frac{\delta}{2}.$$

The correction $\sec(\delta/2)$ is 1.16, so $|G^*| = 57$ kPa. The large difference between the longitudinal and shear properties deserves comment. Soft tissue, not unlike rubber, has a Poisson's ratio approximating $1/2$ and the shear modulus is much less than the bulk modulus. Since longitudinal waves (§5.7) entail volumetric deformation, their speed depends on the bulk modulus.

§7.10.3 Wood

Wood is a cellular composite of biological origin, based on lignin, which is a natural macromolecule [7.10.22]. In addition to the results adduced in Reference [7.2.7], numerous results ($\tan \delta$ vs. temperature at a constant frequency of 1 Hz or 1 kHz) have been obtained for dry woods, lignin, and delignified wood, as well as coal, amber, and oil shale for comparison [7.10.23]. Woods such as spruce and beech exhibit β peaks corresponding to a loss tangent of 0.02 to 0.03 at a temperature of about 200 K, $\tan \delta \approx 0.02$ at room temperature (about 300 K), and an α peak of $\tan \delta \approx 0.08$ at about 400 K. Spectra for lignin, bituminous coal, and amber have similar overall characteristics. The $\tan \delta$ values increase with moisture content in wood [7.10.23]. Fully wet wood (Fig. 7.17) appears to be thermorheologically simple [7.10.23] as far as the stiffness, and master curves for the modulus have been produced for a frequency range 10^{-10} Hz ($E' = 50$ MPa) to more than 100 kHz ($E = 400$ MPa) based on dynamic studies from 23 to 130°C and from 0.6 to 20 Hz. The properties have been associated with those of the lignin within the wood, which exhibits a glass transition temperature near 100°C.

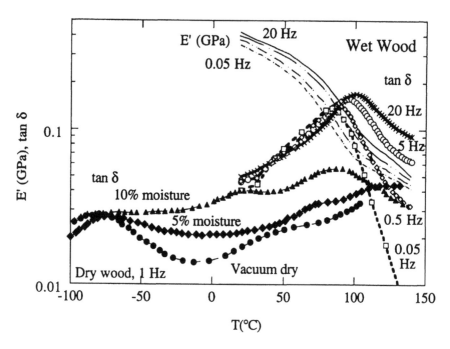

Figure 7.17   Viscoelastic properties of wood. Wet wood (adapted from Salmén [7.10.25]); dry spruce wood, several moisture contents, at 1 Hz in flexure (adapted from Kelley et al. [7.10.24]); vacuum dry spruce wood, torsion, frequency ≈ 1 Hz (adapted from Wert et al. [7.10.23]).

§7.10.4 *Soft plant tissue: apple, potato*

Viscoelastic behavior of many food products has been studied. Considerable creep occurs in such materials. Power law representations ($\varepsilon = a + bt^n$) were useful in representing the creep behavior [7.10.26]. The viscoelastic behavior is nonlinear and is describable by a multiple integral representation of the Green–Rivlin type. Further examples and applications are presented in §10.13.

§7.11   *Porous materials*

Cellular solids may be regarded as composite materials in which one phase is empty space, or fluid. Synthetic cellular solids have been prepared based on polymers, metals, and ceramics as the solid phase. Natural foams include lung, cancellous bone, and wood. A representative stress–strain curve for an open-cell elastomer foam is shown in Fig. 7.18. At small strains the foam deforms by bending of the ribs or struts in the foam structure. At a compressive strain above about 5%, the ribs undergo buckling, which gives rise to a plateau region in which the foam deforms progressively at near constant stress. At a sufficiently high compressive stress, the cell ribs come in contact

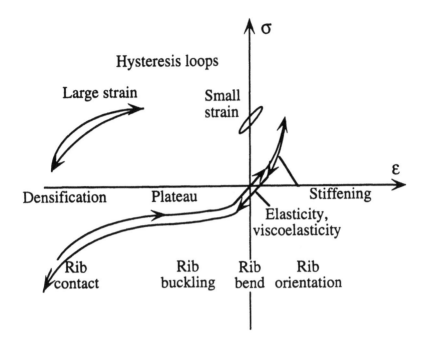

Figure 7.18   Stress–strain curve for an open-cell elastomer foam. Hysteresis loops for dynamic loading at small and large strain amplitudes are shown offset for clarity.

and the foam densifies, which gives rise to a rapid increase in apparent stiffness. Viscoelasticity in the foam manifests itself in creep under transient tests, and energy dissipation in dynamic studies. Foams exhibit linear viscoelasticity, hence elliptical stress–strain diagrams, under small strain amplitude dynamic loading. Under large deformation, foams behave nonlinearly, and the stress–strain diagram becomes distorted, as shown in Fig. 7.18.

Under quasistatic conditions or at low frequency, we may neglect the effect of fluid in the pores, and apply the correspondence principle to the elastic relation between the stiffness E of the foam and the stiffness E_s of the solid from which it is made.

In an open-cell foam [7.11.1],

$$\frac{E}{E_s} = \left[\frac{\rho}{\rho_s}\right]^2, \qquad\qquad (7.11.1)$$

in which ρ is the foam density and ρ_s is the density of the solid from which it is made. Use of the dynamic correspondence principle gives

$$E^* = E_s^* \left[\frac{\rho}{\rho_s}\right]^2, \qquad\qquad (7.11.2)$$

so that under these assumptions the complex dynamic modulus of the foam is proportional to that of the solid from which it is made. In this regime the damping of the foam is governed by the physics and chemistry of the rib material. We have assumed that the foaming process has no chemical effect on the "solid" chemistry which might alter its mechanical properties. In reality, it is possible that the foaming process might alter the chemical and physical properties of the solid phase. Moreover, the correspondence principle applies only to linear behavior.

Foams easily undergo large deformations which cause nonlinear behavior [7.11.1–7.11.3]. Flexible foams, when compressed sufficiently, exhibit buckling of the cell ribs, followed by contact. Frictional contact between ribs provides an additional damping mechanism for heavy load. Elastomeric polymer foams such as those used in seat cushions exhibit incremental properties which depend on the degree of superposed static precompression [7.11.3]. For a representative seat cushion foam, the damping was $\tan \delta \approx 0.2$ and the dynamic Young's modulus was $E \approx 20$ kPa for precompressions 20 to 50% at frequencies of 1 to 10 Hz. For small precompression, the foam was stiffer and had less damping.

Foam materials with interconnected pores and with a viscous fluid in the pores exhibit a viscoelastic loss due to fluid flow, even if the foam skeleton is purely elastic. For open-cell polymer foams, the loss tangent exhibits a peak which can have a magnitude greater than 0.5, depending on the characteristics of the foam and fluid [7.11.4]. One can extract the material constants for the theoretical model from relatively simple experiments [7.11.4, 7.11.5]. Sandstone is a porous material, and the presence of fluid in the pores increases the loss [7.11.6].

At ultrasonic frequencies, one can generate in porous media a second type of compressional wave which is slower than the normal compressional wave. This acoustic slow wave arises from a dynamic interaction between the solid and fluid phases. Experimental results have been obtained for fused glass bead media and for bone [7.11.6–7.11.11]. The attenuations for the fast and slow waves need not be equal.

At sufficiently high frequency the ribs or cell walls in a cellular solid can be set into resonance, which increases the material loss tangent [7.11.12]. For flexible polymer foams, this occurs at about 1 kHz.

§7.12 *Composite materials*

In viscoelastic composite materials, the overall material properties depend on the properties of the constituents; analysis and properties are given in Chapter 9. In fibrous polymer matrix composites, fibers made of a stiff, strong material are embedded in a polymer matrix. Most such fibers, such as boron, graphite, and glass, exhibit much less creep than the matrix polymers [7.12.1]. Consequently, when a tensile stress is applied in a direction in which there are many elastic fibers, the fibers carry most of the load, and creep is minimal. Stress applied in a direction oblique or transverse to the fibers will increase stress in the matrix, hence causing more creep. Kevlar is unusual among fiber materials in that it exhibits significant creep. As for laminates, laminating techniques which minimize stress upon the matrix and which minimize porosity result in the least creep. Laminated composites containing two metal phases have been studied in an effort to achieve simultaneously strength, toughness, and reasonably high damping [7.12.2]. A loss tangent of about 0.001 was observed at 30 Hz, and it increased linearly with frequency.

In some applications such as vibration absorption, the designer wishes to maximize the viscoelastic response of structural members or added layers. One way to do this is to fabricate composites with a controlled amount of slip between the fibers and the matrix [7.12.3]. For steel fibers and a rubber matrix, fibers can be strongly bound via a priming agent or weakly bound by omitting the primer. By varying the surface roughness of the fibers one can control the degree of slip. Large damping and hysteresis are so obtained [7.12.3].

At sufficiently high frequency in the ultrasonic domain, dispersion is observed in composites [7.12.4] as a result of resonances in the fibers or inclusions. This dispersion, observed at ultrasonic frequency, is characterized by a wave speed which decreases with increasing frequency, the opposite of the pattern seen in homogeneous viscoelastic materials.

§7.13   *Inorganic amorphous materials*

Glassy (amorphous) polymers have been discussed above in §7.3. Window glass is also amorphous and exhibits a glass transition similar to that of a glassy polymer, only at a higher temperature, typically above 500°C. The stiffness of window glass in the glassy state is higher than that of a glassy polymer: about 70 GPa compared with 3 GPa. The loss tangent of Pyrex® glass at constant frequency increases with temperature [7.13.1]; at 1 Hz, tan δ goes from 8×10^{-4} at 100°C to about 0.01 at 450°C to 0.1 at 600°C for annealed glass. Window glass also exhibits creep [7.13.1, 7.13.2], and significant creep may occur over long periods of time. Windows in ancient buildings such as cathedrals in Europe are often thicker at the bottom. This may be due to creep; however, glass manufacture in early times did not result in uniformly thick plates, so a judgment on the degree of creep is difficult.

Selenium is an unusual viscoelastic material in that it is a pure element which can be prepared in the glassy state by rapid cooling from the melt; amorphous selenium exhibits dramatic viscoelastic behavior at room temperature and at accessible times. A master curve derived from relaxation experiments was obtained and is representative of polymeric behavior [7.13.3]. In its crystalline form, selenium is piezoelectric as described in §7.9. The flow region can be modified to obtain a rubbery plateau by the addition of arsenic or sulfur and arsenic to the selenium [7.13.3].

As for sulfur, it has long been known that fibers produced by rapidly quenching molten sulfur in cold water exhibit a rubber-like elasticity (see, e.g., [7.3.1, 7.13.3]). Amorphous sulfur becomes transformed to the crystalline form at room temperature. Adding arsenic [7.13.4, 7.13.5] or phosphorus cross-links the chains of polymeric sulfur and retards the crystallization.

§7.14 Common aspects

The phenomenology of viscoelastic response has been presented for many materials. Various materials differ substantially in their viscoelastic response. A common aspect in the viscoelasticity of materials is the role of *temperature* [7.14.1]. If the absolute temperature is small compared with the melting point of a material, it will behave nearly elastically, with minimal viscoelastic response. If the absolute temperature exceeds a significant fraction (0.3 to 0.5) of the melting point (the material is at high homologous temperature), substantial creep, damping, and other manifestations of viscoelasticity may be expected. Many metals have high melting points, and these tend to exhibit little viscoelasticity at room temperature. Most polymers have low melting points, and they tend to exhibit substantial viscoelastic behavior at room temperature.

In many materials, viscoelasticity is governed by a thermally activated process for which

$$\tan \delta = f(v)\exp-\frac{U}{kT}. \tag{7.14.1}$$

with U as the activation energy, k as the Boltzmann constant, and T as the absolute temperature. $k = 1.38 \times 10^{-16}$ erg/K. If U is expressed in calories per mole, k becomes the gas constant $R = 1.986$ cal/molK. This is an Arrhenius equation which arises when the rate-limiting step of the relaxation consists of movement of an atomic or molecular process over an energy barrier. Equation 7.14.1 gives rise to thermorheologically simple behavior with a shift factor a_T as is shown in Example 6.9.

$$\log a_T = \frac{U}{R}\left\{\frac{1}{T_1} - \frac{1}{T_2}\right\}. \tag{7.14.2}$$

Amorphous polymers also depend sensitively on temperature. The time–temperature shift for polymers follows the empirical WLF (after Williams, Landel, and Ferry) equation [1.6.18],

$$\log a_T = -\frac{C_1(T - T_{ref})}{C_2 + (T - T_{ref})}. \tag{7.14.3}$$

The reference temperature T_{ref} can be any temperature but is frequently taken as the glass transition temperature T_g. The constants C_1 and C_2 depend on the particular polymer. Even so, the behavior of different amorphous polymers is sufficiently similar in normalized coordinates that Ferry has proposed the following "universal" constants: for T_{ref} taken as T_g the glass transition temperature is $C_1 = 17.44$ and $C_2 = 51.6$. The effective activation energy associated with the WLF equation is not constant but increases rapidly with decreasing temperature below T_g.

Secondary creep (which is nonlinear) in a wide variety of materials is governed by the following relation for thermal activation [7.14.1]:

$$\frac{d\varepsilon}{dt} = A\sigma_d^n \exp-\frac{U}{kT}, \qquad (7.14.4)$$

in which n is a constant.

In a wide range of crystalline materials such as metals, mechanical damping increases rapidly with temperature. The *high-temperature background* refers to such damping observed at an absolute temperature which is a significant fraction of the melting point. The background appears to be thermally activated, with the following frequency dependence:

$$\tan\delta \propto v^{-n} \qquad (7.14.5)$$

with $n < 1$. Early examples have been reviewed by Nowick and Berry [1.6.23]. Damping in alloys of low melting point were observed to follow Eq. 7.14.5 over a wide range of frequency [7.14.2], with $n \approx 0.2$. By contrast, an elementary dislocation model [7.14.3] predicts $n = 1$. Models are discussed in §8.8.

There are some exceptions to thermally activated temperature dependence. For example, thermoelastic damping, discussed in §8.3, is approximately proportional to absolute temperature. The dependence of creep rate or mechanical loss on temperature and on load level may be depicted in a deformation mechanism map, as discussed in §8.9.

§7.15 Summary

Virtually all materials exhibit some viscoelastic effects. The degree of viscoelasticity as quantified by the loss tangent in known solid materials differs by more than nine orders of magnitude. Some single-crystal materials exhibit the lowest losses, followed by common metals. Polymers in the transition region exhibit the highest losses commonly found in solids: the loss tangent can exceed unity. For common materials, high stiffness is associated with low loss.

Examples

Example 7.1

Is the principle of dissipativity (§2.3) a real physical law? Would it be possible to create a material for which tan δ < 0? If so, what would be required?

Answer

In §7.9 piezoelectric semiconductors such as CdS were discussed. They *amplify* ultrasonic waves, and hence they have tan δ < 0. No thermodynamic principles are violated since the energy required for amplification is derived from an externally applied DC electric field. [7.9.3, 7.9.4].

Example 7.2

Does a stretched rod held at constant extension get fatter or thinner in cross section with time?

Answer

In §7.3, experimental results bearing on the viscoelastic Poisson's ratio of polymers show that Poisson's ratio increases through the α transition but is not monotonic through the β transition. These experimental results (assuming the specimens were in fact isotropic) provide an answer to questions raised in §5.6 and Problem 5.3. A stretched rod held at constant extension gets thinner during the α transition but can get fatter through the β transition.

Example 7.3

For some materials, tan δ > 1. Does that mean they are viscoelastic liquids?

Answer

Polymers in the leathery regime (the α peak) can exhibit tan δ > 1. The distinction between solids and liquids is defined (§1.3.2) in terms of the limiting behavior of the relaxation modulus or creep compliance as time tends to infinity. If the relaxation modulus approaches zero as time tends to infinity (or as frequency tends to zero for the storage modulus), the material is a liquid; otherwise it is a solid. For a liquid, $\lim_{\omega \to 0} \tan \delta = \infty$. So it is not the value of tan δ at a given frequency but the limiting behavior which is relevant to the distinction between solid and liquid.

Example 7.4

Must recovery (at zero stress) following some arbitrary stress history always be monotonic?

Answer

No. If the stress history contains reversals, the strain recovery after removal of the stress can contain reversals. Such reversals have been experimentally demonstrated in polymers [E7.4.1].

Problems

7.1 Are there any materials which do not exhibit fading memory? Discuss your findings in connection with §2.3.

7.2 Plot Eqs. 7.14.2 and 7.14.3 for the shift factor. Is there a regime for which the functional dependence is similar? Discuss.

7.3 Discuss applications of tin and its alloys and how they are influenced by viscoelastic properties.

7.4 In §2.6 the standard linear solid was described. Can you find any materials among the examples given in this chapter which behave in this way? How do real materials differ from the standard linear solid?

7.5 Tuning forks are usually made of a low-loss aluminum alloy. After such a fork is struck, its vibration persists for many cycles. Prepare tuning forks of several available materials such as various polymers and woods. Describe the sounds in connection with free decay of vibration and relate your observations to measured values of tan δ.

References

7.2.1 Braginsky, V. B., Mitrofanov, V. P., and Panov, V. I., *Systems with Small Dissipation*, University of Chicago Press, IL, 1985.

7.2.2 Mason, W. P., "Use of piezoelectric crystals and mechanical resonators in filters and oscillators", in *Physical Acoustics*, Vol. 1A, ed. E. P. Mason, Academic, NY, 1964, pp. 335–416.

7.2.3 Smithells, C. J., *Metals Reference Book*, 5th ed., Butterworths, London and Boston, 1976.

7.2.4 Randall, B. H. and Zener, C., "Internal friction of aluminum", *Phys. Rev.* 58, 472–483, 1940.

7.2.5 Jamieson, R. M. and Kennedy, R., "Internal friction characteristics of iron–vanadium–nitrogen alloys", *J. Iron Steel Inst.* 204, 1208–1210, 1966.

7.2.6 Brennan, B. J. and Stacey, F. D., "Frequency dependence of elasticity of rock-test of seismic velocity dispersion", *Nature* 268, 220–222, 1977.

7.2.7 Knopoff, L., "Attenuation of elastic waves in the earth", in *Physical Acoustics*, Vol. 3B, ed. E. P. Mason, Academic, NY, 1965, pp. 287–324.

7.2.8 Lakes, R. S., Katz, J. L., and Sternstein, S. S., "Viscoelastic properties of wet cortical bone. I. Torsional and biaxial studies", *J. Biomech.* 12, 657–678, 1979.

7.2.9 Koppelmann, V. J., "Über die Bestimmung des dynamischen Elastizitäts-moduls und des dynamischen Schubmoduls im Frequenzbereich von 10^{-5} bis 10^{-1} Hz", *Rheol. Acta* 1, 20–28, 1958.

7.2.10 Koppelmann, V. J., "Über das dynamische elastische Verhalten hochpoly-merer Stoffe", *Kolloid Z.* 144, 12–41, 1955.

7.2.11   Iwayanagi, S. and Hideshima, T., "Low frequency coupled oscillator and its application to high polymer study", *J. Phys. Soc. Jpn.* 8, 365–358, 1953.

7.2.12   Ashby, M. F., "On the engineering properties of materials", *Acta Metall.* 37, 1273–1293, 1989.

7.3.1   Hopkins, I. L. and Kurkjian, C. R., "Relaxation spectra and relaxation processes in solid polymers and glasses", in *Physical Acoustics II*, ed. W. P. Mason, Academic, NY, 1965.

7.3.2   McLoughlin, J. R. and Tobolsky, A. V., "The viscoelastic behavior of polymethyl methacrylate" *J. Colloid Sci.* 7, 555–568, 1952.

7.3.3   Tobolsky, A. V., "Stress relaxation studies of the viscoelastic properties of polymers", *J. Appl. Phys.* 27, 673–685, 1956.

7.3.4   Schmieder, Von K. and Wolf, K., "Über die Temperatur und Frequenzabhängigkeit des mechanischen Verhaltens einiger hochpolymerer Stoffe", *Kolloid Z.* 127, 65–78, 1952.

7.3.5   Havrilak, S., Jr. and Havrilak, S. J., *Dielectric and Mechanical Relaxation in Materials*, Hanser, Munich and Cincinnati, 1997.

7.3.6   Snowdon, J. C., *Vibration and Shock in Damped Mechanical Systems*, J. Wiley, NY, 1968.

7.3.7   Capps, R. N. and Beumel, L. L. "Dynamic mechanical testing, application of polymer development to constrained layer damping", in *Sound and Vibration Damping with Polymers*, ed. R. D. Corsaro and L. H. Sperling, American Chemical Society, Washington, DC, 1990.

7.3.8   Wetton, R. E., "Design of elastomers for damping applications", in *Elastomers: Criteria for Engineering Design*, ed. C. Hepburn, and R. J. W. Reynolds, Applied Science Publishers, Ltd., London, 1979.

7.3.9   Hostettler, F., "Energy attenuating polyurethanes", US Patent 4,722,946 (1988).

7.3.10   Tiao, Wen-Yu and Tiao, Chin-Sheng, "Methods for the manufacture of energy attenuating polyurethanes", US Patent 4,980,386 (1990).

7.3.11   Peoples, W. J., "Shock absorbing device for high heel footwear", US Patent 4,876,805 (1989).

7.3.12   Reader, W. T. and Megill, R. W., "Clorosulfonated polyethylene: a versatile polymer for damping acoustic waves", *Met. Trans.* 22A, 633–640, 1991.

7.3.13   Hypalon®, dupont de Nemours, Wilmington, DE.

7.3.14   Marvin, R. and McKinney, J. E., "Volume relaxations in amorphous polymers", in *Physical Acoustics*, Vol. 2B, ed. E. P. Mason, Academic, NY, 1964, pp. 165–229.

7.3.15   Kono, R., "The dynamic bulk viscosity of polystyrene and polymethyl methacrylate", *J. Phys. Soc. Jpn.* 15, 718–725, 1960.

7.3.16   Sternstein, S. S. and Ho, T. C., "Biaxial stress relaxation in glassy polymers", *J. Appl. Phys.* 43, 4370–4383, 1972.

7.3.17   Sutherland, H. J. and Lingle, R., "An acoustic characterization of polymethyl methacrylate and three epoxy formulations", *J. Appl. Phys.* 43, 4022–4026, 1972.

7.3.18   Sokolnikoff, I. S., *Mathematical Theory of Elasticity*, Krieger, Malabar, FL, 1983.

7.3.19   Flocke, H. A., "Ein Beitrag zum mechanischen Relaxationsverhalten von Polyäthylen, Polypropylen, Gemischen aus diesen und Mischpolymerisaten aus Propylen und Äthylen", *Kolloid Z.* 180, 118–126, 1962.

7.3.20   Aklonis, J. J., MacKnight, W. J., and Shen, M., *Introduction to Polymer Viscoelasticity*, J. Wiley, NY, 1972.

7.3.21   Hodge, I. M., "Physical aging in polymeric glasses", *Science* 267, 1945–1947, 1995.

7.3.22   Struik, L. C. E., *Physical Aging in Amorphous Polymers and Other Materials*, Elsevier Scientific, Amsterdam and New York, 1978.

7.3.23   Kubát, J., "Stress relaxation in solids", *Nature* 205, 378–379, 1965.

7.3.24   Spencer, R. S. and Boyer, R. F., "Thermal expansion and second order transition effects in high polymers. III. Time effects", *J. Appl. Phys.* 17, 398, 1946.

7.3.25   Thiele, J. L. and Cohen, R. E., "Thermal expansion phenomena in filled and unfilled natural rubber vulcanizates", *Rubber Chem. Technol.* 53, 313–320, 1980.

7.4.1   Read, T. A., "The internal friction of single metal crystals", *Phys. Rev.* 58, 371–380, 1940.

7.4.2   Weertman, J., "Internal friction of metal single crystals", *J. Appl. Phys.* 26, 202–210, 1955.

7.4.3   Read, T. A., "Internal friction of single crystals of copper and zinc", *Trans. Am. Inst. Min. Metal. Eng.* 143, 30–44, 1941.

7.4.4   Read, T. A., "Internal friction and plastic extension of zinc single crystals", *J. Appl. Phys.* 17, 713–720, 1946.

7.4.5   Swift, I. H., "Internal friction of zinc single crystals", *J. Appl. Phys.* 18, 417–425, 1947.

7.4.6   Read, T. A., "Internal friction of zinc single crystals", *J. Appl. Phys.* 20, 29–37, 1947.

7.4.7   Hiki, Y., "Internal friction of lead", *J. Phys. Soc. Jpn.* 13, 1138–1144, 1958.

7.4.8   Wert, C. A., "The internal friction of zinc single crystals", *J. Appl. Phys.* 20, 29–37, 1949.

7.4.9   Kê, T. S., Cui, P., and Su, C. M., "Internal friction in high purity aluminium single crystals", *Phys. Status Solidi* 84, 157–164, 1984.

7.4.10   Kê, T. S., "Experimental evidence of the viscous behavior of grain boundaries in metals", *Phys. Rev.* 71, 533–546, 1947.

7.4.11   Kê, T. S. and Zhang, B. S., "Contributions of bamboo boundaries to the internal friction peak in macrocrystalline high purity aluminium", *Phys. Status Solidi* 96, 515–525, 1986.

7.4.12   Kê, T. S., Zhang, D. L., Cheng, B. L., and Zhu, A. W., "On the origin of the macrocrystalline internal friction peak (the bamboo boundary peak)", *Phys. Status Solidi* 108, 569–575, 1988.

7.4.13   Zhu, A. W. and Kê, T. S., "Characteristics of the internal friction peak associated with bamboo grain boundaries", *Phys. Status Solidi* 113, 393–401, 1989.

7.4.14   Firle, T. E., "Amplitude dependence of internal friction and shear modulus of boron fibers", *J. Appl. Phys.* 39, 2839–2845, 1968.

7.4.15   Kê, T. S., "Viscous slip along grain boundaries and diffusion of zinc in alpha brass", *J. Appl. Phys.* 19, 285–290, 1948.

7.4.16   Duffy, W., "Acoustic quality factor of aluminum alloys from 50 mK to 300 K", *J. Appl. Phys.* 68, 5601–5609, 1990.

7.4.17   Wolfenden, A., Harmouche, M. R., and Hartman, J. T., "Mechanical damping at high temperature in aluminides" *J. Phys.* 46, C10-391–C10-394, 1985.

7.4.18   Keene, K. H., Hartman, J. T., Jr., Wolfenden, A., and Ludtka, G. M., "Determination of dynamic Young's modulus, shear modulus, and Poisson's ratio as a function of temperature for depleted uranium–0.75 wt% titanium using the piezoelectric ultrasonic composite oscillator technique", *J. Nuclear Mat.* 149, 218–226, 1987.

7.4.19   Fitzgerald, E., "Mechanical resonance dispersion in metals at audio frequencies", *Phys. Rev.* 108, 690–706, 1957.

7.4.20   Farid, Z. M., Saleh, S., and Mahmoud, S. A., "On the grain boundary internal friction peak of α-brasses", *Mater. Sci. Eng.* A110, L31–L34, 1989.

7.4.21   Bratina, W. J., "Internal friction and basic fatigue mechanisms in body-centered cubic metals, mainly iron and carbon steels", in *Physical Acoustics*, Vol. 3A, ed. E. P. Mason, Academic, NY, 1966, pp. 223–291.

7.4.22   Fast, J. D. and Ferrup, M. B., "Internal friction in lightly deformed pure iron wires", *Philips Res. Rep.* 16, 51–65, 1961.

7.4.23   Fessler, H. and Hyde, T. H., "Creep deformation of metals", *Creep of Engineering Materials*, ed. C. D. Pomeroy, Mechanical Engineering Publications, Ltd., London, 1978.

7.4.24   Tribula, D. and Morris, J. W., Jr., "Creep in shear of experimental solder joints", *J. Electron. Packag.* 112, 87–93, 1990.

7.4.25   Ritchie, I. G., Pan, Z. L., Sprungmann, K. W., Schmidt, H. K., and Dutton, R., "High damping alloys—the metallurgist's cure for unwanted vibration", *Can. Metall. Q.* 26, 239–250, 1987.

7.4.26   James, D. J., "High damping metals for engineering applications", *Mater. Sci. Eng.* 4, 1–8, 1969.

7.4.27   Lazan, B. J., *Damping of Materials and Members in Structural Mechanics*, Pergamon, NY, 1968.

7.4.28   Ritchie, I. G. and Pan, Z. L., "High damping metals and alloys", *Metall. Trans.* 22A, 607–616, 1991.

7.4.29   Cochardt, A. W., "High damping ferromagnetic alloys", *J. Met. (Trans. Am. Inst. Min. Metall. Eng.* 206, 1295–1298, 1956.

7.4.30   Ritchie, I. G., Sprungmann, K. W., and Sahoo, M., "Internal friction in Sonoston — a high damping Mn/Cu-based alloy for marine propeller applications", *J. Phys.* 46, C10-409–C10-412, 1985.

7.4.31   Weissmann, G. F. and Babington, W., "A high damping magnesium alloy for missile applications", *Proc. ASTM* 58, 869–892, 1958.

7.4.32   Schwaneke, A. E. and Nash, R. W., "Effect of preferred orientation on the damping capacity of magnesium alloys", *Metall. Trans.* 2, 3453–3457, 1971.

7.4.33   Sugimoto, K., Niiya, K., Okamoto, T., and Kishitake, K., "A study of damping capacity in magnesium alloys", *Trans. JIM* 18, 277–288, 1977.

7.4.34   Bozorth, R. M., Mason, W. P., and McSkimin, H. J., "Frequency dependence of elastic constants and losses in nickel", *Bell Syst. Tech. J.* 30, 970–989, 1951.

7.4.35   Schneider, W., Schrey, P., Hausch, G., and Török, "Damping capacity of Fe–Cr and Fe–Cr based high damping alloys", *J. Phys.* 42, suppl. 10, C5-635–C5-639, 1981.

7.4.36   Masumoto, H., Sawaya, S., and Hinai, M., "Damping capacity of Gentalloy in the Fe–Co alloys", *Trans. Jpn. Inst. Met.* 19, 312–316, 1978.

7.4.37   Masumoto, H., Sawaya, S., and Hinai, M., "Damping capacity of 'Gentalloy' in the Fe–Mo alloys", *Trans. Jpn. Inst. Met.* 22, 607–613, 1981.

7.4.38   Masumoto, H., Hinai, M., and Sawaya, "Damping capacity and pitting corrosion resistance of Fe–Mo–Cr alloys", *Trans. Jpn. Inst. Met.* 25, 891–899, 1984.

7.4.39   Masumoto, H., Sawaya, S., and Hinai, M., "Damping capacity of Fe–Mo alloys", *Trans. Jpn. Inst. Met.* 18, 581–584, 1977.

7.4.40   Enrietto, J. and Wert, C., "Anelasticity in alloys of Cd and Mg", *Acta Metall.* 6, 130–132, 1958.

7.4.41 Lulay, J. and Wert, C., "Internal friction in alloys of Mg and Cd", *Acta Metall.* 4, 627–631, 1957.

7.4.42 Cook, L. S. and Lakes, R. S., "Viscoelastic spectra of $Cd_{0.67}Mg_{0.33}$ in torsion and bending", *Metall. Trans.* 26A, 2037–2039, 1995.

7.4.43 Sobha, B. and Murti, Y. V. G. S., " Low frequency internal friction spectra of $Cu_{0.81}Pd_{0.19}$ alloy", *Bull. Mater. Sci.* 11, 319–328, 1988.

7.4.44 Sommer, A. W., Motokura, S., Ono, K., and Buck, O., "Relaxation processes in metastable beta titanium alloys", *Acta Metall.* 21, 489–497, 1973.

7.4.45 Kamel, R. "Measurement of the internal friction of solids", *Phys. Rev.* 75, 1606, 1949.

7.4.46 Cook, L. S. and Lakes, R. S., "Damping at high homologous temperature in pure Cd, In, Pb, and Sn", *Scr. Metall. Mater.* 32, 773–777, 1995.

7.4.47 Reed, R. P., McCowan, C. N., Walsh, R. P., Delgado, L. A., and McColskey, "Tensile strength and ductility of indium", *Mater. Sci. Eng. A*, 102, 227–236, 1988.

7.4.48 Gindin, I. A., Lazarev, B. G., Starodubov, Ya. D., and Lebedev, V. P., "Creep of indium in the normal and superconductive states", *Fiz. Met. Metalloved.* 29, 862–868, 1970.

7.4.49 Chen, C. P. and Lakes, R. S., "Viscoelastic behaviour of composite materials with conventional or negative Poisson's ratio foam as one phase", *J. Mater. Sci.* 28, 4288–4298, 1993.

7.4.50 Malhotra, A. K. and Van Aken, D. C., "Experimental and theoretical aspects of the internal friction associated with the melting of embedded particles", *Acta Metall. Mater.* 41, 1337–1346, 1993.

7.4.51 Ritchie, I. G., Pan, Z. L., and Goodwin, F. E., "Characterization of the damping properties of die-cast zinc–aluminum alloys", *Met. Trans.* 22A, 617–622, 1991.

7.4.52 Otani, T., Sakai, T., Hoshino, K., and Kurosawa, T., "Damping capacity of Zn–Al alloy castings", *J. Phys. Colloq.* C10, suppl. 12, 46, c10-417–c10-420, 1985; 8th Int. Conf. on Internal Friction and Ultrasonic Attenuation in Solids, ed. A. V. Granato, G. Mozurkewich, and C. A. Wert, Urbana, IL.

7.4.53 Millett, P., Schaller, R., and Benoit, W., "Internal friction spectrum and damping capacity of grey cast iron", in *Deformation of Multi-Phase and Particle Containing Materials*, Proc. 4th Riso Int. Symp. Metall. Mater. Sci., ed. J. B. Bilde-Sorensen, N. Hansen, A. Horsewell, T. Leffers, and H. Linholt, Riso National Laboratory, Roskilde, Denmark, 1983.

7.4.54 Ashby, M. F. and Abel, C. A., "Materials selection to resist creep", *Phil. Trans. R. Soc. London A* 351, 451–468, 1995.

7.4.55 Pichler, A., Weller, M., and Arzt, E., "High temperature damping in dispersion-strengthened aluminium alloys", *J. Alloys Comp.* 211, 414–418, 1994.

7.4.56 Ashby, M. F. and Jones, D. R. H., *Engineering Materials*, Pergamon, Oxford, 1980.

7.5.1 Gueguen, Y., Woirgard, J., and Darot, M., "Attenuation mechanisms and anelasticity in the upper mantle", in *Anelasticity in the Earth*, ed. F. D. Stacey, M. S. Paterson, and A. Nicholas, American Geophysical Union, Washington, DC, 1981.

7.5.2 Stein, S., Mills, J. M., and Geller, R. J., "Q^{-1} models from data space inversion of fundamental spheroidal mode attenuation measurements", in *Anelasticity in the Earth*, ed. F. D. Stacey, M. S. Paterson, and A. Nicholas, American Geophysical Union, Washington, DC, 1981.

7.5.3   Mason, W. P., "Internal friction at low frequencies due to dislocations: applications to metals and rock mechanics", in *Physical Acoustics*, Vol. 8, ed. E. P. Mason and R. N. Thurston, Academic, NY, 1971, pp. 347–371.

7.5.4   Latham, G. V., Ewing, M., Dorman, J., Press, F., Toksoz, N., Sutton, G., Meissner, R., Duennebier, F., Nakamura, Y., Kovach, R., and Yates, M., "Seismic data from man made impacts on the moon", *Science* 170, 620–626, 1970.

7.5.5   Pandit, B. I. and Tozer, D. C., "Anomalous propagation of elastic energy within the moon", *Science* 226, 335, 1970.

7.5.6   Mason, W. P., "Internal friction in moon and earth rocks", *Nature* 234, 461–463, 1971.

7.5.7   Tittman, B. R., "Internal friction in lunar rocks and terrestrial rocks", 1974 Ultrasonics Symp. Proc. IEEE No.74 CHO 896-ISU, 509–513, 1974.

7.5.8   Tittman, B. R. and Housley, R. M., "High Q (low internal friction) observed in a strongly outgassed terrestrial analog of lunar basalt", *Phys. Status Solidi* 56, K109–K110, 1973.

7.5.9   Pomeroy, C. D., "Time-dependent deformation of rocks", in *Creep of Engineering Materials*, ed. C. D. Pomeroy, Mechanical Engineering Publications, Ltd., London, 1978.

7.5.10 Bassett, R. H., "Time-dependent strains and creep in rock and soil structures", in *Creep of Engineering Materials*, ed. C. D. Pomeroy, Mechanical Engineering Publications, Ltd., London, 1978.

7.5.11 Karato, S. and Wu, P., "Rheology of the upper mantle: a synthesis", *Science* 260, 771–778, 1993.

7.5.12 Poirier, J. P., *Creep of Crystals*, Cambridge University Press, England, 1985.

7.5.13 Wolfenden, A., "Measurement and analysis of elastic and anelastic properties of alumina and silicon carbide", *J. Mater. Sci.* 32, 2275–2282, 1997.

7.5.14 Robinson, W. H. and Birnbaum, H. K., "High temperature internal friction in potassium chloride", *J. Appl. Phys.* 37, 3754–3766, 1966.

7.5.15 Wolfenden, A., Lastrapes, A., Duggan, M. B., and Raj, S. V., "Temperature dependence of the elastic moduli and damping for polycrystalline LiF-22% CaF_2 eutectic salt", *J. Mater. Sci.* 26, 1973–1798, 1991.

7.5.16 Robinson, W. H., "Amplitude-independent mechanical damping in alkali halides", *J. Mater. Sci.* 7, 115–123, 1972.

7.6.1   Branson, D., *Deformation of Concrete Structures*, McGraw-Hill, NY, 1977.

7.6.2   Illston, J. M., "Creep of concrete", in *Creep of Engineering Materials*, ed. C. D. Pomeroy, Mechanical Engineering Publications, Ltd., London, 1978.

7.6.3   Troxell, G. E., Raphael, J. M., and Davis, R. W., "Long time creep and shrinkage tests of plain and reinforced concrete", *ASTM Proc.* 58, 1–20, 1958.

7.6.4   Bazant, Z., *Mathematical Modeling of Creep and Shrinkage of Concrete*, J. Wiley, NY, 1988.

7.6.5   L'Hermite, R., "What do we know about the plastic deformation and creep of concrete?", *Bull. RILEM* 1, 22–51, 1959.

7.6.6   Neville, A., *Creep of Concrete: Plain, Reinforced, and Prestressed*, North-Holland, Amsterdam, 1970.

7.6.7   Zhaoxia, L., "Effective creep Poisson's ratio for damaged concrete", *Int. J. Fracture* 66, 189–196, 1994.

7.7.1   Lesueur, D., Gerard, J. F., Claudy, P., Letoffe, M. M., Planche, J. P., and Martin, D., "A structure related model to describe asphalt linear viscoelasticity", *J. Rheol.* 40, 813–836, 1996.

7.7.2   Sayegh, G., "Variation des modules de quelques bitumes purs et bétons bitumineux", *Cahiers Rhéol.* 2, 51–74, 1966.

7.8.1   Eranti, E. and Lee, G. C., *Cold Region Structural Engineering*, McGraw-Hill, NY, 1986.

7.8.2   Jezek, K. C., Alley, R. B., and Thomas, R. H. "Rheology of glacier ice", *Science* 227, 1335–1337, 1985.

7.8.3   Vassoile, R., Perez, J., Mai, C., and Gobin, P. F., "Internal friction of ice I_h due to crystalline defects", in *Internal Friction and Ultrasonic Attenuation in Solids*, ed. R. R. Hasiguti, and N. Mikoshiba, University of Tokyo Press, 1977 (Proc. 6th Int. Conf. Internal Friction and Ultrasonic Attenuation in Solids, Tokyo, June 4–7, 1977.

7.8.4   Tatiboët, J., Perez, J., and Vassoile, R., "Study of lattice defects in ice Ih by very low frequency internal friction measurements", *J. Phys. Chem.* 87, 4050–4054, 1983.

7.8.5   Nakamura, T. and Abe, O., "Internal friction of snow and ice at low frequency", in *Internal Friction and Ultrasonic Attenuation in Solids*, ed. R. R. Hasiguti and N. Mikoshiba, University of Tokyo Press, 1977.

7.8.6   Hiki, Y. and Tamura, J., "Internal friction in ice crystals", *J. Phys. Chem.* 87, 4054–4059, 1983.

7.8.7   Cole, D. M. and Durell, G. D., "The cyclic loading of saline ice", *Philos. Mag.* A 72, 209–229, 1995.

7.9.1   Leung, W. P. and Yung, K. K., "Internal losses in polyvinylidene fluoride (PVF$_2$) ultrasonic transducers", *J. Appl. Phys.* 50, 8031–8033, 1979.

7.9.2   Royer, D. and Dieulesant, E., "Elastic and piezoelectric constants of trigonal selenium and tellurium crystals", *J. Appl. Phys.* 50, 4042–4045, 1979.

7.9.3   Hutson, A. R., McFee, J. H., and White, D. L., "Ultrasonic amplification in CdS", *Phys. Rev. Lett.* 7, 237–239, 1961.

7.9.4   White, D. L., "Amplification of ultrasonic waves in piezoelectric semiconductors", *J. Appl. Phys.* 33, 2547–2554, 1962.

7.9.5   Holden, A. V., Markus, M., and Othmer, H. G. ed., *Nonlinear Wave Processes in Excitable Media*, Plenum, NY, 1991.

7.9.6   Markus, M., Kloss, G., and Kusch, I., "Disordered waves in a homogeneous, motionless, excitable medium", *Nature* 371, 402–404, 1994.

7.9.7   Berlincourt, D. A., Curran, D. R., and Jaffe, H., "Piezoelectric and piezomagnetic materials and their function in transducers", in *Physical Acoustics*, Vol. 1A, ed. E. P. Mason, Academic, NY, 1964, pp. 169–270.

7.10.1 Lakes, R. S., Katz, J. L., and Sternstein, S. S., "Viscoelastic properties of wet cortical bone I. Torsional and biaxial studies", *J. Biomech.* 12, 657–678, 1979.

7.10.2 Lakes, R. S., "Dynamical study of couple stress effects in human compact bone", *J. Biomech. Eng.* 104, 6–11, 1982.

7.10.3 Rimnac, C. M., Petko, A. A., Santner, T. J., and Wright, T. M., "The effect of temperature, stress, and microstructure on the creep of compact bovine bone", *J. Biomech.* 26, 219–228, 1993.

7.10.4 Caler, W. E. and Carter, D. R., "Bone creep-fatigue damage accumulation", *J. Biomech.* 22, 625–635, 1989.

7.10.5 Mauch, M., Currey, J. D., and Sedman, A. J., "Creep fracture in bones with different stiffnesses", *J. Biomech.* 25, 11–16, 1992.

7.10.6 Bowman, S. M., Keaveny, T. M., Gibson, L. J., Hayes, W. C., and McMahon, T. A., "Compressive creep behavior of bovine trabecular bone", *J. Biomech.* 27, 301–310, 1994.

7.10.7   Fung, Y. C., "Stress strain history relations of soft tissues in simple elongation", in *Biomechanics, its Foundations and Objectives*, Prentice Hall, Englewood Cliffs, NJ, 1970.

7.10.8   Abbott, B. C. and Lowy, J., "Stress relaxation in muscle", *Proc. R. Soc. London
B* 146, 281–288, 1956.

7.10.9   Djerad, S. E., du Burck, F., Naili, S., and Oddou, C., "Analyse du compor-
tement rhéologique instationnaire d'un échantillon de muscle cardiaque",
C. R. Acad. Sci. Paris, Sér. II, 315, 1615–1621, 1992.

7.10.10 Alexander, R. McN., *Elastic Mechanisms in Animal Movement*, Cambridge
University Press, England, 1988.

7.10.11 Ker, R. F., Bennett, M. B., Bibby, S. R., Kester, R. C., and Alexander, R. McN.,
"The spring in the arch of the human foot", *Nature* 325, 147–149, 1987.

7.10.12 Nyquist, G. J., "Rheology of the cornea: experimental techniques and results", *Exp. Eye Res.* 7, 183–188, 1968.

7.10.13 Soergel, F., Jean, B., Seiler, T., Bende, T, Mücke, S., Pechold, W., and Pels, L.,
"Dynamic mechanical spectroscopy of the cornea for measurement of its
viscoelastic properties in vitro", *Ger. J. Ophthalmol.* 4, 151–156, 1995.

7.10.14 Ejiri, M., Thompson, H. E., and O'Neill, W. D., "Dynamic visco-elastic properties of the lens", *Vision Res.* 9, 233–244, 1969.

7.10.15 Iatridis, J. C., Weidenbaum, M., Setton, L. A., and Mow, V. C., "Is the nucleus
pulposus a solid or a fluid? Mechanical behaviors of the nucleus pulposus
of the human intervertebral disc", *Spine* 21, 1174–1184, 1996.

7.10.16 Ohshima, H., Tsuji, H., Hirano, N., Ishihara, H., Katoh, Y., and Yamada, H.,
"Water diffusion pathway, swelling pressure, and biomechanical properties
of the intervertebral disc during compression load", *Spine* 11, 1234–1244,
1989.

7.10.17 Setton, L. A., Mow, V. C., and Howell, D. S., "The mechanical behavior of
articular cartilage in shear is altered by transection of the anterior cruciate
ligament", *J. Orthop. Res.* 11, 228–239, 1993.

7.10.18 Zhu, W. B., Chern, K. Y., and Mow, V. C., "Anisotropic viscoelastic shear
properties of bovine meniscus", *Clin. Orthop.* 306, 34–45, 1994.

7.10.19 Safari, M., Bjelle, A., Gudmundsson, M., and Högfors, G., "Clinical assessment of rheumatic diseases using viscoelastic parameters for synovial fluid",
Biorheology 27, 659–674, 1990.

7.10.20 Bettelheim, F. A. and Wang, T., "Dynamic viscoelastic properties of bovine
vitreous", *Exp. Eye Res.* 23, 435–441, 1976.

7.10.21 Goss, S. A., Johnson, R. L., and Dunn, F., "Comprehensive compilation of
empirical ultrasonic properties of mammalian tissues", *J. Acoust. Soc. Am.*
64, 423–457, 1978.

7.10.22 Thomas, R. J., "Wood: formation and morphology" in *Wood Structure and
Composition*, ed. M. Lewin and I. S. Goldstein, Marcel Dekker, NY, 1991.

7.10.23 Wert, C. A., Weller, M., and Caulfield, D., "Dynamic loss properties of
wood", *J. Appl. Phys.* 56, 2453–2458, 1984.

7.10.24 Kelley, S. S., Rials, T. G., and Glasser, W. G., "Relaxation behaviour of the
amorphous components of wood", *J. Mater. Sci.* 22, 617–624, 1987.

7.10.25 Salmén, L., "Viscoelastic properties of *in situ* lignin under water saturated
conditions", *J. Mater. Sci.* 19, 3090–3096, 1984.

7.10.26 Lu, R. and Puri, V. M., "Characterization of nonlinear creep behavior of two
food products", *J. Rheol.* 35, 1209–1233, 1991.

7.11.1   Gibson, L. J. and Ashby, M. F., *Cellular Solids*, Pergamon, Oxford, 1988.

7.11.2   Hilyard, N. C., "Hysteresis and energy loss in flexible polyurethane foams", in *Low Density Cellular Plastics*, ed. N. C. Hilyard and A. Cunningham, Chapman and Hall, London, 1994.

7.11.3   Cunningham A., Huygens E., and Leenslag J. W., "MDI comfort cushioning for automotive applications", *Cell. Polym.* 13, 461–472, 1994.

7.11.4   Gent, A. N. and Rusch, K. C., "Viscoelastic behavior of open cell foams", *Rubber Chem. Technol.* 39, 388–396, 1966.

7.11.5   Kim, Y. K. and Kingsbury, H. B., "Dynamic characterization of poroelastic materials", *Exp. Mech.* 17, 252–258, 1979.

7.11.6   Fatt, I., "The Biot-Willis elastic coefficients for a sandstone", *J. Appl. Mech.* 26, 296–297, 1959.

7.11.7   Plona, T. J., "Observation of a second bulk compressional wave in a porous medium at ultrasonic frequencies", *Appl. Phys. Lett.* 36, 259–261, 1980.

7.11.8   Berryman, J. G., "Confirmation of Biot's theory", *Appl. Phys. Lett.* 37, 382–384, 1980.

7.11.9   Johnson, D. L., Plona, T. J., Scala, C., Pasierb, F., and Kojima, H., "Tortuosity and acoustic slow waves", *Phys. Rev. Lett.* 49, 1840–1844, 1982.

7.11.10 Johnson, D. L. and Plona, T. J., "Acoustic slow waves and the consolidation transition", *J. Acoust. Soc. Am.* 72, 556–565, 1982.

7.11.11 Lakes, R. S., Yoon, H. S., and Katz, J. L., "Slow compressional wave propagation in wet human and bovine cortical bone", *Science*, 220, 513–515, 1983.

7.11.12 Chen, C. P. and Lakes, R. S., "Dynamic wave dispersion and loss properties of conventional and negative Poisson's ratio polymeric cellular materials", *Cell. Polym.* 8, 343–359, 1989.

7.12.1   Sturgeon, J. B., "Creep of fibre reinforced thermosetting resins", in *Creep of Engineering Materials*, ed. C. D. Pomeroy, Mechanical Engineering Publications, Ltd., London, 1978.

7.12.2   Bonner, B. P., Lesuer, D. R., Syn, C. K., Brown, A. E., and Sherby, O. D., "Damping measurements for ultra-high carbon steel/brass laminates", in *Damping of Multiphase Inorganic Materials*, ASM Materials Week, Chicago, November 2–5, 1992.

7.12.3   Nelson, D. J. and Hancock, J. W., "Interfacial slip and damping in fibre reinforced composites", *J. Mater. Sci.* 13, 2429–2440, 1978.

7.12.4   Sutherland, H. J., "Dispersion of acoustic waves by fiber reinforced viscoelastic materials", *J. Acoust. Soc. Am.* 57, 870–875, 1975.

7.13.1   Irby, P. L., "Kinetics of mechanical relaxation processes in inorganic glasses", in *Non-crystalline Solids*, ed. V. D. Fréchette, J. Wiley, NY, 1960.

7.13.2   Rekhson, S., "Viscoelasticity of glass", in *Glass: Science and Technology*, Vol. 3, Viscosity and Relaxation, ed. D. R. Uhlmann and N. J. Kreidl, Academic, NY, 1986.

7.13.3   Eisenberg, A. and Tobolsky, A. V., "Viscoelastic properties of amorphous selenium", *J. Polym. Sci.* 61, 483–495, 1962.

7.13.4   Tobolsky, A. V. and Eisenberg, A., *J. Am. Chem. Soc.* 81, 780, 1959.

7.13.5   Tobolsky, A. V., Owen. G. D. T., and Eisenberg, A., *J. Colloid Sci.* 17, 717, 1962.

7.14.1   Ashby, M. F. and Jones, D. R. H., *Engineering Materials*, Pergamon, Oxford, 1980.

7.14.2   Lakes, R. S. and Quackenbush, J., "Viscoelastic behaviour in indium tin alloys over a wide range of frequency and time", *Philos. Mag. Lett.* 74, 227–232, 1996.

7.14.3 Schoeck, G., Bisogni, E., and Shyne, J., "The activation energy of high temperature internal friction", *Acta Metall.* 12, 1466–1468, 1964.

E7.4.1 Leaderman, H., "Elastic and creep properties of filamentous materials and other high polymers", Textile Foundation, National Bureau of Standards, Washington, DC, 1943.

chapter eight

Causal mechanisms

§8.1 Introduction

The treatment of viscoelasticity thus far has dealt with phenomena, measurement of material properties, prediction of response to various load histories, and stress analysis. In this section we consider the physical causes of viscoelastic response. Study of mechanisms is motivated by (i) the desire for scientific understanding; (ii) the desire for the ability to choose or tailor materials with specified viscoelastic properties, and (iii) the utility of viscoelastic measurements as a probe of microphysical processes which are causally linked to viscoelasticity. The conceptually simplest causal mechanisms are developed in detail here. The treatment is intended to be introductory, not exhaustive.

§8.2 Survey of viscoelastic mechanisms

There are many causal mechanisms [1.6.22, 1.6.23, 8.2.1] responsible for viscoelastic response; several of these are as follows. Mechanisms indicated by an asterisk (*) are called fundamental mechanisms because they occur even in an ideal perfect single crystal and they are not removable, even in principle.

1. Atomic and molecular processes
   a. Relaxation by electron viscosity (ultrasonic)
   b. Relaxation by point defect motion
   c. Relaxation by motion of solute atoms (Snoek, body centered cubic [bcc] metal; high temperature)
   d. Relaxation by dislocation motion
      (1) Bordoni, in face centered cubic (fcc) metal; low temperature
      (2) High-temperature "background"
      (3) Amplitude dependent, Granato–Lücke

     e. Relaxation by molecular rearrangement (polymers)
     f. Relaxation by atom pair reorientation (Zener, alloys)
     g. Relaxation by diffusion of atoms (high temperature in metals)
     h. Relaxation via phase transition (sharp dependence on temperature)
2. Coupled field effects
     a. Thermoelastic relaxation* (in any material with thermal expansion)
     b. Relaxation by fluid flow (porous materials with fluid in interstices)
     c. Phonon–phonon interactions* (in all materials)
     d. Electron–phonon interactions (in metals)
     e. Piezoelectric relaxation (in piezoelectric materials)
     f. Magnetoelastic relaxation (in magnetic materials)
3. Heterogeneous relaxation
     a. Relaxation in composite materials (multiphase-structured materials)
     b. Relaxation by grain boundary slip (polycrystalline metals or ceramics)
     c. Relaxation by fluid flow (porous materials with fluid in interstices)

A common feature of causal mechanisms is that strain depends not only on macroscopic stress but also on an "internal variable" associated with microscopic processes or coupling to another field variable. Suppose, as we have done in §2.6 that the strain ε depends on the stress σ and on one internal variable β, then [1.6.23]

$$\varepsilon(\sigma, \beta) = J_0 \sigma + c\beta, \qquad (8.2.1)$$

with J_0 as an initial or unrelaxed compliance and c as a constant. Assume further that the internal variable, when perturbed, approaches its equilibrium value β_e (assumed to be zero for zero stress; $\beta_e = \sigma b$ with b characteristic of the material) according to the first-order equation

$$\frac{d\beta}{dt} = -\frac{(\beta - \beta_e)}{\tau}, \qquad (8.2.2)$$

with τ as the relaxation time. By combining these relations to eliminate β, we obtain, following Nowick and Berry [1.6.23],

$$\varepsilon + \tau \frac{d\varepsilon}{dt} = \tau J_0 \frac{d\sigma}{dt} + \sigma(J_0 + bc). \qquad (8.2.3)$$

This describes behavior of a standard linear solid. The foregoing is relevant to the large class of mechanisms for exponential relaxation. Not all mechanisms have this character.

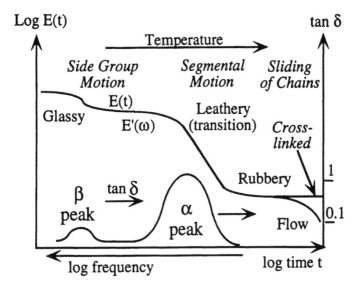

Figure 8.1 Representative viscoelastic behavior due to molecular motions in polymers.

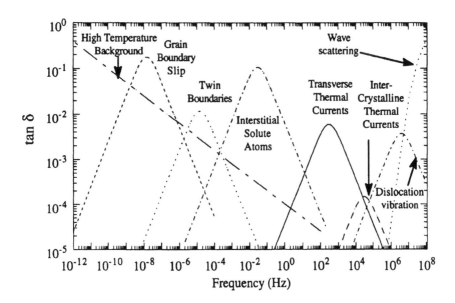

Figure 8.2 Representative viscoelastic behavior due to various processes in metals. In general, not all processes are active.

An overview of representative behavior and causal mechanisms is given for polymers in Fig. 8.1 and for metals in Fig. 8.2.

§8.3 *Thermoelastic relaxation*

In this section let us consider in some detail the relaxation due to stress-induced heat flow, since it is one of the easiest to understand of the viscoelastic mechanisms. It is a coupled-field effect in which strain is related to another field variable, temperature in this case, as well as to stress. The damping depends on the coefficient of thermal expansion which is a *continuum* property having its ultimate origin in a slight nonlinearity of the interatomic force.

§8.3.1 *Thermoelasticity*

As a result of thermoelastic coupling, the stiffness of a material is different depending on whether loading is accomplished slowly (isothermal case) or rapidly (adiabatic case). The difference in stiffness governs the relaxation strength. It is found to depend on the coefficient of thermal expansion.

Consider [8.3.1] a unit volume of *elastic* material with strain ε_{ij} and entropy S as dependent variables, and stress σ_{ij} and absolute temperature T as independent variables:

$$d\varepsilon_{ij} = \frac{\partial \varepsilon_{ij}}{\partial \sigma_{kl}}\bigg|_{T} d\sigma_{kl} + \frac{\partial \varepsilon_{ij}}{\partial T}\bigg|_{\sigma} dT, \tag{8.3.1}$$

$$dS = \frac{\partial S}{\partial \sigma_{kl}}\bigg|_{T} d\sigma_{kl} + \frac{\partial S}{\partial T}\bigg|_{\sigma} dT, \tag{8.3.2}$$

in which $(\partial \varepsilon_{ij}/\partial \sigma_{kl})$ represents elasticity, $(\partial \varepsilon_{ij}/\partial T)$ represents thermal expansion, $(\partial S/\partial \sigma_{kl})$ represents the piezocaloric effect in which heat is generated in response to stress, and $(\partial S/\partial T)$ represents heat capacity. In linear materials, the elasticity equations allowing for temperature changes become [8.3.1]

$$\varepsilon_{ij} = S^{T}_{ijkl}\sigma_{kl} + \alpha_{ij}\Delta T, \tag{8.3.3}$$

$$\Delta S = \alpha_{ij}\sigma_{ij} + \frac{C^{\sigma}}{T}\Delta T, \tag{8.3.4}$$

in which S^{T}_{ijkl} is the elastic compliance tensor at constant temperature, α_{ij} is the thermal expansion tensor, and C^{σ} is the heat capacity per unit volume at constant stress. In the isotropic case, $\alpha_{ij} = \alpha \delta_{ij}$ and Eq. 8.3.3 is equivalent to the following elementary form [5.2.1] of Hooke's law, in engineering notation, allowing thermal expansion:

$$\varepsilon_{xx} = \frac{1}{E}\left\{\sigma_{xx} - v\sigma_{yy} - v\sigma_{zz}\right\} + \alpha\Delta T,$$

$$\varepsilon_{yy} = \frac{1}{E}\left\{\sigma_{yy} - v\sigma_{xx} - v\sigma_{zz}\right\} + \alpha\Delta T,$$

$$\varepsilon_{zz} = \frac{1}{E}\left\{\sigma_{zz} - v\sigma_{xx} - v\sigma_{yy}\right\} + \alpha\Delta T.$$

The compliance S^T_{ijkl} is the isothermal compliance and is the compliance actually measured in an elastic material under deformation which is slow enough that any heat generated via the piezocaloric effect has time to flow, equalizing the temperature. The elastic compliance is different under deformation which is sufficiently fast that this heat has no time to diffuse (adiabatic condition, dS = 0). To calculate the adiabatic compliance, set dS = 0 in Eq. 8.3.2 and combine Eqs. 8.3.1 and 8.3.2 to eliminate dT.

$$d\varepsilon_{ij} = \left.\frac{\partial\varepsilon_{ij}}{\partial\sigma_{kl}}\right|_T d\sigma_{kl} - \frac{\left[\left.\frac{\partial\varepsilon_{ij}}{\partial T}\right|_\sigma \left.\frac{\partial S}{\partial\sigma_{kl}}\right|_T\right]}{\left.\frac{\partial S}{\partial T}\right|_\sigma} d\sigma_{kl}. \qquad (8.3.5)$$

Now $(\partial\varepsilon_{ij}/\partial T)_\sigma = (\partial S/\partial\sigma_{kl})_T$ since these derivatives can be expressed in terms of a thermodynamic potential function by virtue of the first and second laws of thermodynamics [8.3.1]. Divide both sides of Eq. 8.3.5 by $d\sigma_{kl}$ to obtain the adiabatic compliance $(\partial\varepsilon_{ij}/\partial\sigma_{kl})_S$,

$$\left.\frac{\partial\varepsilon_{ij}}{\partial\sigma_{kl}}\right|_S - \left.\frac{\partial\varepsilon_{ij}}{\partial\sigma_{kl}}\right|_T = -\left.\frac{\partial\varepsilon_{ij}}{\partial T}\right|_\sigma \left.\frac{\partial\varepsilon_{kl}}{\partial T}\right|_\sigma \left.\frac{\partial T}{\partial S}\right|_\sigma. \qquad (8.3.6)$$

If the material is linear, Eq. 8.3.6 becomes

$$S^S_{ijkl} - S^T_{ijkl} = -\alpha_{ij}\alpha_{kl}\frac{T}{C^\sigma}, \qquad (8.3.7)$$

giving a relaxation strength

$$\Delta_{ijkl} = \frac{S^T_{ijkl} - S^S_{ijkl}}{S^S_{ijkl}} = \frac{\alpha_{ij}\alpha_{kl}}{S^S_{ijkl}}\frac{T}{C^\sigma}. \qquad (8.3.8)$$

The adiabatic compliance S^S_{ijkl} differs from the isothermal compliance S^T_{ijkl}, and the difference depends on the thermal expansion and on the heat capacity. The heat capacity and absolute temperature are always positive. If, as is usual, the thermal expansion coefficient is also positive, all the adiabatic compliances are smaller and hence the adiabatic stiffnesses are larger than the isothermal ones.

The above deals with elastic behavior only. However, there is a difference in stiffness between fast and slow processes. The adiabatic compliance is determined in fast processes since in a material deformed rapidly there is insufficient time for heat generated by stress to flow. The isothermal compliance is determined in slow processes such that heat generated by deformation has sufficient time to flow, equalizing the temperature throughout the material. We know from the Kramers–Kronig relations discussed in §3.3 that such dispersion of the stiffness is always associated with viscoelastic loss.

§8.3.2 Thermoelastic relaxation kinetics

In this section we consider dynamics of thermoelastic relaxation. Dissipation of mechanical energy in a cyclic load history with heat flow is illustrated in Fig. 8.3. The history consists of three portions. First, the object is loaded slowly at constant temperature (isothermally). Then it is unloaded adiabatically, too rapidly for heat flow to occur, with the slope of the stress–strain curve as the adiabatic modulus which differs from the isothermal modulus as shown in the above section. Next the object is held at constant stress, and it exchanges heat with the environment. Thermal expansion occurs, so the strain changes. Mechanical energy is dissipated in this cycle, since there is a nonzero area enclosed by the load history. The loss tangent developed below refers to sinusoidal loading which differs from the above; however, the general principles are similar.

The differential equation for relaxation due to thermoelastic coupling is derived, adapted from Zener [1.6.22, 8.3.2, 8.3.3]. The thermoelastic Eq. 8.3.3 is specialized to one dimension, with the isothermal compliance tensor replaced by a number: $S^T_{ijkl} = J_T$.

$$\varepsilon = J_T\sigma + \alpha\Delta T. \tag{8.3.9}$$

Thermal diffusion is governed by the following, with τ as the diffusion time. This is homogeneous thermal diffusion, from the specimen to its environment.

$$\left.\frac{d\Delta T}{dt}\right|_{\text{diffusion}} = -\frac{\Delta T}{\tau}. \tag{8.3.10}$$

Under adiabatic conditions, an increase in length results in a decrease in temperature, for $\alpha > 0$.

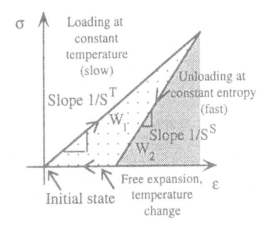

Figure 8.3 Cyclic stress–strain (σ–ε) history with temperature changes allowed. Conversion of mechanical energy into thermal energy via the thermal expansion and piezocaloric effects. Compare with a similar cyclic history in piezoelectric materials in Fig. 8.12. Energy densities W are shown as shaded areas. Material is loaded slowly at constant temperature, then unloaded rapidly at constant entropy (adiabatically), which causes a temperature change, and finally allowed to thermally equilibrate at zero stress. Compliances S are referred to as S or J in the text.

$$\left.\frac{d\Delta T}{dt}\right|_{adiabatic} = -\gamma\frac{d\varepsilon}{dt}, \tag{8.3.11}$$

with

$$\gamma = \partial T/\partial\varepsilon|_{ad}.$$

By combining Eqs. 8.3.10 and 8.3.11, since there are two independent sources for rate of temperature change,

$$\frac{d\Delta T}{dt} = -\frac{\Delta T}{\tau} - \gamma\frac{d\varepsilon}{dt}. \tag{8.3.12}$$

By eliminating ΔT from Eqs. 8.3.9 and 8.3.12, and passing to the frequency domain by substituting a sinusoidal signal of angular frequency ω in the differential equation,

$$i\omega\frac{1}{\alpha}\left(\varepsilon - J_T\sigma\right) = -\frac{1}{\alpha\tau}\left(\varepsilon - J_T\sigma\right) - i\omega\gamma\varepsilon. \tag{8.3.13}$$

Calculate the ratio of stress to strain to obtain the dynamic modulus,

$$E^* = \frac{1}{J_T} \frac{1 + i\omega\tau\alpha\gamma + \omega^2\tau^2\alpha\gamma + \omega^2\tau^2}{1 + \omega^2\tau^2}. \tag{8.3.14}$$

But

$$\tan\delta_E = \frac{\mathrm{Im}\{E^*\}}{\mathrm{Re}\{E^*\}},$$

so the mechanical damping due to thermoelastic effects is

$$\tan\delta_E = \frac{\omega\tau\gamma\alpha}{1 + \omega^2\tau^2(1 + \alpha\gamma)}. \tag{8.3.15}$$

The differential Eq. 8.3.13 is that of a standard linear solid. Recall from §2.6 that the relaxation strength is defined as the change in stiffness during relaxation divided by the stiffness at long time. Since $J_T = J(\infty)$ and $J_S = J(0)$, the relaxation strength Δ is, in terms of the adiabatic compliance J_S (at constant entropy S),

$$J_S = \frac{J_T}{(1 + \alpha\gamma)}. \tag{8.3.16}$$

$$\Delta = \frac{J_T - J_S}{J_S}. \tag{8.3.17}$$

This last expression is the one-dimensional equivalent of the first part of Eq. 8.3.8.

For a Debye peak (due to a single relaxation time process), assuming a small relaxation strength,

$$\tan\delta = \Delta\frac{v/v_0}{1 + (v/v_0)^2} \tag{8.3.18}$$

so that the peak loss at the characteristic frequency v_0 (associated with a relaxation process, not resonance) is

$$\tan\delta_{peak} = \Delta/2.$$

The relaxation strength is given by $\Delta = \alpha\gamma$. By thermodynamic arguments, γ can be eliminated.

Table 8.1   Properties of Some Materials: Young's Modulus E, Density ρ,
Thermal Conductivity k, Heat Capacity per Unit Mass c,
Thermal Expansion α at Room Temperature

Solid	E (GPa)	$\rho\left[\dfrac{10^3\,\mathrm{kg}}{\mathrm{m}^3}\right]$	$k\left[\dfrac{\mathrm{J}}{\mathrm{s}\cdot\mathrm{m}\cdot\mathrm{K}}\right]$	$c\left[\dfrac{\mathrm{J}}{\mathrm{kg}\cdot\mathrm{K}}\right]$	$\alpha\left[\dfrac{10^{-6}}{\mathrm{K}}\right]$	$\Delta = E\alpha^2 T/c\rho$
Al	70	2.7	222	900	23.6	0.0048
Cu	110	8.96	394	380	16.5	0.0026
In	11	7.51	23.9	240	33.0	0.0020
Fe	197	7.87	75.4	460	11.8	0.0023
Mg	44	1.74	154	1030	27.1	0.0054
Ni	207	8.9	92	440	13.3	0.0028
Sn	43	7.3	628	230	23.0	0.0041
Ti	116	4.51	38.9	519	8.41	0.0011
Zn	103	7.13	113	383	39.7	0.0178
Al_2O_3	350	3.8	29.3	840	9.0	0.0027
C	379	2.25	23.9	691	−0.90	6×10^{-5}
SiC	460	3.26	90	1330	4.30	0.0006
TiC	350	4.5	30.3	840	7.0	0.0009

Note: Calculated relaxation strength Δ for thermoelastic damping.

Adapted from Milligan and Kinra [8.3.5].

$$\Delta = \frac{\alpha^2 T}{C_v J^S},$$

(8.3.19)

with T as the absolute temperature and C_v as the heat capacity at constant volume. This is the one-dimensional version of Eq. 8.3.8. In treatments of thermoelastic relaxation, the thermal expansion coefficient itself has been considered not to relax.

For aluminum the relaxation strength is $\Delta = 0.0046$, for iron $\Delta = 0.0024$, for magnesium $\Delta = 0.005$, and for most metals $\Delta < 0.01$, so the loss due to thermoelastic effects is usually small [8.3.3]. However, the total loss in most metals used for structural purposes is also usually small. Therefore, in some frequency ranges the thermoelastic effect can account for most of the loss in common structural metals, as has been demonstrated experimentally [8.3.4]. Table 8.1 shows a longer list of thermoelastic properties of some representative materials after Milligan and Kinra [8.3.5].

Thermoelastic relaxation is present whenever there is heterogeneity of dilatational stress [8.3.3]. The heterogeneity can arise in the type of vibration, as in bending vibration of reeds (Fig. 8.4). Heterogeneity of stress also is present if the material has cavities, discrete phases, or anisotropic crystallites with random orientation. There is also a homogeneous thermoelastic relaxation governed by heat flow between the specimen and the environment.

For a reed of thickness d, vibrating in bending [8.3.2, 8.3.3], the critical frequency v_0 is

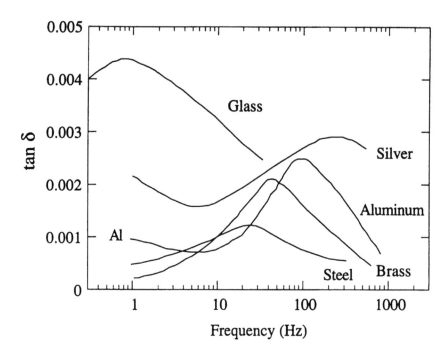

Figure 8.4 Tan δ in several materials in flexure experiments by Bennewitz and
Rötger [8.3.15], adapted from Zener [1.6.22]. The peaks correspond to the predicted
peaks for thermoelastic damping. The frequency for maximum damping depends
on the specimen thickness.

$$v_0 = \frac{\pi}{2} D d^{-2}. \tag{8.3.20}$$

For a vibrating circular rod of radius r,

$$v_0 = 0.539 \, D r^{-2}, \tag{8.3.21}$$

with D as the thermal diffusion coefficient, $D = k/C_v$, with k as the thermal
conductivity and C_v as the heat capacity per unit volume. More sophisticated
analysis of reed vibration discloses a set of characteristic frequencies, but
most of the damping is in the lowest (first) mode given above; the second
mode has about 0.012 the damping of the first. The one-dimensional analysis
of Zener has been extended to three dimensions by Alblas [8.3.6, 8.3.7]. The
specimen *shape* will affect the damping since specimen boundaries influence
the rate of heat flow.

    As for cavities [8.3.3], the heterogeneous stress around them gives rise
to a thermoelastic damping of small magnitude, over a distribution of frequency. For cavities of volume fraction v, in a material with Poisson's ratio

v, the fraction R of strain energy associated with fluctuations in dilatation multiplies the available relaxation strength and is given by

$$R = \frac{10v}{1764} \left\{ \frac{(1-2v)(1+v)}{\left(1-\frac{5}{7}v\right)^2} \right\}. \tag{8.3.22}$$

For representative Poisson's ratios, the bracketed expression is approximately 1. The damping due to thermoelastic damping associated with cavities is small, since the volume fraction v of cavities must be less than 1 and is often substantially less than 1. Cracks [8.3.8] also give rise to thermoelastic damping due to heterogeneous stress.

As for thermal currents between randomly oriented anisotropic crystals in a polycrystalline material, the corresponding values of R have been calculated from crystal anisotropy of several cubic crystals. Results are 0.031 for Cu, 0.0009 for Al, 0.022 for Fe, and 0.065 for Pb; therefore, the associated damping is small. Even so, damping due to this source can comprise most of the total damping in some materials such as brass (69%Cu, 31%Zn), as shown in Fig. 8.5.

Thermoelastic damping in *composite* materials arises due to the heterogeneity of the thermal and mechanical properties of such materials, leading to heat flow between constituents and hence mechanical energy dissipation. The damping depends on the specific phase geometry as well as the constituents involved. The reason is that damping depends on heterogeneity of dilatational stress, and on the nature of the boundary value problem under consideration. Composites modeled as one-dimensional inclusions [8.3.5], laminates [8.3.9], laminates with perfect and imperfect thermal interfaces [8.3.10], and composites with particulate inclusions, have been studied via the second law of thermodynamics. Predicted damping for one-dimensional inclusions is proportional to $(\alpha_1/\rho_1 c_1 - \alpha_2/\rho_2 c_2)^2$ with ρ as density and c as specific heat per unit mass. Selected results are shown in Table 8.2.

Thermoelastic damping has been examined in connection with the theory of thermodynamics [8.3.12–8.3.14]. Some minor corrections to the bending analysis of Zener were given as a result [8.3.12].

§8.4   Relaxation by fluid motion

In a porous elastic material under stress, motion of fluid (such as air or water) within the pores generates viscous drag which gives rise to relaxation. This is a coupled-field type of mechanism in which the stress and strain in the solid phase are coupled to the fluid pressure and fluid volume change. Biot and co-workers [8.4.1–8.4.6] developed a substantial understanding of this process. The relations between stress σ and strain ε including the effect of

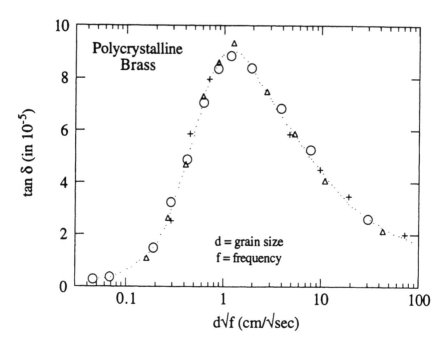

Figure 8.5 Tan δ in polycrystalline brass, adapted from Randall et al. [8.3.16]. The damping arises from intercrystalline thermal currents. Frequency is denoted by f. Symbols: o, f = 6 kHz; Δ, f = 12 kHz; and +, f = 36 kHz.

Table 8.2   Predicted Peak Thermoelastic tan δ for Inclusions in a Matrix

One-dimensional inclusion[a]	Matrix					
	Mg	Al	In	Sn	Zn	Fe
Al_2O_3	0.0077	0.0031	0.0047	0.0093	0.0073	1.6×10^{-5}
SiC	0.0091	0.0049	0.0038	0.014	0.0089	0.0002
Graphite	0.0091	0.0046	0.0063	0.010	0.0087	0.0005

Laminate, uniform stress inclusion[b]	Matrix	
	Mg	Al
Al_2O_3	8×10^{-5}	0.0004
SiC	0.0012	0.0008

Spherical particulate inclusion[c]	Matrix	
	Mg	Al
Al_2O_3	0.0035	0.0018
SiC	0.0067	0.0045

[a] Adapted from Milligan and Kinra [8.3.5].
[b] Bishop and Kinra [8.3.10].
[c] Bishop and Kinra [8.3.11].

fluid pressure P may be written [8.4.1] as follows as a generalization of the elementary isotropic form for Hooke's law. The solid phase is in this treatment assumed to be elastic.

$$\varepsilon_{xx} = \frac{1}{E}\left\{\sigma_{xx} - v\sigma_{yy} - v\sigma_{zz}\right\} + \frac{1}{3H}P,$$

$$\varepsilon_{yy} = \frac{1}{E}\left\{\sigma_{yy} - v\sigma_{xx} - v\sigma_{zz}\right\} + \frac{1}{3H}P,$$

$$\varepsilon_{zz} = \frac{1}{E}\left\{\sigma_{zz} - v\sigma_{yy} - v\sigma_{xx}\right\} + \frac{1}{3H}P,$$

$$2\varepsilon_{xy} = \frac{\sigma_{xy}}{G}, \ 2\varepsilon_{xz} = \frac{\sigma_{xz}}{G}, \ 2\varepsilon_{yz} = \frac{\sigma_{yz}}{G}.$$

(8.4.1)

Here E is Young's modulus, G is the shear modulus, v is Poisson's ratio, and H is a physical constant with dimensions of stress. H^{-1} is a measure of the compressibility of the porous solid for a change in fluid pressure. One may write the above as follows in the tensorial formulation, with S_{ijkl} as the compliance tensor.

$$\varepsilon_{ij} = S_{ijkl}\sigma_{kl} + \frac{1}{3H}\delta_{ij}P. \tag{8.4.2}$$

The volume change Θ of fluid content is given, with σ_{kk} as the trace of the stress tensor, by

$$\Theta = \frac{1}{3H}\left\{\sigma_{kk}\right\} + \frac{1}{R}P. \tag{8.4.3}$$

R^{-1} is a physical constant representing the change in fluid content for a given change in fluid pressure. These have the same form as Eqs. 8.3.3 and 8.3.4 which describe thermoelastic coupling. Moreover, the treatment of damping due to fluid–solid interaction is analogous to the treatment of damping due to thermoelastic coupling, as indicated schematically in Fig. 8.6.

The stress–strain relations can also be written in the modulus formulation as

$$\sigma_{xx} = 2G\left\{\varepsilon_{xx} + \frac{v}{1-2v}\left(\varepsilon_{xx} + \varepsilon_{yy} + \varepsilon_{zz}\right)\right\} - AP, \tag{8.4.4}$$

or, in the Lamé formulation for isotropic materials,

$$\sigma_{ij} = 2G\varepsilon_{ij} + \left(\lambda + A^2Q\right)\varepsilon_{kk}\delta_{ij} - AQ\zeta\delta_{ij}, \tag{8.4.5}$$

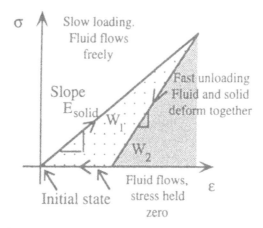

Figure 8.6 Cyclic stress–strain (σ–ε) history with fluid flow, showing energy dissipation. Compare with a similar cyclic history in thermoelastic materials in Fig. 8.3 and in piezoelectric materials in Fig. 8.12.

with ζ as the fluid content, and

$$A = \frac{2(1+v)}{3(1-2v)}\frac{G}{H}.$$ (8.4.6)

Since the bulk modulus B of the solid skeleton without fluid is

$$B = \frac{2G(1+v)}{3(1-2v)},$$ (8.4.7)

one can write

$$A = \frac{B}{H}.$$ (8.4.8)

The coefficient Q^{-1} is a measure of how much fluid can be forced into the material if it is at constant volume.

$$\frac{1}{Q} = \frac{1}{R} - \frac{A}{H}.$$ (8.4.9)

The above coefficients have been related [8.4.2] to the following observable parameters. The *porosity* f is the volume fraction of void space. The *jacketed compressibility* κ is the ratio of strain in the solid to the fluid pressure applied externally, with internal fluid pressure held constant. To measure κ, a specimen is surrounded by a thin impermeable jacket and fluid pressure is then applied. So the compressibility κ is the inverse of the bulk modulus B, of a dry specimen. The *unjacketed compressibility* δ is the ratio of strain in the solid to fluid pressure when the fluid completely penetrates the pores. The coefficient γ of fluid content is given by an experiment in which fluid is injected into a fluid-filled chamber with and without the porous specimen. The difference in volume ΔV is given by ΔV = δ + γ - c, with c as the fluid compressibility. If the porous material is isotropic, macroscopically homogeneous, and fully saturated, then

$$\gamma = f(c - \delta). \tag{8.4.10}$$

The interrelations are as follows [8.4.2, 8.4.6]:

$$Q = \frac{f\left(1 - f - \dfrac{\delta}{\kappa}\right)}{\gamma + \delta - \dfrac{\delta^2}{\kappa}}, \tag{8.4.11}$$

$$R = \frac{f^2}{\gamma + \delta - \dfrac{\delta^2}{\kappa}}, \tag{8.4.12}$$

$$\delta = (1 - A)\kappa, \tag{8.4.13}$$

$$H^{-1} = \kappa - \delta. \tag{8.4.14}$$

The treatment is entirely elastic thus far. Time-dependent behavior is demonstrated by considering the flow of fluid in response to stress, as follows.

The flow rate vector **F** of fluid in response to a gradient of pressure P is governed by Darcy's law,

$$F = -k\nabla P, \tag{8.4.15}$$

with k as the permeability.

Suppose the fluid content Θ is related to the flow by

$$\frac{\partial\Theta}{\partial t} = -\text{div}F. \tag{8.4.16}$$

Then by combining with the generalized three-dimensional Hooke's law in Eq. 8.4.1,

$$k\nabla^2 P = A \frac{\partial}{\partial t}\left\{\varepsilon_{xx} + \varepsilon_{yy} + \varepsilon_{zz}\right\} + \frac{1}{Q}\frac{\partial}{\partial t}P. \tag{8.4.17}$$

Relaxation due to fluid flow depends on the geometry of the object and the surface boundary conditions. As a particular case [8.4.1], consider a slab of thickness h of material confined on the edges, with fluid free to escape from one flat surface. To that end, the above poroelastic equations (in the modulus formulation) may be converted with the aid of the equilibrium conditions $\dfrac{\partial \sigma_{ij}}{\partial x_j} = 0$, into the following differential equation involving the displacement field u_j:

$$G\nabla^2 u_j + \frac{G}{1-2v}\frac{\partial \varepsilon_{kk}}{\partial x_j} - A\frac{\partial P}{\partial x_j} = 0, \tag{8.4.18}$$

which, with Eq. 8.4.17, may be specialized to one spatial dimension,

$$\frac{1}{a}\frac{\partial^2 u_z}{\partial z^2} - A\frac{\partial u_z}{\partial z} = 0, \tag{8.4.19}$$

$$k\frac{\partial^2 P}{\partial z^2} = A\frac{\partial^2 u_z}{\partial z \partial t} + \frac{1}{Q}\frac{\partial P}{\partial t}, \tag{8.4.20}$$

with a as the final compressibility

$$a = \frac{1-2v}{2G(1-v)}. \tag{8.4.21}$$

This is the inverse of C_{1111} for the solid phase. Physically, C_{1111} is the constrained modulus for the solid phase under compression, with the Poisson effect laterally constrained; see Example 5.9.

An initial compressibility is defined (8.4.1),

$$a_i = \frac{a}{1 + A^2 aQ}. \tag{8.4.22}$$

So the relaxation strength, from the following definition (see Eq. 2.6.29):

$$\Delta = \frac{J(\infty) - J(0)}{J(0)}, \tag{8.4.23}$$

with the compliance J as a compressibility, is

$$\Delta = A^2 aQ. \tag{8.4.24}$$

If a step load of magnitude p_0 is applied, the displacement as a function of time t is

$$u_z = hp_0\left\{a - \frac{8}{\pi^2}(a - a_i)\sum_{n=0}^{\infty}\frac{1}{(2n+1)^2}\exp\left\{-(2n+1)^2\frac{t}{\tau}\right\}\right\}, \tag{8.4.25}$$

with

$$\tau = \frac{4h^2}{\pi^2 k}\left\{A^2 a + \frac{1}{Q}\right\}. \tag{8.4.26}$$

The time constant τ for creep depends inversely as the permeability k and increases as the square of the slab thickness h. The first term in the summation is much larger than subsequent terms, so the creep is approximately of single exponential form. Therefore, the corresponding tan δ response is dominated by a Debye peak. In this example, creep results from squeezing of fluid through a surface of a compressed slab. Results will differ for other boundary conditions. For example, a uniform specimen in torsion experiences no volume change, hence no driving force for fluid flow and therefore no creep due to this mechanism. This example is analogous to the case of thermoelastic damping due to exchange of heat with the environment, considered in §8.3.2. If the porous medium is heterogeneous on a scale larger than that of the porosity, there can be stress-induced flow of fluid between regions in the material, analogous to the intercrystalline thermal currents associated with thermoelastic coupling.

The Biot equations assume a particularly simple form for low-density polymer foams for which $f \approx 1$ and the unjacketed compressibility δ (of the solid phase) is much less than the jacketed compressibility κ (of the foam). Then the relaxation strength for compression of a constrained layer is expressed in terms of the bulk moduli,

$$\Delta = \frac{B_{fluid}}{B_{foam}}\frac{1}{3}\frac{1+v}{1-v}. \tag{8.4.27}$$

Poisson's ratio v of the foam appears since the constrained layer compression is not a pure bulk deformation.

Introduction of anisotropy [8.4.2, 8.4.7] in the analysis permits application to problems involving bedded rock or cartilage. More sophisticated theory incorporates viscoelasticity of the solid phase, bubbles in the fluid, and better modeling of the porous structure [8.4.5, 8.4.6]. As for elastic waves, there is the possibility of dynamic relative motion between the solid and fluid phases, and this results in two kinds of compressional waves [8.4.3, 8.4.4] with different speed and different attenuation. Such waves have been observed in synthetic porous media [8.4.8, 8.4.9] and in bone [8.4.10].

Fluid flow in porous materials is important in civil engineering in the understanding consolidation of soils [8.4.1, 8.4.11]. Observe that the time scale of the creep or relaxation, expressed as the time constant τ, increases as the square of the specimen size. Therefore, the coupled-field equations are used directly in such problems. Fluid flow also is important in determining the mechanical and transport properties of biological materials such as cartilage, muscle, and bone, as well as of synthetic polymer foams. The size dependence of the time constant was used to infer an important role for fluid flow in the viscoelasticity of muscle, as described in §7.10.

Relaxation due to fluid flow has been studied [8.4.12, 8.4.13] in the context of polymer foams. The analysis, in common with that of Biot, predicts a peak in the loss tangent at a frequency which depends on the foam slab thickness. For materials which can be idealized as an array of tubes of diameter d, the permeability is

$$K = \frac{d^2}{32}.$$

The predicted peak in the loss tangent and transition from low to high stiffness in E' covers little more than one decade of frequency scale. Large loss tangents of unity or greater are possible via this mechanism. The analysis can be improved, e.g., by allowing for the compressibility of the fluid as required to properly treat gases such as air. The theoretical formulation was used successfully in modeling a damping peak of magnitude 0.6 due to air flow in a polyurethane foam.

In some polymer foams used for wheelchair cushions, bed pads, and spacecraft seats, the foam is made viscoelastic and the orifice between cells in the foam is made narrow [8.4.14]. In large blocks of foam under compression, the slow escape of air contributes significantly to the viscoelastic properties. In foams, flow of air can give rise to large values of tan δ.

§8.5   Relaxation in magnetic media

Magnetoelastic relaxation is also a coupled-field process. Loss tangents from various magnetoelastic processes can be large; exceeding 0.1, as discussed in §7.4. A magnetic field which changes with time as a result of a varying strain gives rise to circulating electric currents in a conducting medium; these are called eddy currents [1.6.23, 8.5.1]. Eddy currents may occupy the entire volume of a magnetized specimen (macroeddy currents) or may occur locally as a result of changes in the magnetization or orientation of domains within the material (microeddy currents). All such electric currents (in a material of finite electric conductivity) involve the dissipation of energy. The damping due to macroeddy currents arises from energy dissipation from electric currents caused by changes in magnetization of the entire specimen. It depends on stress amplitude as well as on frequency and applied magnetic field. Microeddy currents can give rise to damping even if the metal as a whole is not magnetized. Motion of domain walls in the material gives rise to a change in magnetic moment. This contribution depends on frequency but not on amplitude of applied stress. The above effects are linear ones. There is also hysteresis damping, associated with magnetic hysteresis, which is inherently nonlinear. Such damping depends on stress amplitude but not on frequency. Magnetoelastic mechanisms were explored experimentally in a study of Armco iron [8.5.2] at 22.5 kHz, as a function of stress amplitude, applied magnetic field, permanent deformation, and time following vibration at high amplitude. Damping increases with vibration strain but decreases with permanent deformation. The hysteresis damping appears to be governed by bowing of the domain walls, a dynamic process which can occur at high frequency and is linked to internal stress.

§8.6   Relaxation by molecular rearrangement

This mechanism is the most relevant to polymers [1.6.18, 8.6.1], which consist of macromolecules; refer to observed behavior of polymers given in §7.3. The region of short times and low temperatures corresponds to the *glassy* region of behavior of amorphous polymers (Fig. 8.1). The polymer backbones have little freedom of movement; nevertheless, peaks in tan δ may be observed. The peak at the lowest temperature (or highest frequency), called the γ peak, is thought to be a result of flexing or twisting of segments of the main polymer chain. At a higher temperature, a β peak may be observed. Viscoelasticity in this region arises from limited local molecular motion of side groups. Small peaks superposed on a nearly constant loss tangent in this region are called secondary maxima. In the *transition* region, large relaxation of a factor 100 to 10,000 occurs in the stiffness as a result of configurational rearrangement of the polymer chains. In the transition region the loss tangent exhibits a very large peak, called the α peak, of magnitude from 1 to 10. Cross-links play a minor role in this region. In the *plateau* region of rubbery behavior, associated with long times or high temperature, the behavior is governed by entanglements of the long-chain molecules. Rearrangements on this scale occur comparatively slowly. The molecular weight of the polymer chains is important in this region as seen in the plots shown in §7.3. The asymptotic or equilibrium modulus in the *terminal* region depends on the density of cross-links and on the molecular weight. Polymers without cross-links behave as liquids for sufficiently long times, since the molecular chains can then slide over each other given enough time or thermal activation by high temperature.

In recent studies of molecular mechanisms in polymers, regions of elevated energy states or "kinks" in long-chain molecules are considered to [8.6.2, 8.6.3] oscillate at frequency v_0 in Brownian motion and can rearrange to overcome an energy barrier, of height Δ_e. The probability of movement depends on temperature so that the transition rate is $\lambda_0 = e^{-\Delta_e/kT}$. Power law relaxation and fractional derivative representations can arise from this analysis [8.6.3].

The occurrence of crystallinity in polymers requires a sufficiently regular chemical structure to allow chain molecules to pack regularly in a lattice [8.6.4]. The lateral forces between the oriented molecular chains are weak, facilitating torsional oscillation of the molecules [1.6.18]. Macroscopic deformation alters the distribution of oscillations, following a time lag. This gives rise to viscoelastic loss. There are domains of both ordered and amorphous material in crystalline polymers. Moreover, the viscoelastic properties within lamellae and at lamellar interfaces can differ. The polymer can therefore be regarded as a composite. Polyethylene and polytetrafluoroethylene are examples of such semicrystalline polymers. Overall, one major effect of crystallinity in polymers is to broaden the glass transition and reduce the maximum loss tangent.

Power law and other nonexponential relaxation relations have been attributed to quantum mechanical "hopping" processes [8.6.5–8.6.7]. Relaxation following the "stretched exponential" (discussed in §2.6.5) is thought to arise from interactions among a hierarchy of energy levels [8.6.8].

Many amorphous polymers are found to be thermorheologically simple; that is, a change in temperature is equivalent to a shift of the viscoelastic property curves along a log time or log frequency axis. This behavior implies that in similar deformations at different temperatures, the same sequence of molecular events (leading to viscoelasticity) occurs; only the speed changes with temperature [8.6.9]. The Williams, Landel, and Ferry (WLF) equation (Eq. 2.7.4), which describes the temperature dependence, may be obtained from concepts of free volume [1.6.21, 2.7.3]. The plot of volume per unit mass vs. temperature for a typical amorphous polymer exhibits a discontinuity in *slope* at the glass transition temperature T_g, as shown in Fig. 8.7. The additional volume within the polymer above the glass transition is considered to be free volume, or voids on the molecular scale. The fractional free volume $f = v_f/v$ may be written as follows:

$$f = f_g + \alpha_f\left(T - T_g\right), \qquad (8.6.1)$$

in which T is temperature, f_g is the fractional free volume at the glass transition temperature T_g, and α_f is the coefficient of expansion of the free volume. Suppose that each relaxation mode has a time constant $\tau = \eta/E$, in which the viscosity η gives rise to a shift factor

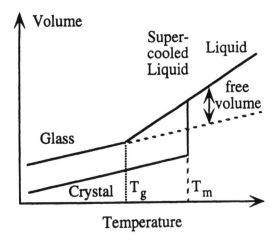

Figure 8.7 Glass transition T_g and melting T_m in polymers.

$$a_T(T) = \frac{\eta_T}{\eta_{T_g}}. \tag{8.6.2}$$

Suppose, moreover, that the viscosity is governed by Doolittle's equation for the viscosity of monomeric liquids, supported by experiment,

$$\eta = a\exp\{bv/v_f\}, \tag{8.6.3}$$

with a and b as constants. By combining Eqs. 8.6.2 and 8.6.3,

$$\ln a_T(T) = b\left\{\frac{1}{f} - \frac{1}{f_g}\right\}. \tag{8.6.4}$$

By incorporating Eq. 8.6.1,

$$\log a_T = \frac{\dfrac{b}{2.303f_g}\left(T - T_g\right)}{\dfrac{f_g}{\alpha_f} + \left(T - T_g\right)}, \tag{8.6.5}$$

which is the WLF equation.

The glass transition temperature T_g is fixed for a given polymer and frequency; nevertheless, one can control T_g by adding various constituents, e.g., a plasticizer may be added to reduce T_g. Plasticizers contain small molecules which penetrate the polymer and allow molecular chain segments greater mobility. For example [8.6.10], polyvinyl chloride (PVC) is a hard plastic with $T_g \approx 80°C$; it is commonly used for water pipes. Plasticized PVC is flexible, has T_g just below room temperature, and is used for shower cushions and seat covers. Particulate inclusions or "fillers" tend to raise T_g slightly since free volume is reduced in the vicinity of the interface. This effect falls off with distance from the interface; therefore, the transition is broadened as well as elevated in temperature. Broad transitions can be achieved by combining immiscible polymers [8.6.10] with different transition temperatures. If the free energy of mixing is close to zero, the heterogeneous structure is on a very small distance scale on the order 10 nm. In that case much of the material is very close to an interface, and the result is a single broad transition rather than two individual transitions.

Materials of biological origin generally contain macromolecules which give rise to relaxations via mechanisms of the types observed in synthetic polymers. Several examples are given by Wert [8.2.1]. Biological materials also tend to have a complex composite structure (§9.5) in which the macro-molecules are highly ordered.

Similarities in the shape (not the magnitude) of the relaxation functions of metals and polymers have been interpreted by a cooperative theory [8.6.11] in which the solid is made up of structural units which behave as two level systems. Structural units interact by exchanging phonons (quanta of sound) giving rise to cooperative behavior. Molecular dynamics simulations via this theory, of stress relaxation, compare well with experiments.

§8.7   Relaxation by interface motion

§8.7.1   Grain boundary slip in metals

In polycrystalline metals, the boundaries between crystals or grains (Fig. 8.8) exhibit viscous-like behavior under prolonged load [1.6.22, 1.6.23]. The material in the grain boundaries is amorphous and appears to be associated with viscous flow. Grain boundary slip is responsible for significant creep in metals at high temperatures and at long times. The creep or relaxation time varies directly with the grain size. At sufficiently high temperature (a significant fraction of the melting temperature), slip can occur sufficiently rapidly that it manifests itself as a peak in the loss tangent [8.7.1]; the loss at the peak can be on the order of 0.1; see §7.4. The effect of grain boundary slip has been studied [8.7.1] in comparisons between polycrystalline metal and the same metal in single crystal form (Fig. 7.9). If the grain size is comparable to the specimen size (macrocrystalline materials), the grain boundaries are called "bamboo boundaries" [7.4.10–7.4.12]. In macrocrys-talline materials, the loss peak height is proportional to the number of grains; by contrast if the grains are small, the loss peak height is independent of grain size. One can show theoretically that grain boundary slip can cause a total relaxation in Young's modulus of 50 to 75% of its unrelaxed value

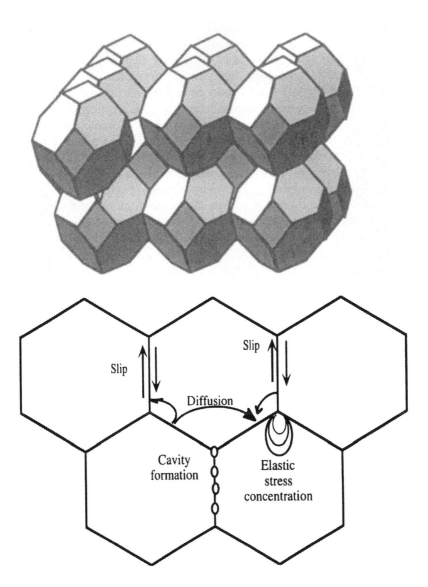

Figure 8.8 Schematic diagram of crystals (grains) in a polycrystalline material. Top: idealized tetrakaidecahedral structure of the grains. Bottom: slip motion at the grain boundaries in cross section and how that motion can be accommodated. (Adapted from Crossman and Ashby [8.7.3].)

[8.7.2]. More recent analysis models creep in metals in view of viscous processes at the grain boundaries combined with power law creep in the grain interior, leading to deformation mechanism maps [8.7.3] discussed in §8.9.

The slip between grains can be directly observed by scribing a line across a polished metal surface and then microscopically observing offsets of the line at the grain boundaries as deformation proceeds [8.7.4]. Grain boundary slip has been examined experimentally in eutectic Pb–62wt%Sn alloy by straining polished specimens within a scanning electron microscope [8.7.5]. As in earlier studies, grain boundary slip was determined by measuring the offset of scribed marker lines. Grain boundary slip along shear surfaces occurred in an heterogeneous manner, and the rate of grain boundary slip changed with strain (at large strains) in a nonmonotonic manner. These observations were consistent with a model in which grain boundary slip is considered a consequence of grain boundary dislocations. Grain boundary motion at high temperature in a magnesium alloy containing fine and coarse grains has been observed using interference optical microscopy [8.7.6]. These observations disclosed grain boundary slip to predominate in the fine grains as a mechanism to accommodate large deformations of 7%. Cooperative movement of groups of grains has also been observed [8.7.7].

The grain boundaries are also important in plastic deformation of metals and in the damage which occurs under prolonged creep at high loads. Under such conditions, grain boundary slip can be accommodated by elastic distortion of the grains, by plastic flow of the grains, and by diffusion [8.7.3]. Diffusion of vacancies can lead to the formation of *cavities* at the boundaries or their junctures [8.7.3, 8.7.8]. In materials subjected to long-term high stress, the cavities can grow and coalesce, leading to fracture.

§8.7.2   Interface motion in composites

In composite materials, slip at viscous interfaces between constituents can give rise to mechanical damping. Specifically, in synthetic composites, fiber–matrix slip can cause damping. In experiments, a weak fiber matrix bond was deliberately formed to encourage such slip [8.7.9]. In compact bone, which is a natural composite, fibers or laminae about 0.2 mm thick are separated by a layer of "cement substance", most likely mucopolysaccharides. Viscous motion at the cement lines gives rise to significant long-term creep in bone [8.7.10].

§8.7.3   Structural interface motion

Interface motion can also occur in rivets, bolted joints, and other interfaces in structures [8.7.11]. Effective "loss tangents" associated with riveted or bolted thin sheet metal structures such as those used in automobiles are on the order of 0.01, even though the loss tangent of the metal itself may be 0.001 or smaller. Welded structures of thick plates such as those used in ships exhibit effective loss tangents on the order of 0.001. Motion at joints may be accompanied by friction, but the amplitude of motion is so small that theories for dry friction do not apply. Much of the damping associated with joints in structures is thought to arise from the viscosity of air which is squeezed out of crevices as the structure vibrates.

§8.8   Other relaxation processes in crystalline materials

We consider here briefly several processes which can give rise to viscoelastic loss in crystalline materials, particularly metals [1.6.23].

§8.8.1   Snoek relaxation

The *Snoek relaxation* [8.8.1, 8.8.2] refers to viscoelastic damping due to motion of interstitial solutes in metals with a body centered cubic (bcc) crystal structure. As an example, carbon or nitrogen dissolved in iron can give rise to significant relaxation, up to tan δ ≈ 0.1. Study of other metals and solutes disclosed that the relaxation kinetics depend strongly on temperature and on the particular alloy. Relaxation times of 0.1 to 1000 sec are possible for temperatures between room temperature and 350°C.

§8.8.2   Zener relaxation in alloys

The *Zener relaxation* in solid solutions refers to reorientation of pairs of atoms in the alloy under stress [1.6.22, 8.8.3]. A classic example of the Zener relaxation in an alloy is that of α-brass (Fig. 7.9) which is a solid solution of copper and zinc. The theory assumes a relatively simple form only for a dilute concentration of solute; in that case tan δ goes as the square of the concentration. The loss peak due to the Zener relaxation is absent in pure elements. Concentrated solid solutions are more complex to understand, but experiments show a peak in tan δ as a function of concentration. The maximum loss tangent is usually less than 0.01; however, some alloys such as AgCd and AgZn exhibit a greater loss [8.8.4]. Moreover, brass exhibits a loss peak of magnitude tan δ = 0.012 near 400°C at 620 Hz. A large tan δ peak is seen in CdMg alloy described in §7.4; this alloy exhibits a loss maximum of about 0.14 at room temperature and near 1 Hz.

Solid solutions under some conditions may exhibit *lower* damping than the corresponding pure metals. Brass (CuZn) alloys with 0 to 30 atomic % zinc, at 12 kHz, near room temperature, exhibit tan δ which is approximately inversely proportional to the concentration of zinc [8.8.5]. Here the temperature is too low for the Zener mechanism to be operative. The reduction in damping is attributed to the fact that dislocation mobility [8.8.6] depends on concentration.

§8.8.3   Relaxations due to dislocations

Dislocations are line defects in crystals. They experience drag as they move under stress, so they are responsible for relaxation or damping. The concept of a pinned dislocation loop oscillating under the influence of a dynamic applied stress leads to two components of loss, according to Granato and Lücke [8.8.7]. One of these depends on frequency and attains a peak in the megahertz range, and the other is a hysteresis type which is independent of frequency but is dependent on strain. In the Granato–Lücke model, the material is assumed to contain a network of dislocations. There are two characteristic lengths in the model (Fig. 8.9), the network length L_n determined by the intersection of dislocation loops and the length L_c determined by impurities. Under a small stress the dislocation loops bow out until a break-away stress is reached at which dislocation strain increases substantially at constant stress. Under further increase of stress, new dislocation loops are created, giving rise to irreversible (plastic) strain. The frequency-dependent dislocation damping process depends on oscillation of the dislocation loops. This is a resonant process, damped by drag on the dislocations. Refinements of the Granato-Lücke theory include effects of point defect drag [8.8.8]. Even so, many aspects of damping in crystalline materials cannot be fully understood by this theory [8.8.9].

The *Bordoni relaxation* is a dislocation-type relaxation associated with dislocation motion in metals with face centered cubic (fcc) crystal structure. The relaxation is most prominent in metals which have been subjected to cold work. The relaxation peak occurs at low temperature, usually one third the Debye temperature (defined as $h\nu_m/k$ with h as Planck's constant, ν_m as the maximum frequency of lattice vibration, and k as the Boltzmann constant). For example, in copper, a loss peak of width several times larger than a single relaxation time process, of magnitude 3×10^{-3} occurred at a temperature of 72 K and at 1.9 kHz [8.8.10]. However, dislocation motion can contribute significantly to damping at room temperature, too. Relaxations appear in metals following cold work, and disappear after annealing; cold work related peaks in damping at audio frequencies have also been observed in metals.

The *Hasiguti relaxation* is attributed to interaction between point defects and dislocations [8.8.11, 8.8.12]. This relaxation occurs in metals which have been cold-worked. Loss peaks due to dislocation processes can be from 0.001 to 0.1.

The *high-temperature background* refers to damping observed in polycrystalline materials such as metals at an absolute temperature T which is a significant fraction of the melting point T_m. The homologous temperature is $T_H = T/T_m$; if it exceeds 0.5, there is usually considerable viscoelasticity, manifested as creep, relaxation, or damping. The viscoelasticity is due to a variety of mechanisms, including grain boundary slip, but at sufficiently high temperature, the background predominates. Early results on the background are summarized in the work of Nowick and Berry [1.6.23]. The background depends on structure: it is smaller in single crystals than in polycrystals; it is smaller in coarse-grained polycrystals than in fine-grained polycrystals; it is enhanced in deformed and partially recovered or polygonized samples; and it is reduced by annealing treatments at successively higher temperatures. It is generally agreed, however, that the background is caused by a combination of thermally activated dislocation mechanisms.

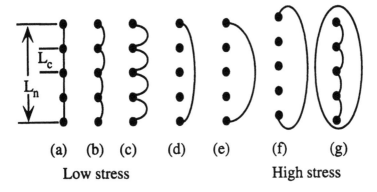

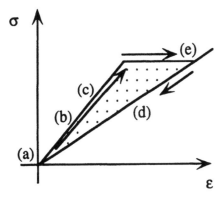

Figure 8.9 Diagram of dislocation loops (curved lines) impeded by impurity particles (small circles) (adapted from Granato and Lücke [8.8.7]). The curve of stress σ vs. strain ε shows the dislocation strain (with the elastic strain subtracted out) for a cycle of loading and unloading. If there is a distribution of loop lengths, the curve for loading becomes a smooth curve. If one continues to apply stress until new dislocation loops are formed, as in (g) in the upper diagram, the dislocation strain is irreversible and represents plastic deformation.

The dependence of the background loss on temperature and frequency has been discussed by Schoeck et al. [8.8.13]. They considered a thermally activated dislocation–point defect mechanism. If the dislocation experiences a restoring force represented by q, the damping follows a Debye peak in angular frequency $\omega = 2\pi\nu$, with ν as frequency:

$$\tan\delta = \frac{\gamma\Lambda Gb}{q}\frac{\omega\tau}{1+\omega^2\tau^2},$$

(8.8.1)

in which G is the shear modulus. The dislocation has length Λ and Burgers vector magnitude b; γ is a geometrical orientation factor of order of magnitude 0.1. The time constant is $\tau = \dfrac{1}{pq}\exp\dfrac{U_0}{kT}$ in which U_0 is an acti-

vation energy, T is the absolute temperature (in K), k is the Boltzmann constant, and p depends on temperature only. If there is no restoring force q,

$$\tan \delta \propto v^{-1}. \tag{8.8.2}$$

Cagnoli et al. [8.8.14] think that such a model cannot account for damping at low frequency. Their recent theoretical development assuming self-organized criticality of stick–slip dislocation processes gives

$$\tan \delta \propto v^{-2}. \tag{8.8.3}$$

The theory is nonlinear, with frequency-dependent and strain-dependent expressions in multiplicative form. Cagnoli et al. point out that the Granato–Lücke theory, while successful in explaining damping at high frequency in metals, fails to explain the low-frequency damping despite all later improvements in the theory.

For a *distribution* of activation energies, the following can be obtained for the case of no restoring force on the dislocations (Schoeck et al. [8.8.13]):

$$\tan \delta \propto v^{-n} \tag{8.8.4}$$

Such a form has been observed in several alloys at high homologous temperature. Since it seems fortuitous that similar distributions occur in many metals, it is natural to consider mechanisms which give rise to the observed behavior without assuming special distributions. The form $\tan \delta \propto v^{-n}$ follows from a stretched exponential relaxation, which arises naturally in many complex materials with strongly interacting constituents; general aspects of such systems are discussed in §8.11. Metals of low melting point have been experimentally observed to follow the form $\tan \delta \propto v^{-n}$ with n in the range 0.2 to 0.3, over many decades of frequency [8.8.15, 8.8.16]. Similar frequency dependence was inferred from *temperature-dependent* damping results [8.8.17] for a nickel–aluminum intermetallic single crystal. Data were recorded at several discrete frequencies and at high temperature but they were fitted with $n \approx 0.6$ at frequencies above 0.1 Hz. This dependence is inconsistent with the self-organized criticality dislocation model which predicts $n = 2$. It is also inconsistent with the simple model [8.8.13], which for a single activation energy, gives $n = 1$. A modified self-organized criticality model [8.8.18] is able to account for fractional values of the exponent n.

§8.8.4 Nonremovable relaxations

Several relaxation processes are fundamental and are not removable in principle [8.8.19]. One may envisage a single crystal without grain boundaries to slip. If the crystal is of a single constituent, processes associated with diffusion in alloys cannot occur. Crystals can be made nearly perfect, with few dislocations or point defects, hence minimal relaxation due to these causes. Therefore, many of the viscoelastic mechanisms can be made inoperative, and the associated losses can be removed.

There remains *thermoelastic relaxation* described above, which occurs in all materials, unless the coefficient of thermal expansion is zero. In pure shear, there is no macroscopic thermoelastic effect, although there may be loss due to intercrystalline thermal currents. In addition, *phonon–phonon interaction* gives rise to damping. A phonon is a quantum of sound. Phonons associated with the applied oscillatory stress can interact with thermal phonons, giving rise to damping. This damping tends to zero as temperature approaches absolute zero. In metals, which conduct electricity, there is also *phonon–electron interaction* which gives rise to damping. The underlying cause for these processes in the anharmonicity of the crystal lattice, which means that the interatomic force–displacement relation, hence the stress–strain relation, is slightly nonlinear. In the above processes [8.8.19] $\tan \delta \propto \omega$. For quartz, $\tan \delta \approx \omega/(6 \times 10^{16})$ due to thermoelasticity and $\tan \delta \approx \omega/10^{14}$ due to phonon–phonon interactions at room temperature (with ω in sec^{-1}). Preparing crystals of sufficient perfection to approach the fundamental limits is challenging.

§8.8.5 Damping due to wave scattering

At very high frequency, attenuation of ultrasonic waves occurs due to scattering of the waves from heterogeneities in the material [8.8.20]. Polycrystalline materials consist of grains of the constituent material, and each grain may be anisotropic and oriented differently from nearby grains. Materials may contain inclusions, voids, or other heterogeneities; and composite materials are heterogeneous by design. If the wavelength of the ultrasonic wave is much larger than the average heterogeneity size d, the attenuation α is given by

$$\alpha = v^4 A d^3,$$

in which v is frequency and the scattering coefficient A depends on the anisotropy of the grains. This regime is known as *Rayleigh scattering* of sound waves. Rayleigh scattering of light in the atmosphere is responsible for the blue color of the sky. Scattering coefficients A have been calculated for a variety of polycrystalline materials: for longitudinal waves, A = 40.2 for aluminum, 3.07×10^3 for copper, 700 for iron, and 5.4×10^3 for lead, in units of decibels (dB)/(cm (MHz)^{4}cm^3).

If the wavelength of the ultrasonic wave is smaller than the heterogeneity size d, the attenuation α is given by

$$\alpha = v^2 Bd,$$

with B as a constant. For typical polycrystalline metals the transition between v^4 dependence and v^2 dependence occurs at frequencies between 1 and 10 MHz. Attenuations between 1 and 10 dB/μsec at 10 MHz are representative. As for interpretation in terms of tan δ, recall from Eq. 3.7.12, that, with c as the wave speed,

$$\alpha \approx \frac{\omega}{2c} \tan \delta.$$

Damping due to scattering cannot be shifted to lower frequencies by cooling the specimen since the physical process involved is geometrical in nature and depends on the ratio of wavelength to heterogeneity size.

§8.9   *Multiple causes: Deformation mechanism maps*

Deformation mechanism maps are diagrams (Fig. 8.10) which display the regions of stress and temperature in which a particular mechanism of *secondary* creep or plastic flow dominates [8.9.1, 8.9.2]. In such a region, the particular mechanism supplies a greater strain rate in creep than any other. The stress axis is normalized to the shear modulus, and the temperature axis is normalized to the melting point. Maps may be constructed from actual creep data or from theoretical relations for each process. In the latter case, the relevant diffusion parameters must be known. The boundaries of the regions in the diagram are determined by equating strain rates associated with pairs of the constitutive equations for particular creep mechanisms.

At least six deformation mechanisms have been identified in connection with the maps. They are (i) flow in the absence of defects at a stress above the theoretical shear stress, (ii) flow by glide of dislocations, (iii) creep by climb of dislocations, (iv) Nabarro–Herring creep from flow of point defects through grains, (v) Coble creep from flow of point defects along grain boundaries, and (vi) twinning of crystals. Constitutive equations are as follows.

Dislocation glide exhibits a threshold effect.

$$\frac{d\varepsilon_2}{dt} = \frac{d\varepsilon}{dt}\bigg|_0 \exp\left\{-\frac{S-\sigma}{kT}ba\right\} \qquad \text{for } \sigma \geq \sigma_0,$$

$$\frac{d\varepsilon_2}{dt} = 0 \qquad\qquad\qquad\qquad \text{for } \sigma \leq \sigma_0.$$

$$(8.9.1)$$

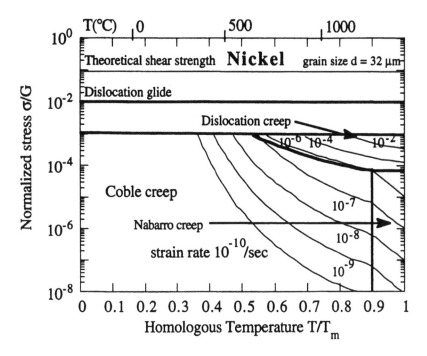

Figure 8.10   Deformation mechanism map for nickel. (Adapted from Ashby [8.9.1] and Frost and Ashby [8.9.2].)

with S as a flow stress, σ as the stress, k as the Boltzmann constant, T as the absolute temperature, a as the activation area, and σ_0 as a cutoff stress.

In the following, diffusional creep due to bulk transport, Nabarro-Herring creep, is represented with the subscript 3. Diffusional creep due to transport along grain boundaries, Coble creep, is represented with the subscript 4.

$$\frac{d\varepsilon_{3;4}}{dt} = 14\frac{\sigma\Omega}{kT}\frac{1}{d^2}D_v\left\{1+\frac{\pi\delta}{d}\frac{D_B}{D_v}\right\}, \qquad (8.9.2)$$

in which Ω is the atomic volume, d is the grain size, D_v is the bulk self-diffusion coefficient, D_B is the boundary self-diffusion coefficient, and δ is the effective cross section of a boundary. These mechanisms may be distinguished physically.

Dislocation creep is nonlinear in stress:

$$\frac{d\varepsilon_s}{dt} = AD_v\frac{Gb}{kT}\left\{\frac{\sigma}{G}\right\}^n,$$

in which G is the shear modulus. This process is due to aggregation of dislocations at temperatures greater than half the melting point. It is diffusion controlled in contrast to dislocation glide.

These maps were presented initially for metals (Fig. 8.10); by explicitly plotting the stress level, nonlinear behavior is allowed for, but detailed history dependence is not. Deformation mechanism maps can be useful in the design of experimental conditions for tests of particular creep processes. They are also useful in design problems involving creep. For example, a strengthening mechanism such as dispersion hardening of metals will slow dislocation creep. Map boundaries will then shift, so that another creep mechanism dominates. Further inhibition of dislocation creep is not productive, since the metal now deforms mostly by diffusional creep, which is not inhibited by the same mechanism [8.9.1]. Ceramics [8.9.3] at sufficiently high temperature exhibit time-dependent deformation as a result of many mechanisms which operate in metals.

A further example is that multiple viscoelastic mechanisms cause flow in the "solid" interior of the earth [8.9.4]. At relatively low stress, or for small grain size, linearly viscoelastic behavior occurs via diffusion of matter between grain boundaries. At a higher stress level, or larger grain size, nonlinearly viscoelastic behavior occurs via motion of dislocations. Creep due to dislocation motion can give rise to preferred orientation in a material. In the context of geology, this gives rise to anisotropy of seismic wave velocities. Relaxation mechanisms in the earth have implications in understanding continental drift.

The activation energy for secondary creep at high temperature in metals is closely related to the activation energy for self-diffusion, [8.9.5] as shown in Fig. 8.11. In pure metals [8.9.5] and in solid solution alloys [8.9.6], the stress dependence of steady-state creep consists of three regions: a linear region, a strongly stress-dependent region, and a breakdown region. In alloys there is an additional possibility of enhanced creep at intermediate stresses due to dragging of solute atoms by dislocations.

Deformation mechanism maps are useful for secondary (steady-state) creep. As currently constituted they do not deal with primary creep or, in the frequency domain, with peaks in the loss tangent.

§8.10 Relaxation in piezoelectric materials

Piezoelectricity is a coupled-field effect as is thermoelasticity. In piezoelectric materials stress and strain are coupled to electrical field and polarization. Not all materials are piezoelectric; only those materials lacking a center of symmetry on the atomic scale can be piezoelectric. Examples of piezoelectric materials include quartz, Rochelle salt, and lead titanate zirconate ceramics.

The piezoelectric contribution to the mechanical loss tangent of a piezoelectric solid is here derived from its complex piezoelectric and dielectric coefficients [8.10.1]. This loss depends on specimen geometry as a result of

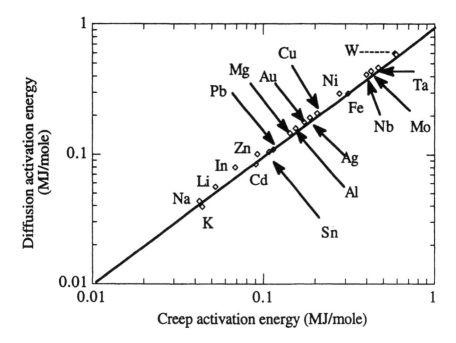

Figure 8.11 Relation between the activation energy for self-diffusion and the activation energy for high-temperature secondary creep of pure metals. (Adapted from Karato and Wu [8.9.4].)

differences in effects related to the electrical boundary conditions. Including a positive out-of-phase piezoelectric modulus results in reduced values of the predicted loss, which constitutes an improvement over earlier theories that predict losses exceeding measured losses by a factor greater than two.

Piezoelectric relaxation has been observed in many materials, including ceramics [8.10.2], composites [8.10.3], and bone [8.10.4]. Such relaxation can be represented with complex piezoelectric coefficients or by a piezoelectric loss tangent [8.10.2, 8.10.5]. Mechanical relaxation also occurs in piezoelectric materials and is important in applications: large damping is considered desirable in materials used to generate short acoustic pulses for flaw detection [8.10.6]; small damping (high mechanical Q) is desirable in stable resonators and high-power transducers.

In this segment the connection between the dielectric, mechanical, and piezoelectric coefficients of a material which exhibits relaxation is considered. Clearly, for an ideal solid which does not relax, this connection, in the form of a piezoelectric contribution to the compliance, has been established. For such a material, in the linear domain, the constitutive equations are [8.3.1]

$$D_i = \left[d_{ijk} \right]_T \sigma_{jk} + \left[K_{ij} \right]_{\sigma,T} E_j + \left[p_i \right]_\sigma \Delta T, \qquad (8.10.1)$$

$$\varepsilon_{ij} = \left[S_{ijkl}\right]_{E,T} \sigma_{kl} + \left[d_{kij}\right]_T E_k + \left[\alpha_{ij}\right]_E \Delta T. \qquad (8.10.2)$$

Here D is the electric displacement vector, d is the piezoelectric modulus tensor at constant temperature T, K is the dielectric tensor at constant stress σ and temperature T, E is the electric field vector, p is the pyroelectric coefficient at constant stress, ε is the strain, S is the elastic compliance tensor at constant electric field, and α is the thermal expansion tensor. The usual Einstein summation convention over repeated subscripts is used. In a material described by these equations, the isothermal compliance measured at constant electric displacement $[S_{ijkl}]_D$ differs [8.3.1] from the compliance measured at constant field $[S_{ijkl}]_E$:

$$\left[S_{ijkl}\right]_D - \left[S_{ijkl}\right]_E = -d_{mij}d_{nkl}\left[K_{mn}\right]_\sigma^{-1}. \qquad (8.10.3)$$

The derivation of this expression follows from the constitutive equations by a similar procedure to that used in obtaining the difference between adiabatic and isothermal compliances in the thermoelastic case. This difference may be regarded as the piezoelectric contribution to the compliance. Compare with the thermoelastic Eq. 8.3.7. Piezoelectric reactions also influence the apparent stiffness of a solid, under conditions in which neither E nor D is constant [8.10.7]. For materials which do relax, the coefficients in Eqs. 8.10.1 and 8.10.2 are complex. The sign is negative since these are compliances.

$$d_{ijk}^* = d_{ijk}' - id_{ijk}'',$$

$$K_{ij}^* = K_{ij}' - iK_{ij}'', \qquad (8.10.4)$$

$$S_{ijkl}^* = S_{ijkl}' - iS_{ijkl}''.$$

One can consider a piezoelectric contribution to mechanical relaxation: experimental evidence for such a contribution in quartz under quasistatic loading has appeared at least as early as 1915 [8.10.8]. A connection between dielectric loss and mechanical loss in piezoelectric solids is to be expected on heuristic grounds in that dielectric relaxation entails dissipation of electrical energy; if this energy has come from the piezoelectric conversion of mechanical energy, then mechanical relaxation or anelasticity must also occur.

The relation between electric field and electric displacement is given by the constitutive equation (8.10.1). For a nonpolar solid under isothermal conditions, with $K_{ij} = k_{ij} e_0$ (e_0 is the permittivity of free space), and with $k^* = k' - ik''$ and $d^* = d' - id''$ to describe dielectric and piezoelectric relaxation, Eqs. 8.10.1 and 8.10.2 become

$$D_i = d^*_{ijk}\sigma_{jk} + \left[k^*_{ij}\right]_\sigma e_0 E_j.$$ (8.10.5)

$$\varepsilon_{ij} = \left[S^*_{ijkl}\right]_E \sigma_{kl} + d^*_{kij} E_k.$$ (8.10.6)

Suppose that the specimen in question is electrically isolated, i.e., free of any attached circuit element of finite impedance and that it is subjected to a stress $\sigma_{11} \neq 0$ which is uniform within the specimen and varies sinusoidally with time: $\sigma_{11}(t) = \sigma_{11}(0)\, e^{i\omega t}$, $\sigma_{ij} = 0$ if $i \neq 1$ or $j \neq 1$. The supposition of uniform stress entails loading below any mechanical resonance. With these assumptions we shall find that the field and the displacement are related in a way which depends on the electrical boundary conditions; therefore, several specimen geometries are considered. In all cases we assume that $d_{311} \neq 0$ and $d_{i\,11} = 0$ if $i \neq 3$ and that the 3 direction is a principal axis of k_{ij}, so that both the field and displacement will be in the 3 direction.

Consider a thin plate of piezoelectric material such that the flat surfaces are perpendicular to the 3 axis. From Gauss's law, the boundary condition on the electric displacement is $D_{in}^{[normal]} - D_{out}^{[normal]} = \Sigma_{free}$; Σ_{free} is the density of free charge on the surface. For the thin plate geometry, $D_{out}^{[normal]} = 0$. Now the charge which accumulates on the surface as a result of conductivity or modes of dielectric relaxation that do not involve dipole rotation may be regarded as free charge. Such a view, while conceptually reasonable, leads to difficulties in gedankenexperiments involving measurement of the electric field in a plate with constant d' and finite conductivity. As a result of such experiments, we conclude that in the present setting, free charge must include only charge that is not associated with processes included in the definition of D. So if k" contains contributions from DC conductivity as well as dielectric relaxation, the free charge is zero. Then inside the plate,

$$\dot{D}_3 = 0,$$ (8.10.7)

as in the case of ideal crystals with zero conductivity [8.10.5]. The relation between electric field and electric displacement depends on boundary conditions. For example, for plates, cylinders, and spheres within the solid, the electric field and displacement are uniform, parallel, and related by the following:

$$D_3 = -E_3 e_0 \Lambda.$$ (8.10.8)

The quantity Λ is zero for the thin plate. For the cylinder, $\Lambda = 1$ and for the sphere, $\Lambda = 2$ [8.10.1].

In this section we develop an expression for the piezoelectric contribution to the anelastic loss tangent in a solid subjected to subresonant, sinusoidal loading. Let the solid obey the constitutive equations (8.10.5) and

(8.10.6), i.e., let it exhibit dielectric and piezoelectric relaxation. The mechanical, dielectric, and piezoelectric loss tangents are defined as follows:

$$\tan \delta_{ijkl}^{(s)} = \frac{S''_{ijkl}}{S'_{ijkl}}, \tag{8.10.9}$$

$$\tan \delta_{ijk}^{(d)} = \frac{d''_{ijk}}{d'_{ijk}}, \tag{8.10.10}$$

$$\tan \delta_{ij}^{(k)} = \frac{k''_{ij}}{k'_{ij}}, \tag{8.10.11}$$

with no summation on the repeated indices.

The piezoelectric contribution to the storage compliance is formally similar to the effect of the temperature field in thermoelasticity, and the effect of the fluid pressure field in poroelasticity. The presence of piezoelectric coupling reduces the compliance and causes the material to appear stiffer. In general, the stiffening effect depends also on geometry and piezoelectric phase angles. However, if we consider a flat plate specimen, $\Lambda = 0$; and if there is no dielectric or piezoelectric relaxation, $\tan \delta^k{}_{33} = 0$, $d''{}_{311} = 0$. Since $\Lambda = 0$ implies that the electric displacement is zero, we obtain

$$S'^{D}_{1111} - S'^{E}_{1111} = -\frac{d^2_{311}}{e_0 k'_{33}}, \tag{8.10.12}$$

which is equivalent to Eq. 8.10.3 in which the sum has collapsed into a single term as a result of the assumptions made. The results, therefore, reduce to those of the classical theory in which no losses or piezoelectric phase angles occur.

We remark that in the field of piezoelectric materials, dimensionless coupling factors K are defined for the purpose of understanding the conversion of mechanical and electrical energy [8.10.10] without damping. The range for K is from zero to one. The relevant energies are shown in Fig. 8.12.

$$K^2 = \frac{W_1}{W_1 + W_2}. \tag{8.10.13}$$

These coupling factors depend on elastic boundary conditions at the material surface. For example, for uniaxial stress [8.10.10],

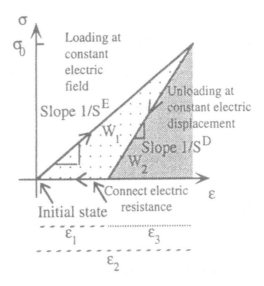

Figure 8.12 Conversion of mechanical energy to electrical energy by a piezoelectric material. (Adapted from IEEE Standard on Piezoelectricity [8.10.10].) Energy densities W are shown as shaded areas. Material is loaded at constant electric field, unloaded at constant electric displacement, and connected to an electric load at zero stress. W_1 represents the work per volume of material done by the material on the electrical load, while $W_1 + W_2$ is the maximum energy per unit volume stored in the material at maximum stress.

$$K_{31} = \frac{d_{31}}{\sqrt{k_{33}S_{11}^E}},$$ (8.10.14)

in which S_{11} is the elastic compliance in the "reduced" notation (it is equivalent to S_{1111}), k_{33} is the dielectric constant and d_{31} is the piezoelectric modulus in the reduced notation. Compare with Eqs. 8.3.8 and 8.3.19 for the thermoelastic case. In piezoelectric ceramics, coupling factors as large as 0.7 are possible. This is much stronger than thermoelastic coupling.

The piezoelectric contribution to the loss tangent of the thin plate is calculated as

$$\tan \delta_{1111} = \frac{\left(d_{311}'^2 - d_{311}''^2\right)\tan \delta_{33}^k - 2d_{311}'d_{311}''}{\varepsilon_0 k_{33}' S_{1111}'\left(1 + \tan^2 \delta_{33}^k\right)}.$$ (8.10.15)

S' in this expression is the storage compliance for the geometry in question; this differs from the constant-field compliance $(S')_E$ which appears in the constitutive equation. For weak coupling, the difference between these compliances is small.

The effect of the piezoelectric relaxation term d" is much more pronounced in the contribution to the anelastic relaxation than in the contribution to the storage compliance. This may explain why the classical theory of linear piezoelectricity [8.3.1, 8.10.7, 8.10.9], which addresses elastic effects in the absence of dissipation, is accurate for this type of problem (prediction of piezoelectric stiffening) despite the neglect of piezoelectric phase angles.

The piezoelectric contribution to the mechanical loss of a solid in the form of a thin plate was obtained. For a different specimen *shape*, the damping may be different as a result of the electrical boundary conditions [8.10.1]. A similar shape dependence of damping was also studied in connection with thermoelastic damping [8.3.6, 8.3.7]. Recently, phase angles in the thermal expansion coefficient [8.10.11] have been shown to affect the thermoelastic damping in a manner similar to the effect of piezoelectric phase angles.

§8.11 *Nonexponential relaxation*

The stretched exponential relaxation function described in §2.6,

$$E(t) = \left(E_0 - E_\infty\right)e^{-(t/\tau_r)^\beta} + E_\infty, \tag{8.11.1}$$

arises naturally in many complex materials with strongly interacting constituents. One may formally express such a function as a superposition of exponentials using the concept of relaxation spectrum developed in §4.2. In view of the Boltzmann superposition principle, linear macroscopic experimental measurements cannot distinguish between causes which involve a true distribution of exponential processes each with its own relaxation time, and causes which are intrinsically nonexponential.

In intrinsically nonexponential mechanisms, the stretched exponential relaxation arises from relaxation in hierarchical stages such that the constraint imposed by a faster degree of freedom must relax before a slower degree of freedom can relax [8.11.1]. The underlying feature of theories giving rise to such relaxation is the generation of a scale invariant distribution of relaxation times [8.11.2]. Stretched exponentials in slow relaxation are so widespread as to be considered "universal" [8.11.3, 8.11.4], perhaps because they represent a probability limit distribution [8.11.5]. In the frequency domain, for frequencies v well above the damping peak,

$$\tan\delta \propto v^{-\beta}, \tag{8.11.2}$$

with $\beta < 1$. The high-temperature background in metals, as discussed above in §7.14 and in §8.8.3, follows this form. However, a copper–beryllium alloy was observed to exhibit damping of the form $\tan\delta \propto v^{-2}$ at low frequency [8.11.6], and that form was modeled theoretically under the assumption of self-organized criticality.

Andrade creep is a particular form of power law creep for which $J(t) = J_0 + At^{1/3}$. Since Andrade creep is ubiquitous in metals at large strain, it must be caused by very general effects and is not limited by specific atomic processes of flow. A mechanism based on correlation volumes has been proposed [8.11.7].

Since macroscopic linear experiments do not allow one to distinguish whether nonexponential relaxation arises intrinsically or from a superposition of exponentials, there has been some debate on the nature of the causal mechanisms [8.11.8]. Recently, a nonlinear technique was used to demonstrate that a broad relaxation spectrum in a dielectric system was the result of a distribution of relaxation times [8.11.9]. The method involved brief exposure of the specimen to a sinusoidal "pump" stimulus followed by a probe stimulus of step function form. If the response to a probe stimulus was altered by the prior presence of a pump stimulus at its characteristic time, that result was interpreted as an excitation of a single exponential mode. Further experiments of this type are called for in viscoelastic materials in addition to dielectric materials.

§8.12   *Damping mechanisms in high-loss alloys*

In known high-damping alloys several viscoelastic mechanisms have been identified. In zinc–aluminum alloys, thermoelastic damping contributes a noticeable peak [8.12.1] due to the high concentration of zinc which has a favorable figure of merit. The remaining damping follows a $\tan \delta \propto \nu^{-n}$ form at high frequency and is attributed to dislocation movement. Damping in CuMn alloys [8.12.2, 8.12.3] is attributed to the movement of twin boundaries and is dependent on the antiferromagnetic nature of the crystals in the metal. This magneto-mechanical mechanism gives rise to strain dependence of damping since at sufficiently large strain, the domains become fully aligned, giving rise to a saturation of the mechanism. Damping drops abruptly above the Néel temperature at which the material changes phase from antiferromagnetic to paramagnetic. High damping in magnesium and some of its alloys is attributed to dislocation movement [8.12.2].

§8.13   *Creep mechanisms in creep-resistant alloys*

As temperature is increased, creep from a variety of mechanisms tends to increase. Creep-resistant alloys have been developed based on an understanding of several viscoelastic mechanisms [1.6.27], as follows:

1. Metals of high melting point are favored since many creep processes depend on the homologous temperature. Nickel alloys are favored for turbine engine applications in which the temperature and stress are high. Cobalt alloys are used in parts for which temperature is very high but stress is not as high. Tungsten has a higher melting point

than these alloys. Tungsten is too dense for structural applications but finds use in light bulb filaments.

2. Elements such as Co, Cr, W, Mo, and V are incorporated in a nickel matrix to form a solid solution. As dislocations move in such a material, they drag an "atmosphere" of solute atoms. The drag reduces the rate of creep. Moreover, the irregular atomic structure tends to resist the movement of dislocations in comparison with the case of a pure metal. This effect of solid solution is in contrast with the *Zener relaxation* in concentrated alloys refers to reorientation of pairs of atoms in the alloy under stress [8.8.1] discussed in §8.8.2. The Zener relaxation provides a relatively small relaxation strength and so is not responsible for large amounts of creep.

3. Refractory particles such as Al_2O_3 and MgO are incorporated to retard dislocation motion. This process is called *dispersion strengthening*. The particles reduce creep indirectly by suppression of grain boundary sliding during the service of the material and by suppression of grain growth during the preparation of the material.

4. Particles of intermetallic compounds such as Ni_3Ti are incorporated. The particles exhibit the remarkable characteristic of a large reversible increase in flow stress as temperature increases.

5. Creep due to grain boundary slip can be reduced by precipitating carbides containing Ti, Cr, W, Mo, and Zr in the grain boundaries.

6. Creep due to grain boundary slip can be reduced by a directional solidification process which produces elongated grains aligned with the expected stress.

7. Creep due to grain boundary slip can be eliminated by a directional solidification process which produces single crystals of metal. Then there are no grain boundaries because there is only one grain. Such crystals are used for turbine blades for jet engines. In single-crystal blades, the grain boundary strengthening elements B, C, Zr, and Hf are omitted since they are no longer needed and since they reduce the melting point of some alloy components.

§8.14 *Relaxation at very long times*

Over a sufficiently long period of time, all matter is fluid, even at zero temperature, as a result of the wave nature of matter [8.14.1]. The time scale for barrier penetration by a mass m behaving as a quantum mechanical matter wave, following Gamow's analysis of radioactive decay is

$$\tau = \tau_0 e^s,$$

with τ_0 as the natural vibration period and S as the action integral

$$S = \frac{2}{\hbar} \int \sqrt{2mU(x)}dx \approx \sqrt{8mUd^2/\hbar^2}, \qquad (8.14.1)$$

with m as the particle mass, U(x) as the potential barrier of thickness d, and $\hbar$ as the normalized Planck's constant $h/2\pi$. For atoms in a solid, $U \approx e^4 m_e / 20\, \hbar^2$, and $d \approx \hbar^2 / m_e e^2$. So with A as the atomic weight of the atom, m_e as the electron mass, and m_p as the proton mass, $S = \sqrt{2 A m_p / 5 m_e} = 27 \sqrt{A}$. For iron, A = 56. For atomic vibration at absolute zero temperature, $\tau_0 \approx 10^{-14}$ sec. The time scale τ is on the order of 10^{65} years. Since this time scale substantially exceeds estimates of the age of the universe (about 10^{10} years) and of the remaining life of the sun (about 5×10^9 years), it has no significance in the practical affairs of humanity. Therefore, most engineers are not concerned about such processes. These arguments nevertheless indicate that there are no solids.

§8.15 Summary

Viscoelastic relaxation can occur whenever there is a delayed rearrangement of the internal structure of the material under stress. There are many ways in which this rearrangement can occur. Understanding viscoelastic mechanisms can aid one in anticipating situations in which viscoelasticity can be expected. Moreover, by understanding the mechanisms, one can at times tailor various materials to exhibit large or small viscoelastic behavior in selected regions of time, temperature, or frequency, as required for various applications.

Examples

Example 8.1

Polymers have higher thermal expansion coefficients than metals. Is thermoelastic damping an important mechanism in polymers?

Solution

Consider the relaxation strength, Eq. 8.3.19, $\Delta = \alpha^2 T / C_v J^S$, and take numerical data from Ashby's review article [E8.1.1]. Suppose further that the relaxation is of the Debye form so the maximum $\tan \delta$ is approximately $\Delta/2$. Properties of high-density polyethylene are

> Thermal expansion coefficient $\alpha = 1.5 \times 10^{-4}/°C$.
> Young's modulus E = 0.6 GPa.
> Heat capacity [E8.1.2], C = 0.55 cal/g°C.
> Density $\rho = 0.95$ g/cm³.

By converting the heat capacity to SI units, and from a mass basis to a volume basis,

$$C = 0.55\,\text{cal/g}°C \times 0.95\,\text{g/cm}^3 \times 4.19\,\text{J/cal} \times 10^6\,\text{cm}^3/\text{m}^3 = 2.19 \times 10^6\,\text{J/}°\text{Cm}^3.$$

Then

$$\Delta = 1.85 \times 10^{-3}\ (\text{dimensionless}),$$

so the maximum tan δ is, for small Δ,

$$\tan \delta_{max} \approx \Delta/2 = 0.93 \times 10^{-3}.$$

This is much smaller than damping observed in polymers. Most of the observed damping is due to molecular motion. Although the thermal expansion coefficient in this polymer is about ten times as large as that of steel, the modulus is some 300 times smaller, so the thermoelastic contribution to the damping in polymers is small. By contrast in most metals, the overall damping is small; and in some frequency ranges, thermoelastic effects, though of small magnitude, can be responsible for most of the damping.

Example 8.2

Negative values of the coefficient of thermal expansion are possible. What are the consequences in connection with thermoelastic damping?

Solution

The relaxation strength is given in Eq. 2.6.29. By using the tensorial form, Eq. 8.3.8,

$$\Delta_{ijkl} = \frac{S^T_{ijkl} - S^S_{ijkl}}{S^S_{ijkl}} = \frac{\alpha_{ij}\alpha_{kl}}{S^S_{ijkl}}\frac{T}{C^\sigma}.$$

Consider several components.

For example, $\Delta_{1111} > 0$ even if $\alpha_{11} < 0$, since α_{11} is squared, so the relaxation strength for axial deformation is always positive. Observe that S_{1111} is the inverse of Young's modulus for stress in the 1 direction, and must be positive for the material to be stable.

Consider $i, j = 1$; $k, l = 2$. If the material is isotropic, $\alpha_{11} = \alpha_{22}$, so again a product of two negative factors appears on the right with a positive result.

If the material is anisotropic, we can have $\alpha_{11} = -\alpha_{22}$. For common materials, $S_{1122} > 0$ since its physical meaning is $-v/E$, and Poisson's ratio is usually positive. The possibility of tan $\delta_{1122} < 0$ does not present physical difficulties since this "loss tangent" represents a phase between a stress in one direction and a strain in an orthogonal direction. The angle δ_{1122} is therefore a phase in a cross property (not a modulus or compliance) with no energy content.

Similarly Poisson's ratio is a cross property, with no energy content. There-fore, not only can Poisson's ratio be negative but also phase angles in Pois-son's ratio can be positive or negative. Negative damping (or gain) in a modulus or compliance can occur only if an external source of energy is supplied.

Example 8.3

Consider a bar of aluminum vibrating in bending. What is the frequency of the damping peak if the bar is 1 mm thick? What if it is 1 cm thick? How large is the peak loss?

Solution

From Eq. 8.3.20, and using data from Table 8.1,

$$v_0 = \frac{\pi}{2}Dd^{-2} = 1.57\frac{222\,J/s \cdot m \cdot K}{900\,J/kg \cdot K}\frac{1}{2.7 \cdot 10^3\,kg/m^3}\frac{1}{\left(10^{-3}m\right)^2} = 144\ Hz.$$

A bar 1 cm thick would have a peak loss at 1.44 Hz, since the thickness d is squared in Eq. 8.3.20. The peak tan δ is approximately $\Delta/2$. Taking Δ from Table 8.1, tan $\delta \approx 2.4 \times 10^{-3}$, at the peak. Compare with the peak for aluminum in Fig. 8.4.

Example 8.4

Consider thermally activated relaxation in aluminum. By following Kê [8.7.1], the grain boundary slip process gives rise to a peak in loss tangent at 0.8 Hz and 280°C. The activation energy is 34,000 cal/mol. At what frequency will this peak occur at 25°C and at 100°C? What are the corre-sponding relaxation times?

Solution

From the Arrhenius equation discussed in §2.7 and its representation in Eq. 6.9.1,

$$\ln\frac{v_2}{v_1} = \frac{U}{R}\left[\frac{1}{T_1} - \frac{1}{T_2}\right].$$

The temperatures are absolute. By substituting $T_2 = 280 + 273 = 553$ K, $T_1 = 25 + 273 = 298$ K, $R = 1.98$ cal/moleK, $\ln\frac{v_2}{v_1} = 26.6$, so $\frac{v_2}{v_1} = 3.47 \times 10^{11}$, so for v_1 at 25°C and v_2 at 280°C,

$$\nu_1 = 2.31 \times 10^{-12}\,\text{Hz}.$$

The time constant is

$$\tau = (2\pi\nu_1)^{-1} = 6.89 \times 10^{10}\ \text{sec, or } 2.19 \times 10^3\ \text{years.}$$

For $T_1 = 100°C$ or 373 K,

$$\ln\frac{\nu_2}{\nu_1} = \frac{34000}{1.98} 8.73 \times 10^{-4} = 14.98,$$

so

$$\nu_1 = 2.49 \times 10^{-7}\,\text{Hz}.$$

The time constant is

$$\tau = (2\pi\nu_1)^{-1} = 6.41 \times 10^5\ \text{sec,}$$

or 7.41 days, a rather more accessible time than at 25°C.

Example 8.5

Does a stretched rod held at constant extension get fatter or thinner with time? Suggest the use of viscoelastic mechanisms to control the rate and direction of lateral expansion or contraction.

Solution

Recall in Example 7.2 and §7.3 that Poisson's ratio increases through the α transition of polymers. In that case the rod gets thinner. Consider a case in which the bulk modulus relaxes but the shear modulus does not relax much. Envisage a porous material with fluid in the pores. Squeezing the fluid out of the pores and through free surfaces during a volume change causes much more relaxation than does a shear deformation. Consider the relation between Poisson's ratio, shear modulus, and bulk modulus for an elastic material.

$$\nu = \frac{3B - 2G}{6B + 2G}.$$

By considering a bulk compliance $\kappa = 1/B$, observing that for $\nu = 1/3$, $G\kappa = 0.462$, and using a Taylor expansion of the denominator to express Poisson's ratio in terms of products,

$$v = \frac{\frac{1}{2}-\frac{1}{3}G\kappa}{1+\frac{1}{3}G\kappa} \approx \left\{\frac{1}{2}-\frac{1}{3}G\kappa\right\}\left\{1-\frac{1}{3}G\kappa\right\} = \frac{1}{2}-\frac{1}{6}G\kappa+\frac{1}{9}G^2\kappa^2.$$

By applying the correspondence principle with G assumed constant,

$$s v(s) \approx \frac{1}{2}-\frac{1}{6}Gs\kappa(s)+\frac{1}{9}G^2 s^2 \kappa^2(s).$$

By dividing by s and transforming back,

$$v(t) \approx \frac{1}{2}-\frac{1}{6}G\kappa(t)+\frac{1}{9}G^2 \int \kappa(t-\tau)\frac{d\kappa(\tau)}{d\tau}\,d\tau.$$

The third term is small compared to the second unless the bulk creep is very fast. Since $\kappa(t)$ is an increasing function, the second term causes Poisson's ratio to decrease with time, so the stretched rod in this case gets fatter.

Example 8.6

Suppose a piezoelectric plate is provided with an electrically resistive path so that stress-generated charge can flow with time. Relate the coupling coefficient K with the relaxation strength Δ. For a Debye peak, what is the relationship between the loss tangent and the coupling coefficient?

Solution

By definition, in terms of the energies W in Fig. 8.12, $K^2 = \dfrac{W_1}{W_1+W_2}$. If stress is applied to the plate slowly, its stiffness is $E_s = 1/S^E$, referring to Fig. 8.12, since for a sufficiently long time, current flow through the resistance forces the electric field to zero, a constant. Recall that S^E is compliance at constant electric field. Similarly for fast loading, $E_f = 1/S^D$. Express the energies in terms of the stiffness values and the strains as shown in Fig. 8.12.

$$W_2 = \frac{1}{2}E_f \varepsilon_3^2, \tag{E8.6.1}$$

$$W_1 + W_2 = \frac{1}{2}E_s \varepsilon_2^2, \tag{E8.6.2}$$

but

$$E_s = \sigma_0/\varepsilon_2 \text{ and } E_f = \sigma_0/\varepsilon_3, \qquad (E8.6.3)$$

so

$$K^2 = \frac{\dfrac{1}{E_s} - \dfrac{1}{E_f}}{\dfrac{1}{E_s}} = 1 - \frac{E_s}{E_f}. \qquad (E8.6.4)$$

Consider the definition of *relaxation strength* as the change in stiffness during relaxation divided by the stiffness at long time, referring to Eqs. 2.6.17 and 2.6.26, $\Delta = \dfrac{E_1}{E_2}$. This may also be written in terms of the stiffness values for fast and slow loading, $E_f = E_1 + E_2$ and $E_s = E_2$. Then,

$$\Delta = \frac{E_f - E_s}{E_s}. \qquad (E8.6.5)$$

By combining,

$$\Delta = \frac{K^2}{1 - K^2}. \qquad \textbf{(E8.6.6)}$$

Observe that the range for the coupling coefficient K is from zero to one, while the range for the relaxation strength Δ is from zero to infinity.

If we have a Debye peak, referring to Eq. 3.2.30,

$$\tan\delta_{max} = \frac{1}{2}\frac{\Delta}{\sqrt{1+\Delta}}, \qquad (E8.6.7)$$

then

$$\tan\delta_{max} = \frac{1}{2}\frac{K^2}{\sqrt{1-K^2}}. \qquad \textbf{(E8.6.8)}$$

Here the limiting aspects of fast and slow loading have been considered, not the time scale of relaxation. As for a comparison with Eq. 8.10.15, we observe that an external resistor was not considered in the development of that equation, which takes into account phase angles in the piezoelectric coefficients.

Problems

8.1 Consider fluid flow based relaxation as a coupled-field process. What fields other than stress and strain are involved? Prepare a diagram analogous to Fig. 8.3. Show the slopes in terms of the Biot coefficients.

8.2 The Zener relaxation in brass occurs at a relatively high temperature. Under what circumstances would the Zener relaxation occur at room temperature? Can you find an example? What might such a material be used for?

8.3 How may an understanding of viscoelastic mechanisms be used in the selection of materials to resist creep?

8.4 Demonstrate the piezocaloric effect as follows. Grip a thick rubber band at both ends. Subjectively evaluate the temperature of the band by briefly touching it to a free area of skin such as the bottom of your nose. Suddenly stretch the band and again evaluate the temperature. Does it get warmer or cooler? Keep it stretched for about a minute so it comes into equilibrium with the environment. Evaluate its temperature, suddenly release the tension, and repeat. This time, does it get warmer or cooler? Discuss.

8.5 What is the effect a phase angle in thermal expansion [8.10.11] on thermoelastic damping of a material? Consider the thermal expansion coefficient to be complex: $\alpha^* = \alpha'(1 + i \tan \delta_\alpha)$.

8.6 There are currently no standard materials for calibration of viscoelastic instrumentation, since the physical processes which give rise to viscoelastic behavior tend to depend on specimen preparation methods as well as on environmental variables such as temperature and humidity. Moreover, for many materials there are multiple relaxation mechanisms simultaneously active in a given frequency range. Develop a candidate standard material based on known relaxation mechanisms discussed in Chapter 8.

References

8.2.1 Wert, C. A., "Internal friction in solids", *J. Appl. Phys.* 1888–1895, 1986.

8.3.1 Nye, J. F., *Physical Properties of Crystals*, Oxford University Press, NY, 1976.

8.3.2 Zener, C., "Internal friction in solids I — Theory of internal friction in reeds", *Phys. Rev.* 52, 230–235, 1937.

8.3.3 Zener, C., "Internal friction in solids II. General theory of thermoelastic internal friction", *Phys. Rev.* 53, 90–99, 1938.

8.3.4 Zener, C., Otis, W., and Nuckolls, R., "Internal friction in solids III. Experimental demonstration of thermoelastic internal friction", *Phys. Rev.* 53, 100–101, 1938.

8.3.5 Milligan, K. B. and Kinra, V. K., "On the thermoelastic damping of a one-dimensional inclusion in a uniaxial bar", *Mech. Res. Commun.* 20, 137–142, 1993.

8.3.6 Alblas, J. B., "On the general theory of thermoelastic damping", *Appl. Sci. Res.* 10, 349–362, 1961.

8.3.7   Alblas, J. B., "A note on the theory of thermoelastic damping", *J. Thermal Stresses* 4, 333–335, 1981.

8.3.8   Kinra, V. K. and Bishop, J. E., "Elastothermodynamic analysis of a Griffith crack", *J. Mech. Phys. Solids* 44, 1305–1336, 1996.

8.3.9   Bishop, J. E. and Kinra, V. K., "Thermoelastic damping of a laminated beam in flexure and extension", *J. Reinforced Plast. and Compos.* 12, 210–226, 1993.

8.3.10 Bishop, J. E. and Kinra, V. K., "Elastothermodynamic damping in composite materials", *Mech. Compos. Mater. Struct.* 1, 75–93, 1994.

8.3.11 Bishop, J. E. and Kinra, V. K., "Analysis of elastothermodynamic damping in particle-reinforced metal-matrix composites", *Metall. Mater. Trans.* 26A, 2773–2783, 1995.

8.3.12 Kinra, V. K. and Milligan, K. B., "A second law analysis of thermoelastic damping", *J. Appl. Mech.* 61, 71–76, 1994.

8.3.13 Biot, M. A., "Thermoelasticity and irreversible thermodynamics", *J. Appl. Phys.* 27, 240–253, 1956.

8.3.14 Bishop, J. E. and Kinra, V. K., "Equivalence of the mechanical and entropic descriptions of elastothermodynamic damping in composite materials", *Mech. Compos. Mater. Struct.* 3, 83–95, 1996.

8.3.15 Bennewitz, K. and Rötger, H., *Phys. Z.* 37, 578, 1936.

8.3.16 Randall, R. H., Rose, F. C., and Zener, C., "Intercrystalline thermal currents as a source of internal friction", *Phys. Rev.* 53, 343–348, 1939.

8.4.1   Biot, M. A., "General theory of three-dimensional consolidation", *J. Appl. Phys.* 12, 155–164, 1941.

8.4.2   Biot, M. A. and Willis, D. G., "The elastic coefficients of the theory of consolidation", *J. Appl. Mech.* 594–601, 1957.

8.4.3   Biot, M. A., "Theory of propagation of elastic waves in a fluid saturated porous solid. I. Low frequency range", *J. Acoust. Soc. Am.* 28, 168–178, 1956.

8.4.4   Biot, M. A., "Theory of propagation of elastic waves in a fluid saturated porous solid. II. Higher frequency range", *J. Acoust. Soc. Am.* 28, 179–191, 1956.

8.4.5   Biot, M. A., "Generalized theory of acoustic propagation in porous dissipative media", *J. Acoust. Soc. Am.* 34, 1254–1264, 1962.

8.4.6   Biot, M. A., "Mechanics of deformation and acoustic propagation in porous media", *J. Appl. Phys.* 33, 1482–1498, 1962.

8.4.7   Kenyon, D. E., "Consolidation in transversely isotropic solids", *J. Appl. Mech.* 46, 65–70, 1979.

8.4.8   Plona, T. J., "Observation of a second bulk compressional wave in a porous medium at ultrasonic frequencies", *Appl. Phys. Lett.* 36, 259–261, 1980.

8.4.9   Berryman, J. G., "Confirmation of Biot's theory", *Appl. Phys. Lett.* 37, 382–384, 1980.

8.4.10 Lakes, R. S., Yoon, H. S., and Katz, J. L., "Slow compressional wave propagation in wet human and bovine cortical bone", *Science* 220, 513–515, 1983.

8.4.11 Zeevaert, L., *Foundation Engineering for Difficult Subsoil Conditions*, Van Nostrand, NY, 1972.

8.4.12 Gent, A. N. and Rusch, K. C., "Viscoelastic behavior of open cell foams", *Rubber Chem. Technol.* 39, 388–396, 1966.

8.4.13 Hilyard, N. C., *Mechanics of Cellular Plastics*, Macmillan, NY, 1982.

8.4.14 AliMed®, Inc., Dedham, MA.

8.5.1   Zener, C., "Internal friction in solids. V. General theory of macroscopic eddy currents", *Phys. Rev.* 53, 1010–1013, 1938.

8.5.2 Coronel, V. F. and Beshers, D. N., "Magnetomechanical damping in iron", *J. Appl. Phys.* 64, 2006–2015, 1988.

8.6.1 Ferry, J. D., Landel, R. F., and Williams, M. L., "Extensions of the Rouse theory of viscoelastic properties to undiluted linear polymers", *J. Appl. Phys.* 26, 359–362, 1955.

8.6.2 Bagley, R. L., "The thermorheologically complex material", *Int. J. Eng. Sci.* 29, 797–806, 1991.

8.6.3 Bendler, J. T. and Schlesinger, M. F., "Generalized Vogel law for glass forming liquids", *J. Stat. Phys.* 53, 531–541, 1988.

8.6.4 Baer, E., Hiltner, A., and Keith, H. D., "Hierarchical structure in polymeric materials", *Science* 235, 1015–1022, 1987.

8.6.5 Jonscher, A. K., "A new understanding of the dielectric relaxation of solids", *J. Mater. Sci.* 16, 2037–2060, 1981.

8.6.6 Jonscher, A. K., "The 'universal' dielectric response", *Nature* 267, 673–679, 1977.

8.6.7 Dissado, L. A. and Hill, R. M., "Non-exponential decay in dielectrics and dynamics of correlated systems", *Nature* 279, 685–689, 1979.

8.6.8 Palmer, R. G., Stein, D. L., Abrahams, E., and Anderson, P. W., "Models of hierarchically constrained dynamics for glassy relaxation", *Phys. Rev. Lett.* 53, 958–961, 1984.

8.6.9 Schwarzl, F. and Staverman, A. J., "Time-temperature dependence of linear viscoelastic behavior", *J. Appl. Phys.* 23, 838–843, 1952.

8.6.10 Sperling, L. H., "Sound and vibration damping with polymers", in *Sound and Vibration Damping with Polymers*, ed. R. D. Corsaro and L. H. Sperling, American Chemical Society, Washington, DC, 1990.

8.6.11 Blonski, S., Brostow, W., and Kubát, J., "Molecular dynamics simulations of stress relaxation in metals and polymers", *Phys. Rev. B* 49, 6494–6500, 1994.

8.7.1 Kê, T. S., "Experimental evidence of the viscous behavior of grain boundaries in metals", *Phys. Rev.* 71, 533–546, 1947.

8.7.2 Zener, C., "Theory of the elasticity of polycrystals with viscous grain boundaries", *Phys. Rev.* 60, 906–908, 1941.

8.7.3 Crossman, F. W. and Ashby, M. F., "The non-uniform flow of polycrystals by grain-boundary sliding accommodated by power law creep", *Acta Metall.* 23, 425–440, 1975.

8.7.4 Bell, R. L. and Langdon, T. G., "An investigation of grain boundary sliding during creep", *J. Mater. Sci.* 2, 313–323, 1967.

8.7.5 Zelin, M. G. and Mukherjee, A. K., "Deformation strengthening of grain-boundary sliding in a Pb–62wt%Sn alloy", *Philos. Mag. Lett.* 68, 201–206, 1993.

8.7.6 Zelin, M. G., Yang, H., Valiev, R. Z., and Mukherjee, A. K., "Interaction of high-temperature deformation mechanisms in a magnesium alloy with mixed fine and coarse grains", *Met. Trans.* 23A, 3135–3140, 1992.

8.7.7 Zelin, M. G. and Mukherjee, A. K., "Common features of intragranular dislocation slip and intergranular sliding", *Philos. Mag. Lett.* 68, 207–214, 1993.

8.7.8 Gittus, J., ed., *Cavities and Cracks in Creep and Fatigue*, Applied Science Publishers, London, 1981.

8.7.9 Nelson, D. J. and Hancock, J. W., "Interfacial slip and damping in fibre reinforced composites", *J. Mater. Sci.* 13, 2429–2440, 1978.

8.7.10 Lakes, R. S. and Saha, S., "Cement line motion in bone," *Science* 204, 501–503, 1979.

8.7.11 Cremer, L. and Heckl, M., *Structure-Borne Sound*, 2nd ed., transl. E. E. Ungar, Springer Verlag, Berlin, 1988.

8.8.1 Snoek, J., "Mechanical after-effect and chemical constitution", *Physica* 6, 591–592, 1939.

8.8.2 Snoek, J., "Effect of small quantities of carbon and nitrogen on the elastic and plastic properties of iron", *Physica* 8, 711–733, 1941.

8.8.3 Zener, C., "Stress-induced preferential ordering of pairs of solute atoms in metallic solid solution", *Phys. Rev.* 71, 34–38, 1947.

8.8.4 Childs, B. G. and Le Claire, A. D., "Relaxation effects in solid solutions arising from changes in local order. I. Experimental", *Acta Metall.* 2, 718–726, 1954.

8.8.5 Spears, C. J. and Feltham, P., "On the amplitude-independent internal friction in crystalline solids", *J. Mater. Sci.* 7, 969–971, 1972.

8.8.6 Feltham, P., "Internal friction in concentrated solid solutions", *Philos. Mag. A* 50, L35–L38, 1984.

8.8.7 Granato, A. and Lücke, K., "Theory of mechanical damping due to dislocations", *J. Appl. Phys.* 27, 583–593, 1956.

8.8.8 Lücke, K. and Granato, A. V., "Simplified theory of dislocation damping including point defect drag. I. Theory of drag by equidistant point defects", *Phys. Rev. B* 24, 6991–7006, 1981.

8.8.9 Robinson, W. H., "Amplitude-independent mechanical damping in alkali halides", *J. Mater. Sci.* 7, 115–123, 1972.

8.8.10 Bordoni, P. G., Nuovo, M., and Verdini, L., "Relaxation of dislocations in copper", *Nuovo Cimento* 14, 273–314, 1959.

8.8.11 Hasiguti, R., "The structure of defects in solids", *Ann. Rev. Mater. Sci.* 69–92, 1972.

8.8.12 Hasiguti, R., Igata, M., and Kamoshita, G., "Internal friction peaks in cold-worked metals", *Acta Metall.* 10, 442–447, 1962.

8.8.13 Schoeck, G., Bisogni, E., and Shyne, J., "The activation energy of high temperature internal friction", *Acta Metall.* 12, 1466–1468, 1964.

8.8.14 Cagnoli, G., Gammaitoni, L., Marchesoni, F., and Segoloni, D., "On dislocation damping at low frequency", *Philos. Mag. A* 68, 865–870, 1993.

8.8.15 Brodt, M., Cook, L. S., and Lakes, R. S., "Apparatus for determining the properties of materials over ten decades of frequency and time: refinements", *Rev. Sci. Instr.* 66, 5292–5297, 1995.

8.8.16 Lakes, R. S. and Quackenbush, J., "Viscoelastic behaviour in indium tin alloys over a wide range of frequency and time", *Philos. Mag. Lett.* 74, 227–232, 1996.

8.8.17 Hirscher, M., Schweitzer, E., Weller, M., and Kronmüller, H., "Internal friction in NiAl single crystals, *Philos. Mag. Lett.* 74, 189–194, 1996.

8.8.18 Marchesoni, F. and Patriarca, M., "Self-organized criticality in dislocation networks", *Phys. Rev. Lett.* 72, 4101–4104, 1994.

8.8.19 Braginsky, V. B., Mitrofanov, V. P., and Panov, V. I., *Systems with Small Dissipation*, University of Chicago Press, IL, 1985.

8.8.20 Papadakis, E. P., "Ultrasonic attenuation caused by scattering in polycrystalline media", in *Phys. Acoust.* IVB, 269–328, 1968.

8.9.1 Ashby, M. F., "A first report on deformation mechanism maps", *Acta Metall.* 20, 887–897, 1972.

8.9.2   Frost, H. J. and Ashby, M. F., *Deformation Mechanism Maps*, Pergamon, Oxford, 1982.

8.9.3   Hynes, A. and Doremus, R., "Theories of creep in ceramics", *Critical Rev. Solid State Mater. Sci.* 21, 129–187, 1996.

8.9.4   Karato, S. and Wu, P., "Rheology of the upper mantle: a synthesis", *Science* 260, 771–778, 1993.

8.9.5   Martin, J. L., "Creep in pure metals", in *Creep Behaviour of Crystalline Solids*, ed. B. Wiltshire and R. W. Evans, Pineridge Press, Swansea, UK, 1985, pp. 1–33; adduced diffusion results of Sherby and Miller, 1979.

8.9.6   Oikawa, H. and Langdon, T. G., "The creep characteristics of pure metals and metallic solid solution alloys", in *Creep Behaviour of Crystalline Solids*, ed. B. Wiltshire and R. W. Evans, Pineridge Press, Swansea, UK, 1985, pp. 33–82.

8.10.1 Lakes, R. S., "Shape-dependent damping in piezoelectric solids," *IEEE Trans. Sonics Ultrasonics* SU27, 208–213, 1980.

8.10.2 Martin, G. E., "Dielectric, piezoelectric, and elastic losses in longitudinally polarized segmented ceramic tubes," *U.S. Navy J. Underwater Acoust.* 15, 329–332, 1965.

8.10.3 Furukawa, T. and Fukada, E., "Piezoelectric relaxation in composite epoxy-PZT system due to ionic conduction", *Jpn. J. Appl. Phys.* 16, 453–458, 1977.

8.10.4 Bur, A. J., "Measurements of the dynamic piezoelectric properties of bone as a function of temperature and humidity," *J. Biomech.* 9, 495–507, 1976.

8.10.5 Holland, R., "Representation of dielectric, elastic, and piezoelectric losses by complex coefficients," *IEEE Trans. Sonics Ultrasonics.* SU14, 18–20, 1967.

8.10.6 Jaffe, H. and Berlincourt, D. A., "Piezoelectric transducer materials," *Proc. IEEE* 53, 1372–1386, 1965.

8.10.7 Cady, W. G., *Piezoelectricity*, Dover, NY, 1964.

8.10.8 Joffe, A. F., *The Physics of Crystals*, McGraw-Hill, NY, 1928.

8.10.9 "IRE standards on piezoelectric crystals: measurements of piezoelectric ceramics," *Proc. IRE* 49, 1161–1169, 1961.

8.10.10 IEEE Standard on Piezoelectricity, IEEE 176–1978; Inst. Electrical, Electronics Engineers, NY, 1978.

8.10.11 Lakes, R. S., "Thermoelastic damping in materials with a complex coefficient of thermal expansion", *J. Mech. Behavior Mater.* 8, 201–216, 1997.

8.11.1 Palmer, R. G., Stein, D. L., Abrahams, E., and Anderson, P. W., "Models of hierarchically constrained dynamics for glassy relaxation", *Phys. Rev. Lett.* 53, 958–961, 1984.

8.11.2 Klafter, J. and Schlesinger, M. F., "On the relationship among three theories of relaxation in disordered systems", *Proc. Natl. Acad. Sci. USA* 83, 848–851, 1986.

8.11.3 Jonscher, A. K., "The 'universal' dielectric response", *Nature* 267, 673–679, 1977.

8.11.4 Ngai, K. L., "Universality of low-frequency fluctuation, dissipation, and relaxation properties of condensed matter." I, *Comments Solid State Phys.* 9, 127–140, 1979.

8.11.5 Scher, H., Shlesinger, M. F., and Bendler, J. T., "Time scale invariance in transport and relaxation", *Phys. Today* 44, 26–34, 1991.

8.11.6 Quinn, J. J., Speake, C. C., and Brown, L. M., "Materials problems in the construction of long-period pendulums", *Philos. Mag. A.* 65, 261–276, 1992.

8.11.7 Cottrell, A. H., "Andrade Creep", *Philos. Mag. Lett.* 73, 35–37, 1996.

8.11.8 Jonscher, A. K., *Dielectric Relaxation in Solids*, Chelsea Dielectrics Press, London, 1982, p. 294.

8.11.9 Schiener, B., Böhmer, R., Loidl, A., and Chamberlin, R. V., "Nonresonant spectral hole-burning in the slow dielectric response of supercooled liquids", *Science* 274, 752–754, 1996.

8.12.1 Ritchie, I. G., Pan, Z. L., and Goodwin, F. E., "Characterization of the damping properties of die-cast zinc–aluminum alloys", *Met. Trans.* 22A, 617–622, 1991.

8.12.2 Ritchie, I. G. and Pan, Z. L., "High damping metals and alloys", *Met.Trans.* 22A, 607–616, 1991.

8.12.3 Laddha, S. and Van Aken, D. C., "On the application of magnetomechanical models to explain damping in an antiferromagnetic copper–manganese alloy", *Met. Trans.* 26A, 957–964, 1995.

8.14.1 Dyson, F., "Time without end—physics and biology in an open universe", *Rev. Mod. Phys.* 51, 447–460, 1979.

E8.1.1 Ashby, M. F., "On the engineering properties of materials", *Acta Metall.* 37, 1273–1293, 1989.

E8.1.2 *Modern Plastics Encyclopedia*, 51 No. 10A, McGraw-Hill, NY, 1974–1975.

chapter nine

Viscoelastic composite materials

§9.1 Introduction

Composite materials are those which contain two or more distinct constituent materials or phases, on a microscopic or macroscopic size scale larger than the atomic scale, and in which properties such as the elastic modulus are significantly altered in comparison with those of a homogeneous material. In this vein, fiberglass and other fibrous materials are viewed as composites, but alloys such as brass are not. Semicrystalline polymers such as polyethylene have a heterogeneous structure which can be treated via composite theory. Biological materials also have a heterogeneous structure and are known as natural composites.

§9.2 Composite structures

The properties of composites are greatly dependent on *microstructure*. Composites differ from homogeneous materials in that considerable control can be exerted over the larger scale structure, and hence over the desired properties. In particular, the properties of a composite depend on the *shape* of the heterogeneities, on the *volume fraction* occupied by them, and on the *interface* between the constituents. Volume fraction refers to the ratio of the volume of a constituent to the total volume of a composite specimen. The shape of the heterogeneities in a composite is classified as follows. The principal inclusion shape categories (Fig. 9.1) are the particle, with no long dimension; the fiber, with one long dimension; and the platelet or lamina, with two long dimensions. The inclusions may vary in size and shape within a category. For example, particulate inclusions may be spherical, ellipsoidal, polyhedral, or irregular. *Cellular solids* are those in which the "inclusions" are voids or cells, filled with air or liquid. Cellular solids include honeycombs, in which the structure is largely two dimensional; and foams, in which the structure is three dimensional. Foams can be open cell, in which the foam has a structure of "ribs", with no barrier between adjacent cells; or closed cell, in which plate or membrane elements separate adjacent cells. Open-cell foams are of particular interest in the context of viscoelasticity, since viscoelastic damping can arise as a result of the viscosity of fluids (such as water or air) moving through the pore structure as discussed in §8.4. The *coated spheres* morphology has multiple length scales. The entire volume is filled with particles of one phase coated with a layer of a second phase. This morphology is used in theoretical analyses of extremal behavior.

Composites may be isotropic or anisotropic. Properties of anisotropic materials depend on direction; isotropic materials have no preferred direction. Composites containing spherical particulate inclusions distributed randomly in an isotropic matrix are macroscopically isotropic. Unidirectional fibrous composites are highly anisotropic. Laminates containing fibrous layers can be prepared with a controlled degree of anisotropy. Anisotropy can be dealt with using a tensorial stress–strain relation. For elastic materials, Hooke's law of linear elasticity is given by

$$\sigma_{ij} = C_{ijkl} \varepsilon_{kl}. \tag{9.2.1}$$

with C_{ijkl} as the elastic modulus tensor, and the usual Einstein summation convention assumed in which repeated indices are summed over [2.8.1–2.8.3]. There are 81 components of C_{ijkl}, but taking into account the symmetry of the stress and strain tensors, only 36 of them are independent. If the elastic solid is describable by a strain energy function, the number of independent elastic constants is reduced to 21 by virtue of the resulting symmetry $C_{ijkl} = C_{klij}$. An elastic modulus tensor with 21 independent constants describes an anisotropic material with the most general type of anisotropy, triclinic symmetry. Materials with orthotropic symmetry are invariant to reflections in two orthogonal planes and are describable by nine elastic constants. Materials with axisymmetry, also called transverse isotropy or hexagonal symmetry, are invariant to 60° rotations about an axis and are describable by five independent elastic constants. For example, a unidirectional fibrous material may have orthotropic symmetry if the fibers are arranged in a rectangular packing, or hexagonal symmetry if the fibers are packed hexagonally. Wood has orthotropic symmetry while human bone has hexagonal symmetry. Materials with cubic symmetry are invariant to 90° rotations about each of three orthogonal axes. Cubic materials are describable by three elastic constants. Isotropic materials with properties independent of direction are describable by two independent elastic constants.

The elastic properties of composites have been studied extensively. The upper and lower bounds of *stiffness* of two-phase and multiphase composites have been obtained by Hashin [9.2.1] and Hashin and Shtrikman [9.2.2] in

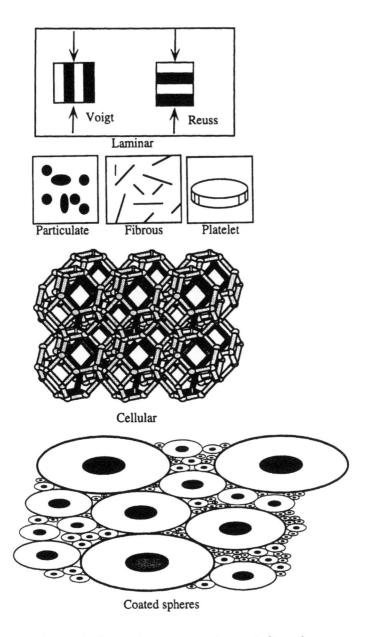

Figure 9.1 Composite structures. Arrows indicate force.

terms of volume fraction of constituents. Much work on composites is reviewed by Hashin [9.2.3].

In linearly viscoelastic materials, the Boltzmann superposition integral has the following form in the tensorial modulus formulation:

$$\sigma_{ij}(t) = \int_0^t C_{ijkl}(t - \tau) \frac{d\varepsilon_{kl}}{d\tau} \, d\tau. \tag{9.2.2}$$

Each component of the relaxation modulus tensor $C_{ijkl}(t)$ can have not only a different value but also a different time dependence. For viscoelastic materials, we may have an asymmetric modulus tensor $C_{ijkl}(t) \neq C_{klij}(t)$ [9.2.4, 9.2.5] since there is no strain energy function for viscoelastic materials. However, $C_{ijkl}(0) = C_{klij}(0)$ and $C_{ijkl}(\infty) = C_{klij}(\infty)$. No effect of asymmetry in the modulus or compliance tensor is seen in isotropic materials; however, in some anisotropic materials, certain equalities of Poisson's ratio terms seen in elastic materials do not appear in viscoelastic materials.

As for viscoelastic composites, if the phases are linearly viscoelastic, the effective relaxation and creep functions of the composite can be calculated from constituent properties by the correspondence principle of the theory of linear viscoelasticity [9.2.6, 9.2.7]. In some cases explicit results in terms of linear viscoelastic matrix properties were given, permitting direct use of experimental information [9.2.8].

§9.3 *Prediction of elastic and viscoelastic properties*

For the simplest case of an *elastic* two-phase composite, the stiffness of Voigt and Reuss composites described below represent rigorous upper and lower bounds on Young's modulus for a given volume fraction of one phase (Fig. 9.2). The Hashin–Shtrikman equations presented below represent upper and lower bounds on the elastic stiffness of *isotropic* composites. Properties of viscoelastic composites containing spherical or platelet inclusions are also presented in terms of constituent properties. Behavior of viscoelastic composites is predicted by the correspondence principle [1.6.1, 9.3.1–9.3.3], by which the relationship between constituent and composite elastic properties can be converted to a steady-state harmonic viscoelastic relation by replacing Young's moduli E by $E^*(i\omega)$ or E^*, in which ω is the angular frequency of the harmonic loading.

In this section we explore the viscoelastic properties of several two-phase (with volume fractions $V_1 + V_2 = 1$) composites of well-defined structure, in terms of the assumed properties of the constituents [9.3.4]. As for the Voigt and Reuss structures, we shall find that curves representing the viscoelastic properties enclose a region in a stiffness–loss map. However, these curves are not proven to be bounds on the viscoelastic behavior.

§9.3.1 *Voigt composite*

Let phase 1 be stiff; let phase 2 be high loss. The composite can contain laminations as shown in Fig. 9.1 or it can be made of continuous unidirectional fibers; in either case the strain in each phase is the same. In a simple

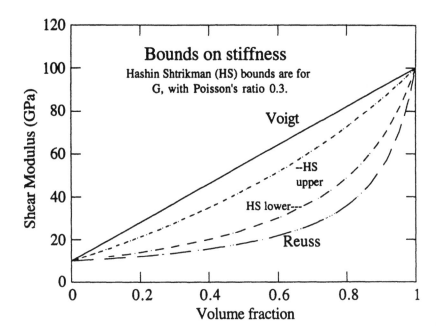

Figure 9.2 Modulus vs. volume fraction of a constituent for several elastic composites. HS refers to the Hashin–Shtrikman bounds for isotropic composites.

one-dimensional view one neglects any complications due to mismatch in the Poisson effect. For elastic materials with no slip between the phases,

$$E_c = E_1 V_1 + E_2 V_2,$$

in which E_c, E_1, and E_2 refer to Young's modulus of the composite, phase 1 and phase 2; and V_1 and V_2 refer to the volume fraction of phase 1 and phase 2 with $V_1 + V_2 = 1$. The dependence of stiffness on volume fraction is shown in Fig. 9.2.

Use of the correspondence principle gives

$$E_c^* = E_1^* V_1 + E_2^* V_2, \tag{9.3.1}$$

with $E^* = E' + i\,E''$ and loss tangent $\tan\delta = E''/E'$. By taking the ratio of real and imaginary parts, the loss tangent of the composite $\tan\delta_c = E_c''/E_c'$ is given by:

$$\tan\delta_c = \frac{V_1 \tan\delta_1 + V_2 \dfrac{E_2'}{E_1'} \tan\delta_2}{V_1 + \dfrac{E_2'}{E_1'} V_2}. \tag{9.3.2}$$

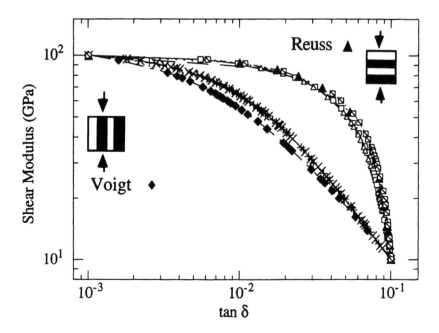

Figure 9.3 Stiffness–loss map. Calculated behavior of several composites. Each point corresponds to a different volume fraction. Reuss: solid triangles, ▲. Voigt: solid diamonds, ◆. Hashin–Shtrikman, different values of Poisson's ratio of compliant phase. H-S lower (top), G_L, Eq. 9.3.8, $v = 0.3$, □; $v = 0.45$, △; $v = 0.3(1 + 0.1i)$, ◺. H-S upper, G_U, Eq. 9.3.9, $v = 0.3$, x; $v = 0.45$, +.

The relation between stiffness and damping is shown in Fig. 9.3.

§9.3.2 *Reuss composite*

The geometry of the Reuss model structure is shown in Fig. 9.1; each phase experiences the same stress. For elastic materials,

$$\frac{1}{E_c} = \frac{V_1}{E_1} + \frac{V_2}{E_2}.$$

The dependence of stiffness on volume fraction is shown in Fig. 9.2. Since the constituents are aligned, this composite is structurally anisotropic. Since the Reuss laminate is identical to the Voigt laminate except for orientation with respect to the stress, the laminate is mechanically anisotropic.

Again by using the correspondence principle, the viscoelastic relation is obtained as

$$\frac{1}{E_c^*} = \frac{V_1}{E_1^*} + \frac{V_2}{E_2^*}. \tag{9.3.3}$$

Again by separating the real and imaginary parts of E_c^*, the loss tangent of the composite $\tan \delta_c$ is obtained:

$$\tan \delta_c =$$

$$\frac{(\tan \delta_1 + \tan \delta_2)\left[V_1 + V_2 \frac{E_1'}{E_2'}\right] - (1 - \tan \delta_1 \tan \delta_2)\left[V_1 \tan \delta_2 + V_2 \tan \delta_1 \frac{E_1'}{E_2'}\right]}{(1 - \tan \delta_1 \tan \delta_2)\left[V_1 + V_2 \frac{E_1'}{E_2'}\right] + (\tan \delta_1 + \tan \delta_2)\left[V_1 \tan \delta_2 + V_2 \tan \delta_1 \frac{E_1'}{E_2'}\right]}.$$

$$(9.3.4)$$

In the compliance formulation, the Reuss relation can be written more simply in terms of the compliances $J^* = 1/E^*$.

$$J_c^* = J_1^* V_1 + J_2^* V_2. \tag{9.3.5}$$

The corresponding analysis of the loss tangent is also simpler.

$$\tan \delta_c = \frac{V_1 \tan \delta_1 + V_2 \frac{J_2'}{J_1'} \tan \delta_2}{V_1 + \frac{J_2'}{J_1'} V_2}. \tag{9.3.6}$$

The relation between stiffness and damping is shown, in comparison with the Voigt composite, in Fig. 9.3.

§9.3.3 Hashin–Shtrikman composite

By allowing for "arbitrary" phase geometry, the upper and lower bounds on the elastic moduli of an *isotropic* composite as a function of composition were developed using variational principles. The lower bound for the elastic shear modulus G_L of the composite was given as [9.2.2]

$$G_L = G_2 + \frac{V_1}{\dfrac{1}{G_1 - G_2} + \dfrac{6(K_2 + 2G_2)V_2}{5(3K_2 + 4G_2)G_2}}, \tag{9.3.7}$$

in which K_1, G_1, and V_1 and K_2, G_2, and V_2 are the bulk modulus, shear modulus, and volume fraction of phases 1 and 2, respectively. Here $G_1 > G_2$, so that G_L represents the lower bound on the shear modulus. Interchanging the numbers 1 and 2 in Eq. 9.3.7 results in the upper bound G_U for the shear modulus. Bounds on the stiffness are shown in Fig. 9.2.

As for viscoelastic materials, the correspondence principle is again applied [9.3.4]. The complex viscoelastic shear moduli of the composite $G_L{}^*$ and $G_U{}^*$ are obtained as

$$G_L^{\bullet} = G_2^{\bullet} + \cfrac{V_1}{\cfrac{1}{G_1^{\bullet} - G_2^{\bullet}} + \cfrac{6\left(K_2^{\bullet} + 2G_2^{\bullet}\right)V_2}{5\left(3K_2^{\bullet} + 4G_2^{\bullet}\right)G_2^{\bullet}}}, \qquad (9.3.8)$$

and

$$G_U^{\bullet} = G_1^{\bullet} + \cfrac{V_2}{\cfrac{1}{G_2^{\bullet} - G_1^{\bullet}} + \cfrac{6\left(K_1^{\bullet} + 2G_1^{\bullet}\right)V_1}{5\left(3K_1^{\bullet} + 4G_1^{\bullet}\right)G_1^{\bullet}}}. \qquad (9.3.9)$$

In these cases the loss tangent is most complicated to write explicitly, so it is more expedient to graphically display computed numerical values in Fig. 9.3.

§9.3.4   Spherical particulate inclusions

For a small volume fraction $V_1 = 1 - V_2$ of spherical elastic inclusions (particles) in a continuous phase of another elastic material, the shear modulus of the composite G_c was given as [9.3.3]

$$\frac{G_c}{G_1} = 1 - \cfrac{15(1 - v_1)\left(1 - \cfrac{G_2}{G_1}\right)V_2}{7 - 5v_1 + 2(4 - 5v_1)\cfrac{G_2}{G_1}}, \qquad (9.3.10)$$

in which v_1 is the Poisson's ratio of phase 1, and phase 1 and phase 2 represent the matrix material and the inclusion material, respectively. The stiffness of such a composite, as a function of volume fraction of inclusions, is close to the Hashin–Shtrikman lower bound for isotropic materials. For larger volume fractions of inclusions, analysis is more complicated as a result of the interaction of stress fields around nearby inclusions (§9.7). Stiff spherical inclusions are less efficient, per volume, in achieving a stiff composite than fiber or particle inclusions. Conversely, soft spherical inclusions have the least effect in reducing the stiffness, in comparison to other inclusion shapes.

By using the correspondence principle again and assuming there is no relaxation in Poisson's ratio, Eq. 9.3.10 becomes [9.3.4]

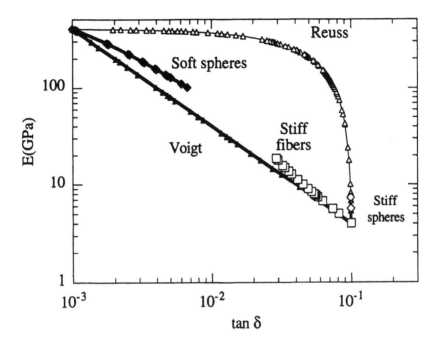

Figure 9.4 Stiffness–loss map. Calculated behavior of several composites with assumed phase properties different from those in Fig. 9.3. Each point corresponds to a different volume fraction. Effect of phase geometry for Voigt (▲), Reuss (△), stiff spheres (◇), soft spheres (◆), and randomly oriented stiff fibers (□).

$$G_c^* = G_1^* - \frac{15(1-v_1)(G_1^* - G_2^*)V_2}{7 - 5v_1 + 2(4 - 5v_1)\dfrac{G_2^*}{G_1^*}},$$

(9.3.11)

for the complex shear modulus of the composite. The loss tangent again is complicated to write explicitly, but can readily be computed numerically, as presented in Fig. 9.4. Behavior of a composite with soft particles approximates the Voigt relation in the stiffness–loss map; by contrast a composite with stiff spherical inclusions behaves more as a Reuss composite.

§9.3.5 Fiber inclusions

For a dilute random suspension of fiber elastic inclusions of phase 2 in a matrix of phase 1, Young's modulus E_c of the composite is as follows [9.3.2], assuming a Poisson's ratio of both fiber and matrix to be 1/4:

$$E_c = \frac{1}{6}E_2 V_2 + E_1 \frac{1 + \dfrac{1}{4}V_2 + \dfrac{1}{6}V_2^2}{1 - V_2}.$$

(9.3.12)

Using the correspondence principle gives

$$E_c^* = \frac{1}{6}E_2^*V_2 + E_1^* \frac{1 + \frac{1}{4}V_2 + \frac{1}{6}V_2^2}{1 - V_2}. \tag{9.3.13}$$

At low volume fraction V_2 of randomly oriented fibers, the stiffening effect of the fibers is 1/6 of the value obtained in the Voigt composite in which the fibers are all aligned. However, for viscoelastic composites, the curve for random fibers in the stiffness–loss map in Fig. 9.4 is close to the curve for the Voigt solid. Corresponding volume fractions are not, however, identical.

§9.3.6 Platelet inclusions

For a dilute random suspension of platelet elastic inclusions of phase 2 in a matrix of phase 1, the shear modulus G_c of the composite was given as [9.3.3]

$$G_c = G_1 + \frac{V_2(G_2 - G_1)}{15} \left[\frac{9K_2 + 4(G_1 + 2G_2)}{K_2 + \frac{4}{3}G_2} + 6\frac{G_1}{G_2} \right]. \tag{9.3.14}$$

Stiff platelets are more efficient, per volume, in achieving a stiff composite than fiber or particle inclusions. The predicted stiffness of a random platelet-reinforced composite is identical to the Hashin–Shtrikman upper bound for isotropic materials [9.3.5]; however, this limiting stiffness will only be achieved if the platelets are infinitely thin [9.3.6].

Again, by using the correspondence principle, Eq. 9.3.14 becomes

$$G_c^* = G_1^* + \frac{V_2(G_2^* - G_1^*)}{15} \left[\frac{9K_2^* + 4(G_1^* + 2G_2^*)}{K_2^* + \frac{4}{3}G_2^*} + 6\frac{G_1^*}{G_2^*} \right], \tag{9.3.15}$$

for the complex shear modulus of the composites. In viscoelastic composites, the curve for platelet inclusions in the stiffness–loss map is close to the Voigt and Hashin–Shtrikman upper bound curves, and corresponds to minimal damping for given composite stiffness.

We may consider "unintentional" platelet structure in materials such as gray cast iron, as described by Millett et al. [9.3.7]. Gray cast iron exhibits a loss tangent of about 0.011 over a range of frequencies, and it contains platelet-shaped inclusions of graphite with a loss tangent of 0.015. The graphite gives rise to most of the loss; the cause is thought to be dislocation loop motion.

§9.3.7   Stiffness–loss maps

Predicted properties of viscoelastic composites are plotted as stiffness–loss maps (plots of $|E^*|$ vs. tan δ) as shown in the following figures.

In the stiffness–loss map in Fig. 9.3, the Reuss curve which had been on the bottom in the plot of stiffness vs. volume fraction in Fig. 9.2 now is on top; the Voigt curve becomes the lower curve in the stiffness–loss map. Therefore, while the Voigt geometry is most favorable in terms of attaining a stiff composite from the least amount of stiff inclusions, it gives the least damping for given stiffness. Moreover, the lower and upper two-phase Hashin–Shtrikman composites behave similarly to the Voigt and Reuss composites, respectively, in the stiffness–loss map even though they differ greatly in a plot of stiffness vs. volume fraction. As for the physical attainment of the Voigt and Reuss composites, simple laminates can be made as in Fig. 9.1, but these are anisotropic. In the Reuss structure each phase carries the full stress, so that a composite of this type will be weak if, as is usual, the soft phase is weak.

The effect of inclusion shape is shown in Fig. 9.4. The composite containing soft spherical inclusions is also found to behave similarly to the Voigt composite in that a small volume fraction of soft, viscoelastic material has a comparatively small effect on the loss tangent [9.3.4]. A small amount of stiff spherical inclusions confers a Reuss-like effect of increasing the stiffness while not changing the damping much. Stiff fibers are more Voigt-like in reducing the damping as they increase the stiffness. In a composite containing soft platelet inclusions, it is found that the results are similar to those of the Reuss structure. A small volume fraction of platelet inclusions as phase 2 results in a large increase in loss tangent without significant stiffness reduction. However, soft platelets resemble penny-shaped cracks in the matrix, so that such a composite would be weaker than the matrix, particularly if the matrix were brittle.

The effect of increasing the stiffness of the stiff phase, from 200 GPa, corresponding to steel, to 400 GPa, corresponding to tungsten, is shown in Fig. 9.5. For both composites, tan δ of the inclusions is assumed to be 0.001, a value representative of results reported in the literature. Observe that if the stiff phase is made stiffer, the resulting composite can be made both stiff and lossy (with a large value of tan δ).

Figure 9.6 shows the effect of changing the properties of the lossy phase. As would be expected, a lossy phase of the highest possible damping gives rise to a composite with both high stiffness and high loss [9.3.8]. Moreover, both the Reuss and Voigt curves are convex to the right when the lossy phase is stiff.

Plotted viscoelastic behavior of the Voigt and Reuss composites encloses a region on the stiffness–loss map, and the properties of the other composites considered lie within that region. However, no proof has been given that these curves constitute bounds on the viscoelastic behavior. That issue is considered in the next section.

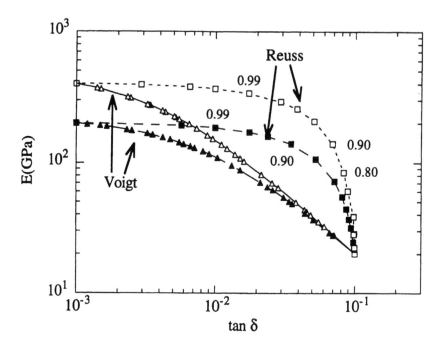

Figure 9.5 Stiffness–loss map showing the effect of the stiffness of the stiff phase [9.3.8]. Each point corresponds to a different volume fraction, indicated by numbers.

Results of this section are summarized as follows:

1. In a stiffness–loss map, the two-phase Hashin composites corresponding to upper and lower elastic bounds behave similarly to the Voigt and Reuss composites, respectively.
2. Reuss laminates and materials filled with stiff, low-loss platelets in a compliant high-loss phase exhibit high stiffness combined with high-loss tangent. However, in the Reuss structure each phase carries the full stress, so that a composite of this type will not be strong if, as is usual, the compliant phase is weak.
3. A composite containing soft lossy spherical inclusions in a stiff matrix behaves similarly to the Voigt composite: low loss and a reduction in stiffness.
4. If one desires a composite with maximal stiffness for given volume fraction, fibrous or platelet geometries are the best. However, if one desires a composite with high stiffness combined with high viscoelastic damping, a Reuss geometry or a concentrated suspension of stiff particles will give better results.

In a related vein, the effect of inclusion shape on viscoelastic response in the time domain has been considered, with the aim of minimizing creep in composites subjected to constant stress [9.3.9]. Composites containing stiff

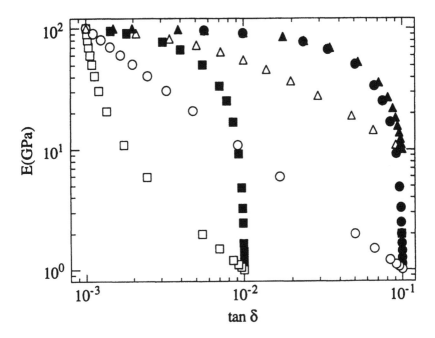

Figure 9.6 Stiffness–loss map showing the effect of the properties of the soft phase [9.3.8].

fibers are predicted to exhibit less creep than composites with stiff particles. Many composites currently in use have a rich and complicated hierarchical structure in which fibrous layers are organized into laminae. Analysis of such materials, even in the elastic case, can be quite complicated as a result of such structure. The transition to the viscoelastic case is facilitated by use of the dynamic correspondence principle in which elastic constants are replaced by complex functions of frequency [9.3.10]. Results obtained in this way do not account for damping due to unintentional cracks, pores, or chemical compounds formed at interfaces. Viscoelastic composites have also been analyzed via the finite element method [9.3.11] and the dynamic correspondence principle, with application to fibrous materials. Complex phase geometries which are not easily amenable to analytical solution may be treated by this approach.

§9.4   Bounds on viscoelastic properties

As for bounds on the properties, the curves for the Voigt and Reuss composites enclose a region in the stiffness–loss map, as do the curves for the upper and lower Hashin–Shtrikman composites. They represent extremes of composites which can be fabricated; however, the issue of bounds on the behavior is distinct. Roscoe [9.4.1] has mathematically established bounds for the real and imaginary parts E′ and E″ of the complex modulus of composites and has shown them to be equivalent to the Voigt and Reuss relations. Therefore, the stiffness $|E^*|$ of the composite is bounded from above by the Voigt limit and cannot exceed the stiffness of the stiff phase. This is not quite the same as establishing bounds for a stiffness–loss map since it is not obvious whether a maximum in $\tan \delta = E''/E'$ could be obtained simultaneously with a maximum in E'. In particular, we can construct $\tan \delta_c = E''_{Voigt}/E'_{Reuss} > E''_{Reuss}/E'_{Reuss}$ and be within the bounds of Roscoe. Bounds on the viscoelastic behavior in bulk deformation have been developed recently for fixed volume fractions of constituents [9.4.2] and for arbitrary volume fractions [9.4.3], incorporating the three-dimensional aspects of deformation. The bounds are in most cases attainable in a composite with the Hashin–Shtrikman coated spheres morphology [9.2.3], illustrated in Fig. 9.1. These expressions are as follows [9.2.3, 9.4.3]:

$$K_1^* = V_1 K_1 + (1 - V_1) K_2 - \frac{V_1 (1 - V_1)(K_1 - K_2)^2}{(1 - V_1) K_1 + V_1 K_2 + \frac{4}{3} G_1}, \qquad (9.4.1)$$

$$K_2^* = V_1 K_1 + (1 - V_1) K_2 - \frac{V_1 (1 - V_1)(K_1 - K_2)^2}{(1 - V_1) K_1 + V_1 K_2 + \frac{4}{3} G_2}. \qquad (9.4.2)$$

Here K_1 and K_2 are the bulk moduli of the two phases, G_1 and G_2 are the shear moduli, and V_1 is the volume fraction of the first phase. In the viscoelastic case, these moduli become complex. Since the shear moduli are also involved, the bounds on the complex bulk moduli depend on the phase Poisson's ratios. Increasing Poisson's ratio of the soft phase increases both the upper and lower bounds on the composite damping. If Poisson's ratio is complex, which is certainly possible, the composite damping can increase. The loss tangent $\tan \delta_B$ for bulk deformation of the composite can exceed the bulk loss tangent of either phase. However, the bulk loss tangent of the composite is bounded from above by the largest of $\tan \delta_B$ and $\tan \delta_G$ of the phases and bounded from below by the smallest of these. If the phases differ greatly in stiffness, the bounds can approach the Voigt and Reuss curves described above.

The bounds for viscoelastic composites, and, in particular, the microstructures identified as extremal, can be used to guide the design and fabrication of composites with specified damping, which may be high or low, as required for specific applications.

§9.5   Biological composite materials

Most composites of biological origin exhibit a rich hierarchical structure [9.5.1]. Hierarchical solids contain structural elements which themselves have structure. Human compact bone is a natural composite which exhibits a complex hierarchical structure [9.5.2, 9.5.3] as shown in Fig. 9.7. In bone, the presence of proteinaceous or polysaccharide phases can give rise to significant viscoelasticity, as discussed in §7.10. Observe in Fig. 7.15 that the loss tangent of compact bone attains a broad *minimum* over the frequency range associated with most bodily activities. The mineral phase of bone is crystalline hydroxyapatite ($Ca_{10}(PO_4)_6(OH)_2$) which is virtually elastic; it provides the stiffness of bone [9.5.4]. On the microstructural level are the osteons [9.5.5], which are large (~200-μm diameter) hollow fibers composed of concentric lamellae and of pores. The lamellae are built of fibers, and the fibers contain fibrils (smaller fibers). At the ultrastructural level (nanoscale) the fibers are a composite of the mineral hydroxyapatite and the protein collagen, which has a triple helix structure. Specific structural features have been associated with properties such as stiffness via the mineral crystallites [9.5.2], creep via the cement lines between osteons [9.5.6], and toughness via osteon pullout at the cement lines [9.5.7]. Lacunae are ellipsoidal pores with dimensions on the order 10 μm which provide spaces for the osteocytes (bone cells) which maintain the bone and allow it to adapt to changing conditions of stress by mediating growth or resorption of bone in response to stress. Haversian canals contain blood vessels which nourish the tissue, and nerves for sensation. Flow of fluid within the pore space in bone is important in the nutrition of bone cells. Stress-generated fluid flow can give rise to mechanical damping in bending or in tension–compression via the Biot mechanism. Two-level hierarchical analytical models involving the osteon as a large fiber have been used to understand anisotropic elasticity [9.5.8] and viscoelasticity [9.5.9] of bone.

Plant tissues such as wood [9.5.10] and bamboo [9.5.11] are cellular solids with complex hierarchical structures. The cell walls themselves contain small fibers. Viscoelastic properties of wood, discussed in §7.10.3, have been associated with those of the lignin within the wood, which exhibits a glass transition temperature near 100°C. Bamboo contains fiber-like structural features known as bundle sheaths [9.5.11] as well as oriented porosity along the stem axis. Bamboo, moreover, has functional gradient properties in which there is a distribution of Young's modulus across the culm (stem) cross section. Plant fibrous materials including jute, bamboo, sisal, and bamboo contain lignin and cellulose, in which cellulose microfibrils are embedded in a matrix of lignin and hemicellulose. In dry bamboo [9.5.12] tan δ was about 0.01 in bending and 0.02 to 0.03 in torsion, with little dependence on frequency in the audio range. Wet bamboo exhibited somewhat greater tan δ: 0.012 to 0.015 in bending and 0.03 to 0.04 in torsion. These figures are comparable to those for wood.

§9.6 Poisson's ratio of viscoelastic composites

Complex Poisson's ratios can occur in viscoelastic composites [9.4.3]. The sign of the imaginary part can be positive or negative. In the time domain,

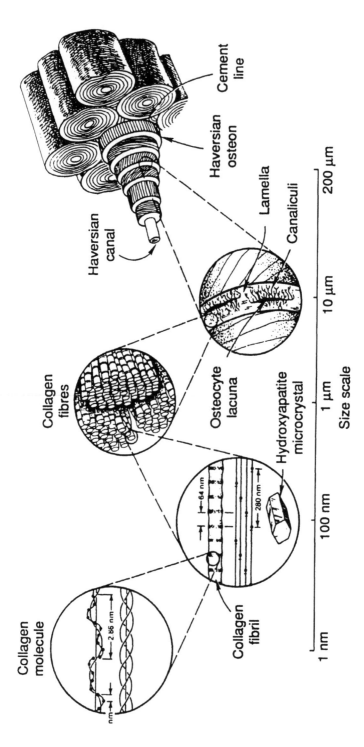

Figure 9.7 Hierarchical structure of human bone [9.5.1].

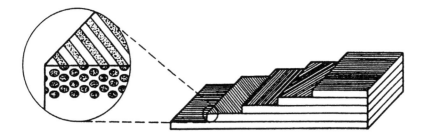

Figure 9.8   Hierarchical structure of a practical fibrous laminate [9.7.3].

the viscoelastic Poisson's ratio can increase or decrease [9.6.1]. This may be demonstrated by envisaging honeycombs or foams which have concave or "re-entrant" cells, and a corresponding negative Poisson's ratio. The interstices may be filled with a honeycomb or with a foam with a positive Poisson's ratio. By making one cellular structure elastic and the other viscoelastic, one can achieve time-dependent Poisson's ratios which increase or decrease with time. Complex Poisson's ratios have been determined for several matrix materials and particulate composites [9.6.2].

§9.7   Behavior of practical composite materials

§9.7.1   Structure

Composites used in engineering applications may have a fibrous [9.7.1, 9.7.2], cellular, or particulate structure. Practical fibrous composites commonly have a low order of hierarchical structure [9.7.3] in which fibers are embedded in a matrix to form an anisotropic sheet or lamina; such laminae are bonded to form a laminate (Fig. 9.8). In the analysis of fibrous composites, the fibers and matrix are regarded as continuous media in the analysis of the lamina; the laminae are then regarded as continuous in the analysis of the laminate. The stacking sequence of laminae and the orientation of fibers within them govern the composite anisotropy.

§9.7.2   Polymer matrix composites

Common fiber materials such as graphite and boron, and particle materials such as silica (SiO_2) tend to exhibit minimal creep at moderate stresses and at ambient temperature. Polymer matrix materials such as epoxy are viscoelastic materials. Viscoelastic behavior of the composite as a whole depends on the stress experienced by the matrix [9.7.4]. Consequently, polymer matrix unidirectional fiber composites creep little if they are loaded along the direction of the fibers, and creep considerably if loaded transversely or in shear (Fig. 9.9). As for laminates of fibrous layers, fiber-dominated graphite–epoxy layups such as $\{0\}_{48}$ and $\{0/45/0/-45\}_{6s}$ exhibit little viscoelastic response and little redistribution of strain due to creep [9.7.5].

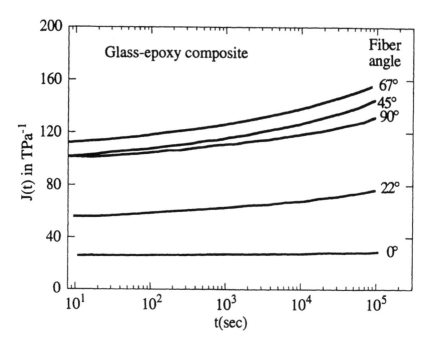

Figure 9.9 Experimental creep curves for various angles of stress with respect to fiber orientation for glass-epoxy composite. (Adapted from Sturgeon [9.7.4].)

The subscripts in the lamination code represent the number of plies, and s represents a laminate symmetrical about its midplane; the quantities in the brackets represent the angle of each ply with respect to a given direction. By contrast, matrix-dominated layups such as $\{90\}_{48}$ and $\{90/-45/90/-45\}_{6s}$ exhibited significant creep and strain redistribution.

   The $\{\pm45\}_s$ stacking sequence of graphite–epoxy laminates is considered matrix dominated since much of the applied load is transmitted through the matrix. Master curves for these composites disclose relaxation approaching zero stiffness at sufficiently long time or sufficiently high temperature. The shift factor depends on moisture content as well as temperature [9.7.6]. In graphite–epoxy composites, fiber spacings of 2 to 7 μm occur. This length scale is comparable to the size of structural features such as shear bands, crazes, or spherulites in polymer matrix materials; therefore, the behavior of the pure matrix (neat resin) may differ from the same matrix material as a constituent of a composite [9.7.7]. In this situation, use of the correspondence principle with data for pure matrix can be expected to generate errors.

   In many particulate composites [9.7.8], a polymer matrix is reinforced by stiff inclusions. The stiffness of a composite with spherical inclusions, as a function of volume fraction of inclusions, is experimentally found to be close to the Hashin–Shtrikman lower bound for isotropic materials. Most particulate inclusions used ordinarily exhibit little viscoelastic response. Dental composite filling resins, for example, contain particles of stiff mineral such as silica in a polymer matrix [9.7.9, 9.7.10]. The silica inclusions confer stiffness and abrasion resistance comparable to that of the tooth. Even though these materials contain a high concentration of filler, significant long-term creep, corresponding to a change of a factor of four in stiffness with time, can occur [9.7.11].

§9.7.3 Metal–matrix composites

Metal–matrix composites [9.7.12] include aluminum stiffened with inclusions of alumina (Al_2O_3) particles; and aluminum reinforced with graphite, boron, or silicon carbide. They are of particular interest in applications involving high temperature, for which a polymer matrix composite would be inappropriate. Such metal–matrix composites exhibit increased stiffness, strength, and wear resistance. Particulate inclusions do not have as much stiffening effect as fibers, as discussed in §9.3, but they are less expensive than fibers, so they are more often used. Materials used for inclusions tend to be stiff and exhibit low damping. Silicon carbide (SiC), for example, exhibits E = 450 GPa and tan $\delta \approx 10^{-4}$ at 20°C [9.7.13] compared with 70 GPa for aluminum. Aluminum itself exhibits low damping, typically below 10^{-3} for aluminum and below 10^{-5} for some aluminum alloys.

As for damping, magnesium–matrix composites may exhibit relatively high damping (to tan $\delta \approx 0.01$) principally as a result of the high damping exhibited by magnesium [9.7.14]. Damping in laminated composites of this type depends on the ply angles. Graphite–aluminum composites can attain tan δ up to 4×10^{-3} due to thermoelastic damping [9.7.15]. For metal–matrix composites with particulate inclusions of ceramic at room temperature, tan $\delta \approx 0.005$ for boron–aluminum, tan $\delta \approx 0.007$ for alumina (Al_2O_3)–aluminum, and tan $\delta \approx 0.004$ for SiC–aluminum [9.7.16]. As for creep at high stress and high temperature (up to 350°C), a high volume fraction of fine particle reinforcement can significantly reduce creep [9.7.17].

Damping in metal–matrix composites can arise due to damping in one or both of the phases. Use of a high-loss phase such as magnesium can give rise to a high-loss composite. In some particulate composites, an excess of damping is attributed to high concentrations of dislocations near the metal–ceramic interface [9.7.18]. Damping due to thermoelastic coupling (§8.3) may be important in some metal–matrix composites. Thermoelastic damping in composite materials arises due to the heterogeneity of the thermal and mechanical properties of such materials, leading to heat flow between constituents, hence, mechanical energy dissipation. Such damping is *not* accounted for in the above treatments (§9.3) via the correspondence principle since the overall actual damping contains a contribution from a coupled-field interaction between the constituents.

Including flake graphite in up to 10% volume fraction into 6061-T6 aluminum alloy by spray deposition gives rise to E = 44 GPa (lower than that of aluminum alloy, 69 GPa) and tan $\delta \approx 0.01$ at low audio frequency [9.7.19]. The damping is attributed to sliding of the planar structure of graphite. Including alumina (Al_2O_3) fibers increases the stiffness of aluminum but reduces the damping at 80 kHz and almost completely suppresses the amplitude-dependent response [9.7.20]. In pure aluminum, tan δ is about 2.5×10^{-4} for strains up to 10^{-5}; the damping increases at higher strains due to dislocation motion. Addition of tungsten fibers to aluminum increases the damping.

§9.7.4   Cellular solids

Cellular solids [9.7.21] are composites in which one phase is empty space or a fluid such as water or air. Cellular solids include honeycombs, foams, and other porous materials. In low-density closed-cell foams, the pressure of air or other gas in the pores contributes to the overall stiffness. In open-cell foams, air in the pores is free to escape as the material is stressed. Under quasistatic conditions or at low frequency, air flow has little effect. The correspondence principle may then be applied to the foam as a composite with one mechanically active phase. So if an elastic open-cell foam of Young's modulus E and density ρ is governed by

$$\frac{E}{E_s} = \left[\frac{\rho}{\rho_s} \right]^2 , \tag{9.7.1}$$

(with ρ_s as the density and E_s as the modulus of the solid material of which the foam is made), then for a corresponding viscoelastic foam,

$$E^* = E_s^* \left[\frac{\rho}{\rho_s} \right]^2 . \tag{9.7.2}$$

Under such circumstances the loss tangent of the foam is the same as that of the solid phase, assuming no chemical changes occur during the foaming process. However, the air has some viscosity. At sufficiently high frequency, significant viscoelastic loss can occur in these materials as a result of the viscosity of the air or water in the pores, as we have examined in §8.4. At sufficiently high frequency it may be possible to set the ribs or walls of the foam into vibration, resulting in additional loss and dispersion [9.7.22]. Moreover, acoustic absorption in negative Poisson's ratio foams is higher than that in foams of conventional structure [9.7.23]. Further examples of viscoelasticity in composites are given in §7.12. Under quasistatic circumstances, below such frequencies, the loss tangent of open-cell cellular solids is that of the solid material of which it is made, following the correspondence principle. In closed-cell foams, the air contained within the cells can support some load. If a steady load is applied in a creep test, air is lost from the cells due to diffusion through the cell walls. This gives rise to creep [9.7.24].

§9.7.5 Piezoelectric composites

Piezoelectric composites find use in electromechanical transducers and in transducers intended for "smart" materials. Piezoelectric composites are available with particles of lead titanate piezoelectric ceramic embedded in a polychloroprene rubber [9.7.25]. The overall damping properties of such composites are, as anticipated in §9.3, dominated by the damping of the polymer matrix which exhibits a large peak due to the glass transition. There is also a peak in the piezoelectric coupling and in the dielectric loss at the glass transition temperature. Piezoelectric composites have also been considered with the aim of achieving high damping in stiff structural composites [9.7.26]. It is considered possible to achieve a peak damping of 0.12 and a longitudinal modulus of 43 GPa with 30% by volume of resistively shunted piezoelectric fibers in a hybrid glass–epoxy composite. Experiments with ceramic–epoxy "composites" containing one ring-shaped inclusion of resistively shunted piezoelectric ceramic disclosed damping exceeding 0.10 over a narrow frequency range [9.7.27].

§9.8 Dispersion in composites

In composites there are several physical processes which give rise to dispersion (frequency dependence of the wave speeds or moduli inferred from them). Viscoelastic damping, as we have seen in §3.3, is linked to dispersion via the Kramers–Kronig relations. Dispersion also can arise due to microscopic resonance or standing wave effects in the structural elements of the composite. This is referred to as geometrical dispersion [9.8.1]. Dispersion due to viscoelasticity always entails increase of wave speed with frequency since the damping in passive materials is positive. Geometrical dispersion can give rise to a decrease of wave speed with frequency [9.8.2]. In typical fibrous composites such phenomena occur at frequencies in the megahertz range and in foams in the kilohertz range or below. If the wavelength becomes comparable to the size of structural elements in the material, extreme dispersion combined with large wave attenuation can occur [9.8.3]. Composites with periodic structure can exhibit cutoff frequencies. If particulate inclusions are distributed randomly, there are no cutoff frequencies [9.8.4]. In strongly scattering media, the group velocity [9.8.5] associated with pulses of waves can differ substantially from the phase velocity associated with individual peaks associated with waves.

§9.9 Summary

The viscoelastic properties of a composite depend on the properties of its phases. For cases of simple geometry, an exact calculation of these properties can be obtained by use of the correspondence principle.

Examples

Example 9.1

Consider a composite containing stiff elastic particles embedded in a viscoelastic matrix. We want to reduce creep. Should we use stiffer particles? What are some applications of such composites?

Solution

At the simplest level one may approximate the behavior as that of a Reuss composite according to Eqs. 9.3.5 and 9.3.6. If $V_1 = 0.5$,

$$\tan \delta_c \approx \frac{\tan \delta_1 + \frac{J_2'}{J_1'} \tan \delta_2}{1 + \frac{J_2'}{J_1'}}.$$

Suppose phase 1 is soft and high loss; then $J_1 \gg J_2$, so $\tan \delta_2 = 0$ (elastic inclusions),

$$\tan \delta_c \approx \tan \delta_1 \left\{ 1 - \frac{J_2'}{J_1'} \right\},$$

so stiffer inclusions slightly reduce the $\tan \delta$; hence, there is a slight reduction in the creep since the slope of the creep curve on a log–log scale is proportional to $\tan \delta$. One may pursue a more sophisticated approximation based on the Hashin–Shtrikman formulae or on a spherical inclusion model. Silica particles in a polymer matrix are used in composite dental fillings. Creep can be a problem but wear is usually more of a problem. The hard mineral particles in a high concentration reduce the wear. A very high concentration of stiff elastic particles would be needed to significantly reduce the creep. It may be impractical to achieve such a high concentration. It may be more promising to seek a polymer which exhibits less creep, and to allow time for aging of the polymer after it is polymerized.

Example 9.2

Derive $E_c = E_1 V_1 + E_2 V_2$, for the elastic Voigt composite microstructure, in one dimension, neglecting Poisson effects.

Solution

Since it is elastic, Hooke's law for phases 1 and 2 is

$$\sigma_1 = E_1\varepsilon_1, \quad \text{and} \quad \sigma_2 = E_2\varepsilon_2.$$

The loads P in terms of the cross-sectional areas A are

$$P_1 = A_1\sigma_1 = E_1\varepsilon_1 A_1, \quad \text{and} \quad P_2 = A_2\sigma_2 = E_2\varepsilon_2 A_2.$$

In view of the geometry, the load P_c on the composite block must be the sum of the loads on the constituents:

$$P_c = P_1 + P_2,$$

$$P_c = \sigma_c A_c = \sigma_1 A_1 + \sigma_2 A_2.$$

So by dividing by A_c, and recognizing the volume fraction for this geometry as the ratio of cross-sectional areas,

$$\sigma_c = \sigma_1 \frac{A_1}{A_c} + \sigma_2 \frac{A_2}{A_c} = \sigma_1 V_1 + \sigma_2 V_2.$$

Finally, divide by the strain, which is the same in both constituents provided they are perfectly bonded:

$$E_c = E_1 V_1 + E_2 V_2,$$

as desired.

The viscoelastic version of this may be obtained without appeal to the correspondence principle by following the same steps assuming the moduli to be complex quantities. The volume fractions are always real quantities based on their definition.

Example 9.3

Consider an isotropic particulate composite containing an epoxy matrix with E = 3 GPa and tan δ = 0.1 and silica inclusions with E = 70 GPa and tan $\delta = 10^{-5}$. Determine and discuss the stiffness and damping of such a composite. For the purpose of calculation assume a Poisson's ratio of 0.3. What are the prospects of achieving minimal viscoelasticity in such a composite?

Solution

Stiffness and damping based on the Hashin–Shtrikman lower bound on stiffness, Eqs. 9.3.8 and 9.3.11, for a small volume fraction of spherical inclu-

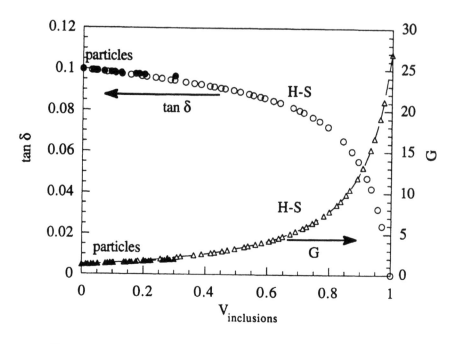

Figure E9.3 Effect of spherical silica inclusions in an epoxy matrix.

sions are plotted in Fig. E9.3. Observe that stiff spherical inclusions are inefficient in creating a stiff composite: the stiffness is close to the Hashin–Shtrikman lower bound on the stiffness. Moreover, for readily attainable volume fractions of inclusions, the damping of the composite is close to that of the matrix itself. If one wishes to make a stiff composite with low damping, the particulate morphology is problematic. A fibrous or platelet morphology is more promising for such a goal.

Example 9.4

Calculate the fluid flow relaxation strength assuming a polymer foam skeleton with G = 100 kPa. Let the fluid in the interstices be air, and consider the surfaces of the foam to freely allow fluid flow. Consider relaxation in shear, bulk deformation, and uniaxial compression.

Solution

Air has a pressure of P = 14.7 psi or 100 kPa under normal sea level conditions. Air as a fluid has a shear modulus of zero, and a bulk modulus governed by the ideal gas equation

$$PV = nRT.$$

So

$$\frac{dP}{dV} = -\frac{nRT}{V^2},$$

so

$$\frac{dP}{\dfrac{dV}{V}} = P.$$

So the bulk modulus is

$$B = \frac{\sigma}{\varepsilon} = \frac{dP}{\dfrac{1}{3}\dfrac{dV}{V}} = 3P.$$

For air at the given pressure, B = 300 kPa. The factor of three arises from the fact that the volumetric deformation depends on the product of deformations in three directions.

The bulk modulus in terms of the shear modulus G and Poisson's ratio v is

$$B = 2G\frac{1+v}{3(1-2v)}.$$

The foam has a Poisson's ratio of about 0.3, so B = 2.167 G = 216.7 kPa.

For an elementary treatment of this problem, neglect the deformation of the solid phase as pressure is increased in the fluid which permeates the pores. Then the bulk stiffness at zero time is the sum of the stiffness of the foam skeleton and that of the air, which at zero time has no time to escape. For an exact solution, the full Biot analysis of §8.4 is called for.

$$B(0) = 217 \text{ kPa} + 300 \text{ kPa}.$$

The relaxation strength for bulk deformation is

$$\Delta_B = \frac{B(0) - B(\infty)}{B(\infty)} = \frac{\{300 + 217\} - 217}{217} = 1.32.$$

The peak damping for bulk deformation, assuming a Debye peak, is

$$\tan \delta = \frac{1}{2} \frac{\Delta}{\sqrt{1+\Delta}} = 0.45.$$

This is a substantial damping.

As for shear, air has zero stiffness in shear. Moreover, the viscosity of air is low so that damping due to global shear is negligible, except at very high frequency. Therefore, the corresponding relaxation strength and damping of the foam in shear due to the air in the pores are zero.

As for damping in axial deformation, observe that Young's modulus is given by

$$E = 3B(1-2v) = \frac{9GB}{3B+G}.$$

So, with $G(0) = G(\infty) = 100$ kPa, $B(0) = 517$ kPa, and $B(\infty) = 217$ kPa, the relaxation strength for axial deformation is

$$\Delta_E = 0.084.$$

This is considerably less than in the case of bulk deformation, since uniaxial tension deformation is considerably influenced by the shear properties, which in this example do not relax.

Problems

9.1 Show how complex Poisson's ratios v^* can be achieved in composites in which each phase has a real Poisson's ratio.

9.2 Why does a unidirectional graphite epoxy fibrous composite have low damping and creep in the fiber direction?

9.3 Derive the composite stiffness in terms of constituent stiffness values for the Reuss composite microstructure. Neglect the Poisson effects.

9.4 Calculate the fluid flow relaxation strength assuming a polymer foam skeleton with $G = 100$ kPa. Let the fluid be water.

References

9.2.1 Hashin, Z., "The elastic moduli of heterogeneous materials", *J. Appl. Mech., Trans. ASME* 84E, 143–150, 1962.

9.2.2 Hashin Z. and Shtrikman, S., "A variational approach to the theory of the elastic behavior of multiphase materials", *J. Mech. Phys. Solids* 11, 127–140, 1963.

9.2.3 Hashin, Z., "Analysis of composite materials—a survey", *J. Appl. Mech.* 50, 481–505, 1983.

9.2.4 Day, W. A., "Restrictions on relaxation functions in linear viscoelasticity", *Q. J. Mech. Appl. Math.* 24, 487–497, 1971.

9.2.5 Rogers, T. G. and Pipkin, A. G., "Asymmetric relaxation and compliance matrices in linear viscoelasticity", *Z. Angew. Math. Phys. (ZAMP)* 14, 334–343, 1963.

9.2.6 Hashin, Z., "Viscoelastic behavior of heterogeneous media", *J. Appl. Mech., Trans. ASME* 32E, 630–636, 1965.

9.2.7 Schapery, R. A., "Stress analysis of viscoelastic composite materials", *J. Compos. Mater.* 1, 228–267, 1967.

9.2.8 Hashin, Z., "Viscoelastic fiber reinforced materials", *AIAA J.* 4, 1411–1417, 1966.

9.3.1 Hashin, Z., "Complex moduli of viscoelastic composites. I. General theory and application to participate composites", *Int. J. Solids, Struct.* 6, 539–552, 1970.

9.3.2 Christensen, R. M., *Mechanics of Composite Materials* , J. Wiley, NY, 1979.

9.3.3 Christensen, R. M., "Viscoelastic properties of heterogeneous media", *J. Mech. Phys. Solids* 17, 23–41, 1969.

9.3.4 Chen, C. P. and Lakes, R. S., "Analysis of high loss viscoelastic composites", *J. Mater. Sci.* 28, 4299–4304, 1993.

9.3.5 Christensen, R. M., "Isotropic properties of platelet reinforced media", *J. Eng. Mater. Technol.* 101, 299–303, 1979.

9.3.6 Norris, A. N., "The mechanical properties of platelet reinforced composites", *Int. J. Solids Struct.* 26, 663–674, 1990.

9.3.7 Millett, P., Schaller, R., and Benoit, W., "Internal friction spectrum and damping capacity of grey cast iron", in *Deformation of Multi-phase and Particle Containing Materials*, Proc. 4th Riso Int. Symp. Metall. Mater. Sci. ed. J. B. Bilde-Sorensen, N. Hansen, A. Horsewell, T. Leffers, and H. Linholt; Riso National Laboratory, Roskilde, Denmark, 1983.

9.3.8 Brodt, M. and Lakes, R. S., "Composite materials which exhibit high stiffness and high viscoelastic damping", *J. Compos. Mater.* 29, 1823–1833, 1995.

9.3.9 Wang, Y. M. and Weng, G. J., "The influence of inclusion shape on the overall viscoelastic behavior of composites", *J. Appl. Mech.* 59, 510–518, 1992.

9.3.10 Zinoviev, P. A. and Ermakov, Y. N., *Energy Dissipation in Composite Materials*, Technomic, Lancaster, PA, 1994.

9.3.11 Brinson, L. C. and Knauss, W. G., "Finite element analysis of multiphase viscoelastic solids", *J. Appl. Mech.* 59, 730–737, 1992.

9.4.1 Roscoe, R., "Bounds for the real and imaginary parts of the dynamic moduli of composite viscoelastic systems", *J. Mech. Phys. Solids* 17, 17–22, 1969.

9.4.2 Gibiansky, L. V. and Milton, G. W., "On the effective viscoelastic moduli of two phase media. I. Rigorous bounds on the complex bulk modulus", *Proc. R. Soc. London* 440, 163–188, 1993.

9.4.3 Gibiansky, L. and Lakes, R. S., "Bounds on the complex bulk modulus of a two phase viscoelastic composite with arbitrary volume fractions of the components", *Mech. Mater.* 16, 317–331, 1993.

9.5.1 Lakes, R. S., "Materials with structural hierarchy", *Nature* 361, 511–515, 1993.

9.5.2 Hancox, N. M., *Biology of Bone*, Cambridge University Press, Cambridge, 1972.

9.5.3 Currey, J., *The Mechanical Adaptations of Bones*, Princeton University Press, Princeton, NJ, 1984.

9.5.4 Katz, J. L., "Hard tissue as a composite material I. Bounds on the elastic behavior", *J. Biomech.* 4, 455–473, 1971.

9.5.5   Frasca, P., Harper, R. A., and Katz, J. L., "Isolation of single osteons and osteon lamellae", *Acta Anat.* 95, 122–129, 1976.

9.5.6   Lakes, R. S. and Saha, S., "Cement line motion in bone," *Science* 204, 501–503, 1979.

9.5.7   Piekarski, K., "Fracture of bone", *J. Appl. Phys.* 41, 215–223, 1970.

9.5.8   Katz, J. L. "Anisotropy of Young's modulus of bone", *Nature* 283, 106–107, 1980.

9.5.9   Gottesman, T. and Hashin, Z., "Analysis of viscoelastic behaviour of bones on the basis of microstructure", *J. Biomech.* 13, 89–96, 1980.

9.5.10 Thomas, R. J., "Wood: formation and morphology" in *Wood Structure and Composition*, ed. M. Lewin and I. S. Goldstein, Marcel Dekker, NY, 1991.

9.5.11 Amada, S., "Hierarchical functionally gradient structures of bamboo, barley, and corn", *MRS Bull.* 20, 35–36, 1995.

9.5.12 Amada, S. and Lakes, R. S., "Viscoelastic properties of bamboo", *J. Mater. Sci.* 32, 2693–2697, 1997.

9.6.1   Lakes, R. S., "The time dependent Poisson's ratio of viscoelastic cellular materials can increase or decrease", *Cell. Polym.* 10, 466–469, 1992.

9.6.2   Agbossou, A., Bergeret, A., Benzarti, K., and Alberola, N., "Modelling of the viscoelastic behaviour of amorphous thermoplastic/glass beads composites based on the evaluation of the complex Poisson's ratio of the polymer matrix", *J. Mater. Sci.* 28, 1963–1972, 1993.

9.7.1   Hashin Z. and Rosen, B. W., "The elastic moduli of fiber-reinforced materials", *J. Appl. Mech., Trans. ASME* 31, 223–232, 1964.

9.7.2   Hashin, Z., "On elastic behaviour of fibre reinforced materials of arbitrary transverse phase geometry", *J. Mech. Phys. Solids* 13, 119–134, 1965.

9.7.3   Lakes, R. S., "Materials with structural hierarchy", *Nature* 361, 511–515, 1993.

9.7.4   Sturgeon, J. B., "Creep of fibre reinforced thermosetting resins", in *Creep of Engineering Materials*, ed. C. D. Pomeroy, Mechanical Engineering Publications, Ltd., London, 1978.

9.7.5   Tuttle, M. E. and Graesser, D. L., "Compression creep of graphite/epoxy laminates monitored using moiré interferometry", *Optics Lasers Eng.* 12, 151–171, 1990.

9.7.6   Flaggs, D. L. and Crossman, F. W., "Analysis of the viscoelastic response of composite laminates during hygrothermal exposure", *J. Compos. Mater.* 15, 21–40, 1981.

9.7.7   Sternstein, S. S., Srinavasan, K., Liu, S. and Yurgartis, S., "Viscoelastic characterization of neat resins and composites", *Polym. Prepr.* 25, 201–202, 1984.

9.7.8   Ahmed, S. and Jones, F. R., "A review of particulate reinforcement theories for polymer composites", *J. Mater. Sci.* 25 , 4933–4942, 1990.

9.7.9   Cannon, M. L., "Composite resins", in *Encyclopedia of Medical Devices and Instrumentation*, ed. J. G. Webster, J. Wiley, NY, 1988.

9.7.10 Craig, R., "Chemistry, composition, and properties of composite resins", in *Dental Clinics of North America*, ed. H. Horn, Saunders, Philadelphia, PA, 1981.

9.7.11 Papadogianis, Y., Boyer, D. B., and Lakes, R. S., "Creep of conventional and microfilled dental composites", *J. Biomed. Mater. Res.* 18, 15–24, 1984.

9.7.12 Taya, M. and Arsenault, R. J., *Metal Matrix Composites*, Pergamon, Oxford, 1989.

9.7.13 Wolfenden, A., Rynn, P. J., and Singh, M., "Measurement of elastic and anelastic properties of reaction-formed silicon carbide materials", *J. Mater. Sci.* 30, 5502–5507, 1995.

9.7.14 Wren, G. G. and Kinra, V. K., "Axial damping in metal-matrix composites. II: A theoretical model and experimental verification", *Exp. Mech.* 32, 172–178, 1992.

9.7.15 Kinra, V. K., Wren, G. G., Rawal, S., and Misra, M., "On the influence of ply-angle on damping and modulus of elasticity of a metal-matrix composite", *Metall. Trans.* 22A, 641–651, 1991.

9.7.16 Wong, C. R. and Holcomb, S., "Damping studies of ceramic reinforced aluminum", in ASTM STP 1169, Mechanics and Mechanisms of Material Damping, DTRC-SME-91/15, David Taylor Research Center, Bethesda, MD, 1991.

9.7.17 Lloyd, D. J., "Particle reinforced aluminium and magnesium matrix composites", *Int. Mater. Rev.* 39, 1–23, 1994.

9.7.18 Ledbetter, H. M. and Datta, S. K., "Young's modulus and the internal friction of an SiC particle reinforced aluminum composite", *Mater. Sci. Eng.* 67, 25–30, 1984.

9.7.19 Zhang, J., Perez, R. J., Gungur, M. N., and Lavernia, E. J., "Damping characterization of graphite particulate reinforced aluminum composite", in *Developments in Ceramic and Metal-Matrix Composites*, ed. K. Upadhya, TMS 1992 Meet. Minerals, Metals, and Materials Society, San Diego, 1992.

9.7.20 Wolfenden, A. and Wolla, J. M., "Mechanical damping and dynamic modulus measurements in alumina and tungsten fibre–reinforced aluminium composites", *J. Mater. Sci.* 24, 3205–3212, 1989.

9.7.21 Gibson, L. J. and Ashby, M. F., *Cellular Solids*, Pergamon, Oxford, 1988.

9.7.22 Chen, C. P. and Lakes, R. S., "Dynamic wave dispersion and loss properties of conventional and negative Poisson's ratio polymeric cellular materials", *Cell. Polym.* 8, 343–359, 1989.

9.7.23 Howell, B., Prendergast, P., and Hansen, L., "Acoustic behavior of negative Poisson's ratio materials", DTRC-SME-91/01, David Taylor Research Center, Bethesda, MD, March 1991.

9.7.24 Mills, N. J., "Time dependence of the compressive response of polypropylene bead foam", *Cell. Polym.* 16, 194–215, 1997.

9.7.25 Rittenmyer, K., "Temperature dependence of the electromechanical properties of 0-3 PbTiO$_3$ polymer piezoelectric composite materials", *J. Acoust. Soc. Am.* 96, 307–318, 1994.

9.7.26 Lesieutre, G. A., Yarlagadda, S., Yoshikawa, S., Kurtz, S. K., and Xu, Q. C., "Passively damped structural composite materials using resistively shunted piezoceramic fibers", *J. Mater. Eng. Performance* 2, 887–892, 1993.

9.7.27 Law, H. H., Rossiter, P. L., Koss, L. L., and Simon, G. P., "Mechanisms in damping of mechanical vibration by piezoelectric ceramic-polymer composite materials", *J. Mater. Sci.* 30, 2648–2655, 1993.

9.8.1 Sutherland, H. J., "On the separation of geometric and viscoelastic dispersion in composite materials", *Int. J. Solids, Struct.* 11, 233–246, 1975.

9.8.2 Sutherland, H. J., "Dispersion of acoustic waves by fiber-reinforced viscoelastic materials", *J. Acoust. Soc. Am.* 57, 870–875, 1975.

9.8.3 Kinra, V. K. and Anand, A., "Wave propagation in a random particulate composite at long and short wavelengths", *Int. J. Solids Struct.* 18, 367–380, 1982.

9.8.4   Kinra, V. K., "Ultrasonic wave propagation in a random particulate composite", *Int. J. Solids Struct.* 16, 301–312, 1980.

9.8.5   Page, J. H., Sheng, P., Schriemer, H. P., Jones, I., Jing, X., and Weitz, D. A., "Group velocity in strongly scattering media", *Science* 271, 634–637, 1996.

chapter ten

Applications and case studies

§10.1   Introduction

Viscoelasticity can enter in the application of materials in many ways. In some applications one must deal with *natural* materials such as stone, earth, or wood in the case of building construction; or bone and soft tissue in the case of biomedical engineering. In these cases the viscoelastic behavior of the natural materials should be known. *Artificial* materials used in engineering applications may exhibit viscoelastic behavior as an unintentional side effect. Finally, one may *deliberately* use the viscoelasticity of certain materials in the design process to achieve a particular goal.

§10.2   A viscoelastic earplug: use of recovery

A foam earplug [10.2.1, 10.2.2] was designed to be easily fitted into the ear by making use of controlled viscoelastic behavior of the polymer from which it is made. The earplug serves to attenuate sound entering the ear to protect the ear from damage from excessive noise, and also to alleviate suffering and reduce fatigue.

To insert the earplug, the user rolls it into a narrow cylindrical shape and then inserts it into the ear canal. Insertion is easier if the outer ear is pulled outward and upward, since that straightens the ear canal. The earplug gradually expands as a result of viscoelastic recovery to fill and contact the ear canal, and it then effectively blocks noise.

The earplug is [10.2.2] cylindrical, somewhat larger than the ear canal, and made of a foamed polymeric material with a recovery from 60% compression to 40% compression occurring in 1 to 60 sec and an equilibrium stiffness at 40% compression from 0.2 to 1.3 psi (1.4 to 9 kPa). In the preferred embodiment of the invention, the recovery occurs in 2 to 20 sec. Detailed specifications for size were a diameter range of 3/8 to 3/4 in. (9.5 to 19 mm) optimally 9/16 to 11/16 in. (14 to 17 mm), and a length from 1/2 to 1 in. (13 to 25 mm).

As for materials, any flexible polymeric material which can be foamed and which has the desired viscoelastic characteristics may be used [10.2.2]. The inventor considers the use of polymers of ethylene, propylene, vinyl chloride, vinyl acetate, diisocyanate, cellulose acetate, or isobutylene. A composition of vinyl chloride homopolymers and copolymers comprising at least 85% by weight of vinyl chloride and up to 15% of other monomers is favored. The viscoelastic behavior is governed by the content of organic plasticizer. The relatively slow recovery behavior confers to the user the ability to initially compress the earplug and provide sufficient time for insertion into the ear canal. Slow recovery allows the earplug to gradually expand and attempt to regain its former shape and thus conform to the ear canal.

Commercially available foam earplugs were found to be about 13 mm in diameter and 18 to 20 mm long. The material was observed to have a Poisson's ratio of zero in compression. Viscoelastic characterization tests in our laboratory on representative earplugs disclosed power law behavior in creep from about 10 to 10^6 sec, typically $J(t) \propto At^{1/8}$, with J^{-1} (10 sec) $\approx$ 0.1 MPa. Different earplugs differed in their mechanical properties. The observed creep behavior did not correspond to a single recovery time described in the patent. Evidently the inventor considered that only a limited aspect of the full viscoelastic behavior was relevant to the application. As for Poisson's ratio, we initially considered the zero Poisson's ratio to be advantageous in that it implies a zero shearing stress on the ear canal during recovery. Any shear stress on the ear canal might be perceived as a tickle, especially if slip occurred. However, there is no mention of this in the patent, so the zero Poisson's ratio is most likely a side effect of the type of foam structure used.

§10.3   Creep and relaxation of materials and structures

§10.3.1   Concrete

Concrete used in structural applications can exhibit significant creep. Excessive deformation of concrete structures, which may in part result from creep, can lead to the following [10.3.1]. Visually objectionable sag may occur; the sag may result in ponding of water on roofs which can result in accelerated deterioration of the building. Sag may also interfere with the operation of sliding doors or other machinery. Partition walls may crack as a result of structural deformation; doors may stick or jam. Excess deformation can cause cracks at joints, permitting the entrance of water into the structure. Therefore, design of concrete structures for adequate rigidity must take into account the creep of concrete. Concrete, moreover, exhibits aging, and this must be taken into account in the analysis and design of concrete structures [10.3.1].

§10.3.2 Wood

Wood used in construction can exhibit significant creep, as shown in Fig. 10.1. Over sufficiently long time intervals, particularly if stresses are high, this creep can lead to visible sag in wooden boats and building structures [10.3.2].

§10.3.3 Leather

Creep can cause stress to be redistributed since the highly stressed parts or regions creep the most. The fact that old shoes are more comfortable than new ones is attributed to this phenomenon [10.3.2].

§10.3.4 Creep-resistant alloys: turbine blades

Creep-resistant alloys are used in high-temperature applications including heat exchangers, furnace linings, boiler baffles, bolts at high temperature, power plants, exhaust systems, and components for gas turbines [1.6.27]. Turbine blades in jet engines are particularly demanding of material creep resistance in that they are subject to large tensile stress of centrifugal origin, and high temperature. The designer of the engine has an incentive to raise the operating temperature to improve the engine's thermodynamic efficiency as well as the power to weight ratio. If the temperature is too high, metals can creep. Significant creep in this application is unacceptable since the end of each blade must closely fit the stationary engine housing, else compressed gas in the engine will escape, lowering engine efficiency. Therefore, materials which resist creep at high temperature are desirable.

Development of creep-resistant alloys [1.6.27, 10.3.3, 10.3.4] has been facilitated by an understanding of viscoelastic mechanisms (§8.13). At high temperature, creep in metals results from several mechanisms including dislocation motion leading to power law creep, and viscous slip at the grain boundaries. Nickel-based superalloys contain significant amounts of other elements, as shown for a representative detailed composition given in §7.4.4. These elements impede the motion of dislocations by the formation of a solid solution as well as by the formation of hard precipitates [10.3.4], e.g., Ni_3Al, Ni_3Ti, and MoC, on the scale of microns. Small inclusions of refractory oxides are also known to improve the strength and creep resistance of metals [1.6.27]. Use of such inclusions is known as *dispersion strengthening*. Dislocation motion is impeded by such inclusions since the dislocation cannot penetrate the particle readily.

Grain boundary slip can also be reduced by improvements in alloy chemistry. Specifically [10.3.3], a high-volume fraction of a second phase (such as nickel aluminide precipitate) improves grain strength, and tailored grain boundary structure containing carbides and borides reduces grain boundary slip. Current superalloys for turbine blades and disks contain nickel, cobalt, chromium, aluminum, titanium, and often tungsten and

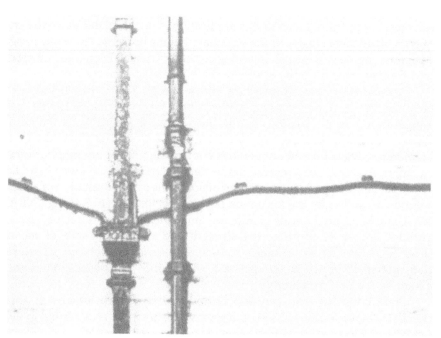

Figure 10.1   Sag under the prolonged action of gravity. (top) Sag of a wood building structure. (bottom) Sag of lead pipes. (After Ashby and Jones [10.3.4]. With
permission.)

molybdenum; trade names for these alloys include Inconel and Mar-M. Despite advances in alloys, creep associated with grain boundaries can be excessive. One solution to the problem is *directional solidification* in which columnar grains are created so that they are aligned with the centrifugal stress in the blade and therefore experience no shear or tensile stress across them. Directional solidification is achieved by solidifying an investment casting of a blade in the presence of a thermal gradient. In current engines, creep of the blades is reduced significantly by forming each blade from a single crystal of metal, eliminating the grain boundaries entirely. The blades are grown by a modified directional solidification method, in which grain reduction and selection procedures are used in the early stages of solidification. The consequence of this control of viscoelastic behavior is that people can travel by air at reduced cost and with increased safety.

§10.3.5 Relaxation of screws

Some machinery which contains bolted connections is expected to perform at elevated temperatures. If it fails to perform properly, people may be injured or lose access to services on which they depend. Examples include engines, pressure vessels, and reactors. Stress in the bolts relaxes with time. At sufficiently high temperature, even steel will creep. Therefore, the bolts must be periodically retightened to provide sufficient fastening force. If the bolt stress is sufficiently high, secondary creep proceeds by a power law in stress,

$$\frac{d\varepsilon}{dt} = B\sigma^n. \tag{10.3.1}$$

The following calculation [10.3.4] can be performed to determine how often one must retighten the bolt. The total strain ε is regarded as the sum of an elastic part $\varepsilon_{el} = \sigma/E$ and a creep part ε_{creep}.

$$\varepsilon = \varepsilon_{el} + \varepsilon_{creep}. \tag{10.3.2}$$

In the bolt, the total strain is constant. By differentiating the above and substituting,

$$\frac{1}{E}\frac{d\sigma}{dt} = -B\sigma^n. \tag{10.3.3}$$

By integrating from an initial stress σ_0 at time $t = 0$ to a final stress σ_f at time t,

$$\frac{1}{\sigma_f^{n-1}} - \frac{1}{\sigma_0^{n-1}} = BE(n-1)t. \tag{10.3.4}$$

The final stress is

$$\sigma_f = \left\{ BE(n-1)t + \frac{1}{\sigma_0^{n-1}} \right\}^{-1/(n-1)}. \tag{10.3.5}$$

Now the bolt must carry a minimum load to hold the machine together. For the sake of doing an example, assume that half the initial stress is sufficient: $\sigma_f = \frac{1}{2}\sigma_0$. Then, the time required for this amount of relaxation is

$$t = \frac{2^{n-1} - 1}{E\sigma_0^{n-1}(n-1)B}. \tag{10.3.6}$$

This would be the minimum time interval between retightening of the bolts by a maintenance worker.

    The above analysis seems simple in view of the fact that the interrelation between creep and relaxation for linear materials (which are simpler than nonlinear ones) involves a convolution integral as studied in §2.4. The simplifying assumption in the above analysis is that the same relation between stress and creep strain rate is valid both under conditions of constant stress (specifically, secondary creep) and constant strain (relaxation). The material may not in fact behave this way. Complex behavior can occur in nonlinear viscoelastic materials. Moreover, the above results for final stress and for the time give rise to indeterminate forms for $n = 1$, which corresponds to a linearly viscoelastic material. Nevertheless, such simple analyses can be of practical use.

§10.3.6 Computer disk drive: case study of relaxation

A computer disk drive was introduced some time ago, and after about six months of use, the drives experienced tracking errors and were returned to the factory [10.3.5]. The vendor examined some of the failed drives and found reduced screw torque holding the interrupter flag to the stepper motor shaft within the drive. At first the problem was thought to be a thermal expansion mismatch between the steel of the shaft and the aluminum thought to be used in the flag, but calculation revealed a deformation of this origin would be too small to be a problem. It was discovered that the flag was in fact made of zinc. Zinc was chosen since it could be die cast cheaply and easily. However, zinc exhibits significant creep and relaxation at room temperature. This was in fact the source of the problem. The following diagnostic test was performed. The interrupter flag set screws were torqued to the design specification value of 3.0 inch pounds. The disk drives were placed in an environmental chamber and the temperature raised to the maximum allowable storage temperature of 60°C at less than the specified maximum ramp rate of 10°C/h. The temperature was then lowered to 40°C, the maximum allowable operating temperature. None of the drives failed after ten days under these conditions; however, that was not unexpected since the drives which failed in service did so after a longer period of time. However, measurement of the set screw torque disclosed that it had relaxed by an average of 54%, so that the clamping force was reduced to less than the design value. It was concluded that the material was inappropriate for the application. The following caveat from the Zinc Institute was quoted:

> There are, however two factors peculiar to zinc which must be borne in mind in the engineering design of threaded connections. First, the relationship between the torque applied to a fastener and the retention or clamping load is erratic, apparently due to the tendency of zinc alloy threads to gall under thread contact pressures. A prescribed tightening torque will develop retention or clamping forces that may vary widely from one joint to the next. Second, zinc alloys tend to cold flow, even at room temperatures, so that the stresses in the threads relax and the clamping force is reduced. The rate of relaxation, which is a function of the elasticity of the joint, diminishes rapidly as the load decreases, tending to level off at a low stress level. At high stresses, such as those developed at 50 percent to 75 percent of the thread stripping load, noticeable relaxation occurs within a few days. The long term retention strength of a zinc alloy is therefore lower than other metals, even though the published value for tensile and shear strength of the zinc alloy may be greater. It is essential to provide for relaxation, especially when converting components to zinc alloys from aluminum, copper, or ferrous alloys.

A calculation of the required clamping force disclosed that after relaxation, the force was less than half the necessary amount. Consequently, the attempt to save money by choosing zinc as a material was unwise.

§10.3.7 Creep and recovery in human tissue

§10.3.7.1 Spinal disks

The disks in the human spinal column are viscoelastic and exhibit creep [10.3.6]. One result of the creep under normal body weight is that most people are about 1 cm taller in the morning than in the evening. For young people, the difference in height can be as much as 2%, which corresponds to a 1.4-in. (3.7-cm) variation in the height of a 6-ft (183-cm) person [10.3.7]. The increase in height after bed rest is due to creep recovery in the disks. Astronauts under microgravity conditions gained 5 cm in height, and then

Figure 10.2 "Droop" in biological and synthetic materials. Apparent droop of the nose is due to growth. Structural metals do not droop at ambient temperature and over humanly accessible time scales since crystal imperfections are largely immobile in metals of high melting point.

suffered backaches. One can measure stature to within 0.4 mm. With refined measurements of this kind, classic creep and recovery curves can be obtained from a living person under gravitational load, following a change in sitting position or following placement of a weight on the shoulders. In one study, 5 min of sitting gave rise to an average of 0.2% reduction in body height for young adults, and 0.4% for people age 60 to 65 [10.3.8]. Such results may be of use in minimizing back pain in ergonomic design, such as design or choice of chairs. Dynamic mechanical damping also occurs in spinal disks [10.3.9] resulting in absorption of energy in cyclic loading.

§10.3.7.2   *The nose*

The noses of older people may appear to droop (Fig. 10.2). This change in appearance is not necessarily due to viscoelasticity. The nose actually continues to grow during adulthood [10.3.10]. Moreover, other soft and hard tissues in the face undergo growth during adulthood. Knowledge of change in these tissues is useful to orthodontists and reconstructive surgeons.

§10.3.7.3   *Skin*

Skin is viscoelastic; therefore, it has a delayed recovery in response to transient deformation. The skin's ability to recover its original shape is important clinically [10.3.11] in the context of plastic surgery. Moreover, many skin diseases result in changes in the mechanical properties of skin, and differential diagnosis may be possible based on mechanical tests [10.3.12]. Quantitative measurements of skin recovery can be made in the living subject. An informal test of recovery may be done by pinching the skin on the back of the hand for several seconds. In a healthy young person, the skin snaps back in less than 1 sec but in older people and in cigarette smokers, recovery following this deformation may require many seconds to several minutes.

Figure 10.3   Tilt of a retaining wall due to creep in the soil.

§10.3.7.4   The head

Babies have deformable heads. The head may be misshapen following the birth process, but the shape returns to normal after about six weeks. If the baby sleeps in a single position, the long-term load component can cause an abnormal head shape [10.3.13] due to creep deformation. Such deformation was observed after it was recommended that infants be positioned to sleep on their backs or sides to reduce risk of sudden infant death syndrome [10.3.14]. Following that publication, many children were observed with abnormal head shapes. Abnormal head shape in infants can be corrected by a helmet-like device worn on the head [10.3.15, 10.3.16]. The device provides a gentle pressure to gradually restore the normal shape of the head. Heads of infants have been deliberately deformed by ancient Egyptians as early as 2000 BC as well as by aboriginal peoples in Africa and in North and South America.

§10.3.8   Earth, rock, and ice

Creep of soils can give rise to the tilt of retaining walls (Fig. 10.3), settlement of foundations, movement of buildings [10.3.17, 10.3.18], or even tilt of large buildings (as in the Leaning Tower of Pisa). Prediction of such creep requires consideration of the fact that the creep depends on the hydration of the soil and its temperature, hence on weather and climatic conditions. Effective Young's moduli for initial deformation of soils is from 1 to 3 MPa for soft clay to about 70 MPa for dense sand [10.3.17]. Time scales for time-dependent settlement can range from months to years. Soil is actually a fluid–solid composite. Creep in soils may be understood in part by Biot-type poroelastic analysis as discussed in §8.4. The time scale of the creep or relaxation, expressed as the time constant τ, increases as the square of the size of the region subject to consolidation. Therefore, the coupled-field equations are used directly in such problems.

As for rock, creep in rock can give rise to redistribution of stress around mine tunnels and is therefore of some interest to mining engineers. Creep at very slow strain rates occurs in continental drift. Movement of the earth's plates is associated with steady-state creep at strain rates of 10^{-20} to 10^{-29} per second [7.5.10].

As for ice, its viscoelasticity (§7.8) is of interest in connection with glaciers and with structures built where there is moving ice.

§10.3.9   Solder

Solders are alloys of low melting point (120 to 320°C) used to join metal parts in plumbing or to achieve electrical connections [10.3.19–10.3.22]. Elements commonly used for electrical solders are tin, lead, silver, bismuth, indium, antimony, and cadmium. Lead–tin alloys are commonly used. In view of the toxicity of lead, other alloys such as tin–antimony solders have been developed. Since solders are used at high homologous temperatures, several relaxation mechanisms are active. Dislocation movement and grain boundary slip are considered to be the most important mechanisms [10.3.21]. Most solders exhibit considerable creep, and this creep can be of concern when it is desired that electrical connections be maintained over long periods of time. Development of alloys which exhibit reduced creep, yet have a low melting point, is challenging. One possibility is dispersion hardening by magnetically distributed iron particles [10.3.22]. Moreover, thermal cycling occurs in electrical equipment which is turned on and off, and this cycling can affect the performance of the connections. Failure of solder joints can occur via creep, via fatigue, or by their interaction.

Some solder alloys and selected properties are as follows [10.3.19].

Table 10.1   Solder Properties

Alloy	Melting range, solidus T (°C)	Melting range, liquidus T (°C)	Creep-resistance rank	Ultimate tensile strength (MPa)
63%Sn–37%Pb	183	183	Moderate	35
60%Sn–40%Pb	183	190	Low	28
95%Sn–5%Sb	235	240	Higher	56
52%In–48%Sn	118	131	n/a	n/a
60%In–40%Sn	122	113	Low	7.6
96.5%Sn–3.5%Ag	221	221	High	58

§10.3.10 Filaments in light bulbs and other devices

Filaments in light bulbs are made of tungsten, a metal with a high melting point (>3300°C) [10.3.4]. The filament can therefore be electrically heated to a temperature high enough for light emission. The filament is made as hot as possible (≈2000°C), to improve efficiency. Although tungsten at room temperature exhibits little creep, it does creep at elevated temperature. This creep is a principal cause of the failure of the light bulb: the filament sags and its coils touch each other [10.3.4] leading to a localized short circuit. The creep resistance of tungsten can be improved by alloying. For example, "nonsag" tungsten contains microscopic bubbles of elemental potassium [10.3.23]. In an illuminated light bulb, the potassium becomes a pressurized gas. The gas bubbles prevent recrystallization of the tungsten, which is responsible for much of its high-temperature creep. Light bulbs can also be made to last longer if they are operated at a lower temperature. The user can apply a lower average voltage by placing a diode in series with the bulb. The light is then redder and dimmer, and conversion of electricity into light is less efficient.

Electrodes in ion lasers such as the argon–ion laser get hot since the power density in the laser is high. The electrodes therefore exhibit creep. Sag of the electrodes under their own weight is a major failure mechanism in these lasers.

§10.3.11 Tires

Tires in a car parked in cold weather are comparatively stiff, and exhibit more noticeable creep as a result of shifting of the time scale of retardation processes due to the low temperature. Consequently, the flat portions of the tires in contact with the road may persist for a while after the car is driven, so the driver perceives repetitive thumps from the tires. This is known as "tire flatspotting", and it is related to the creep behavior of the tire materials, particularly nylon fibers [1.6.18, 10.3.24, 10.3.25]. Nylon fibers also creep considerably during service, so that the tire grows gradually larger [10.3.25]. Deformation due to this creep is large enough that size adjustments are made in the retreading of tires for commercial vehicles.

§10.3.12 Cushions for seats and wheelchairs

Polymer foams used in seat cushions are viscoelastic. Consequences include (i) progressive conformation of the cushion to the body shape, (ii) damping of vibration which may be transmitted to people via seats in automobiles and other vehicles, and (iii) progressive densification of the foam due to long-term creep. The perceived comfort of a foam cushion depends on the compliance of the foam and particulars of its stress–strain behavior, discussed in §7.11. Foams which are too compliant may bottom out (undergo densification). In this nonlinear regime the incremental stiffness is much higher than the initial stiffness near zero strain [10.3.26]. It has been suggested that foams with a bimodal distribution of cell sizes are more comfortable than foams with a single cell size, but the nonlinear stress–strain curves are similar [10.3.27]. Moreover, the damping of the cushion governs the transmission of vibration to the person; and this perception of vibration is important in seats in automobiles, trucks, and heavy equipment.

The properties of cushions are crucially important in reducing illness and suffering in people who are confined to wheelchairs or hospital beds for long periods. Prolonged pressure on any part of the body can obstruct circulation in the capillaries sufficient to cause a sore or ulcer known as a pressure sore, also called a bed sore. In its most severe manifestation, a pressure sore can form a deep crater-like ulcer in which underlying muscle or bone is exposed [10.3.28, 10.3.29]. To reduce the incidence of pressure sores, the maximum pressure on any part of the body surface should not exceed 32 mmHg (4.3 kPa) for long periods. This pressure is comparable to the pressure of blood in capillaries. Too much external pressure prevents blood flow and damages the tissue. Moreover, pressure should be uniformly distributed, adequate air flow should be provided to the skin, and frictional forces should be minimized. Several cushion materials have been tried to minimize the incidence and severity of pressure sores [10.3.30, 10.3.31]. Viscoelastic foam allows the cushion to progressively conform to the body shape. However, foam densification due to creep results in a stiffer cushion. Too stiff a cushion increases the risk of pressure sores; therefore, current foam cushions must be replaced after six months' use. Efforts to develop better materials continue.

§10.3.13 Artificial hip joints

The hip joint has a ball and socket structure which allows considerable mobility of the leg. In severe cases of arthritis, pain in the joint can prevent a person from moving about. To alleviate this pain and disability it has become commonplace to surgically replace the joint with a total hip prosthesis consisting of a ball joint fixed in the upper end of the femur (thigh bone), and a socket of ultrahigh molecular weight polyethylene cemented into the pelvis. Over years of use, the ball migrates slowly into the socket, due to both wear and creep. Most of the migration appears to be due to creep [10.3.32]. Loosening of the ball joint portion in the thigh bone is usually more of a problem clinically.

§10.3.14 Dental fillings

Dental composite tooth fillings, which consist of inorganic mineral particles in a polymeric matrix, offer the cosmetic benefit of resembling tooth structure, and do not release metallic ions into the oral environment. The polymer phase is viscoelastic, and some studies of viscoelastic behavior of dental composites have been reported [10.3.33–10.3.36]. The study of viscoelasticity in these materials is relevant in elucidating polymer properties such as cross-link density, segmental motion, and degree of polymerization; and composite properties such as interfacial bonding. Viscoelastic behavior of dental materials governs some aspects of their performance in that the forces of mastication have a static component which over time can lead to an indentation of the restoration as a result of its creep. Wear processes also cause indentation of the filling. The creep is greatest soon after the composites are prepared; and as polymerization proceeds, the creep becomes less. Silver amalgam fillings [10.3.36] also exhibit creep as well as aging due to continued chemical reactions in the filling after it is prepared. For that reason it is best to avoid chewing stiff foods for a while after receiving a filling.

§10.4   Creep damage and creep rupture

Creep which occurs under the combination of large stress and long time can lead to damage and rupture of the material. During tertiary creep in polycrystalline materials such as metals, voids form along the grain boundaries [10.3.4]. Growth of the voids results in a higher stress for given load, and hence a higher creep strain rate. The strain rate increases until the material breaks. Perhaps the most dramatic case history of creep rupture is the Vajont slide [10.4.1] in which a large mass of rock on a steep slope suddenly collapsed, and slid down the slope into a reservoir. This created a water wave which obliterated the town of Longeroné at the cost of 2500 lives. The tragedy occurred in 1963; however, slow creep of the rock had been observed for three years prior to the creep rupture. Creep rupture is also a matter of concern in metals exposed to high stress and high temperature for extended periods. Optimal design of structural elements in this regime is complicated by the intrinsically nonlinear nature of secondary and tertiary creep [10.4.2].

§10.5   Seals and gaskets

Rubbery materials are commonly used in various kinds of seals and gaskets which prevent the flow of pressurized gas or liquid through the gap between two adjacent solid objects. The gap may change with time, and the seal must accommodate this change. If the gap changes too rapidly for the seal to follow, the pressurized fluid will not be contained. An example is the catastrophic explosion of the space shuttle Challenger in 1986. O-rings were used to prevent hot exhaust gas from escaping through joints in the solid rocket boosters. The shuttle was launched on a cold day. The low temperature caused O-ring material to acquire a more leathery rather than rubbery consistency, with more marked viscoelastic response (see §7.3). The material, therefore, could not rapidly adapt to changes in the spacing of the joint it was intended to seal. Hot gas escaped through the side of the booster causing the spacecraft's fuel tank to explode, resulting in death of the astronauts. The leathery viscoelastic response was described as a loss of resiliency. The change in behavior with temperature was graphically demonstrated by

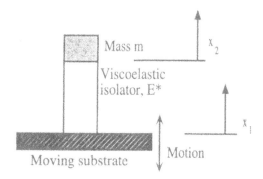

Figure 10.4   Sketch of a mass m isolated from a moving support by a viscoelastic bar of stiffness E* and length ℓ.

physicist Richard Feynman who immersed a segment of O-ring rubber in ice water, squeezed it with a clamp, and observed slow recovery on release of the clamp [10.5.1]. Feynman demonstrated not only viscoelastic behavior but also physical insight in general. Feynman earned his Nobel Prize in particle physics and was not a specialist in viscoelastic materials. Engineers were aware of the role of thermoviscoelasticity in the O-rings and had warned against launching the spacecraft. The disaster was a consequence of management decisions [10.5.1].

§10.6   Vibration control

§10.6.1   Analysis of vibration transmission

An example of the use of compliant viscoelastic materials in vibration isolation is presented. Vibration isolators are used to protect people or machinery from the harmful effects of vibration from engines, road travel, or other sources. Suppose that a substrate (Fig. 10.4) undergoes vibration given by $x_1(t) = a\, e^{i\omega t}$ (in which a is the amplitude of the displacement), and upon the substrate is a block of modulus E* of length ℓ and cross-sectional area A supporting a mass m. In the following, a simple one-dimensional analysis is conducted, with effects due to Poisson's ratio neglected. If such effects were included, an additional geometrical factor would appear multiplying the modulus. The analysis is similar to that in §3.5 pursued for a lumped system in which the dynamic compliance of the system was calculated; here the issue of interest is the transmission of vibratory motion from the substrate to the supported mass. This problem is amenable to the correspondence principle; however, since the solution is simple, it is treated directly in the frequency domain. The motion of mass m is described by $x_2(t) = b^* e^{i\omega t}$, in which the response amplitude b may have a phase and so be describable as a complex number. The system is considered to have one degree of freedom. In applications involving vibration isolation, the system *transmissibility* T is of interest. The transmissibility is defined as follows [10.6.1]:

$$T = \left| \frac{b^*}{a} \right|. \tag{10.6.1}$$

The transmissibility represents the dimensionless ratio of the displacement amplitude of the driven mass to the displacement amplitude of the moving substrate. It is usually desirable to minimize the transmissibility to protect the mass from too much vibration.

For this system, Newton's second law of motion becomes as follows [10.6.1]:

$$m \frac{d^2 x_2(t)}{dt^2} = \frac{A}{\ell} E^* (x_1 - x_2). \tag{10.6.2}$$

By substituting the harmonic time dependence,

$$-\omega^2 m b^* = \frac{A}{\ell} E^* (a - b^*). \tag{10.6.3}$$

So, the complex ratio of response displacement to driving displacement is

$$\frac{b^*}{a} = \frac{\dfrac{A}{\ell} E^*}{\dfrac{A}{\ell} E^* - \omega^2 m}. \tag{10.6.4}$$

Recall that $E^* = E'(1 + i \tan \delta)$ and that both E' and $\tan \delta$ depend on frequency,

$$\frac{b^*}{a} = \frac{1 + i \tan \delta}{1 + i \tan \delta - \dfrac{\omega^2}{\omega_0^2}}, \tag{10.6.5}$$

with

$$\omega_0^2 = \frac{\dfrac{A}{\ell} E'}{m} \tag{10.6.6}$$

as the natural angular frequency. Recall that the angular frequency is given by $\omega = 2\pi v$, with v as frequency.

By separating this into magnitude and phase θ,

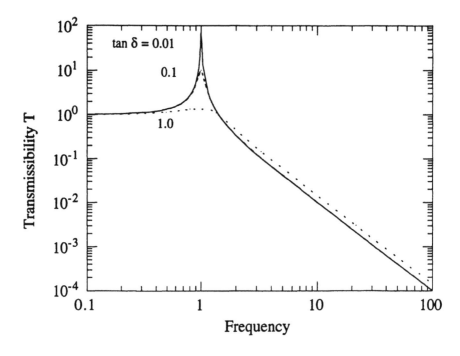

Figure 10.5   Transmissibility of lumped mass–isolator system, of single degree of freedom, for various tan δ vs. normalized angular frequency (ω/ω₀).

$$T = \left|\frac{b^*}{a}\right| = \frac{\sqrt{1 + \tan^2 \delta}}{\sqrt{\left(1 - \dfrac{\omega^2}{\omega_0^2}\right)^2 + \tan^2 \delta}}, \tag{10.6.7}$$

$$\theta = \tan^{-1}\left\{\frac{-\dfrac{\omega^2}{\omega_0^2}\tan\delta}{1 - \dfrac{\omega^2}{\omega_0^2} + \tan^2\delta}\right\}. \tag{10.6.8}$$

The transmissibility T tends to 1 at frequencies far enough below the natural frequency, and it decreases with frequency as $(\omega_0/\omega)^2$ for frequencies well above the natural frequency v_0 with the angular frequency $\omega = 2\pi v$ (Fig. 10.5). Therefore, if the transmissibility is to be minimized at the higher frequencies, the natural frequency should also be minimized. To make a detailed plot of T vs. ω, the frequency dependence of E′ as it influences ω_0 must be taken into account. Although some writers [10.6.1] present plots based on different assumptions about E′(ω), recall that the Kramers–Kronig relations developed in §3.3 prescribe a unique relationship between the real

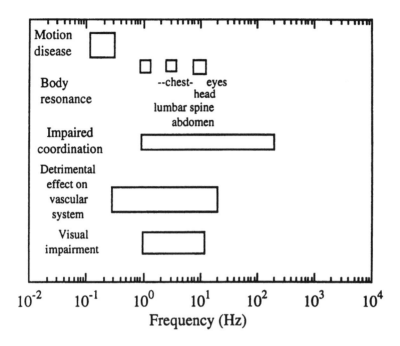

Figure 10.6 Frequency regions for effects of vibration on the human body. (Adapted from Frolov and Furman [10.6.2].)

and imaginary parts of a complex mechanical property function. Further examples and analyses are given in References [10.6.1, 10.6.2].

In practical systems, the isolator, represented by the block of stiffness E^*, is chosen so that the natural frequency is as far as possible below the frequencies of excitation of the substrate, so that the transmissibility is minimized in the frequency range of interest. The natural frequency is reduced by making the isolator more compliant (Eq. 10.6.6). If the isolator is too compliant, any force on the mass itself will cause excessive motion or even bottoming. In some cases it may be unavoidable that excitation motion appears at or near the natural frequency. For such cases, the isolator must be viscoelastic, to reduce the amplitude of the response at resonance. Since vibration can affect the human body (Fig. 10.6), isolation systems including shock absorbers and seat cushions are used to protect the body as described below in §10.6.11.

Resonant damping. If damping is required over a limited range of frequency, resonant (or tuned) damping may be used [10.6.3–10.6.5]. An example of a situation involving narrowband excitation is an electrical transformer. Transformers produce acoustic noise at harmonics of the power line frequency, as a result of magnetostriction of the iron alloys in them. The axial vibration configuration in Fig. 10.4 may be used to achieve a viscous or resistive input impedance at the base. Alternatively, one may use a configuration based on a shearing resonance. Here the objective is to maximize the dissipation of power from the vibrating "base" rather than to minimize the acceleration of the mass as considered above. Therefore, the material requirements differ. The viscoelastic element should have the smallest possible loss tangent if the dissipation is to be maximized at the resonance frequency. Under such conditions the supported mass will undergo high acceleration, but that is immaterial since it is a deadweight rather than a fragile object to be protected.

§10.6.2   Rotating equipment

Audio turntables are ordinarily isolated from outside vibration by suspending them from a relatively complex system of springs. A recent design for a turntable uses a viscoelastic platform for support rather than a costly and complex spring-based system [10.6.6]. The viscoelastic material, referred to as "a dry goo" by a manufacturer's representative, is sandwiched between layers of fiberboard.

Viscoelastic elastomer is available to be used as rubber "feet" [10.2.1] or base material to support scientific apparatus or other equipment which may be perturbed by vibration transmitted through the laboratory floor. Gaskets made of such materials not only can function as seals but also can reduce vibration in structures.

Viscoelastic elastomers are also used in the form of inertial dampers to reduce settling time during rapid angular accelerations of head-positioning motors in computer disk drives [10.2.1], and to reduce resonant vibration in high-speed robots. Nutation motion of spinning rotors mounted upon gimbals (pivot joints) occurs when a perturbing torque is suddenly applied. Nutation is an oscillatory motion of the spin axis. In some gyroscopic, aerospace, and satellite applications it is desirable to damp this nutation motion. Such damping may be achieved by attaching masses to the rotor by a compliant, viscoelastic stalk or annulus [10.6.7, 10.6.8]. The nutation motion causes an oscillatory strain in the viscoelastic material, dissipating energy and damping the motion.

§10.6.3   Large structures

Bridge and building structures are dynamically loaded by the wind and by earthquakes, and the resulting vibration can be problematic [10.6.9]. Sway of buildings may annoy or nauseate the occupants, and oscillation of bridges can cause structural damage. Damping in building arises from viscosity of the air surrounding the building, viscoelasticity of the building materials, friction damping at joints and contact points, hysteresis if the loads in the building or ground exceed the elastic range, and damping due to radiation of elastic waves through the supporting ground [10.6.10].

Viscoelastic dampers are used in some tall buildings to damp out sway oscillations of the building caused by wind. Viscoelastic dampers are used in the World Trade Center in New York [10.6.10], and viscoelastic dampers consisting of steel plates coated with a polymer compound are used in a 76-story building called the Columbia Center in Seattle. The viscoelastic dampers are connected to some of the diagonal bracing members [10.6.11]. Vibration damping may be achieved by means other than viscoelastic solids. For example, the tuned mass damper is used in other buildings such as the City Corp Center in New York and the John Hancock Tower in Boston. The tuned mass damper consists of a large concrete block weighing about 2% of the entire building. It is near the top of the building, upon a smooth concrete surface, and linked by a spring and macroscopic shock absorber to the building structure [10.6.10]. The mass and spring constant are "tuned" so that the resonant frequency of the system corresponds to the principal sway resonance frequency of the building. If building accelerations reach a threshold (such as 0.003 of the acceleration of gravity) due to a windstorm, pumps are activated which float the block upon a layer of oil so that it can oscillate in synchrony with the vibration, and damp it out. This is considered an active damping method.

§10.6.4   Plate and beam vibration

Flexural vibration in thin plate elements of structures can result in noise which is objectionable or harmful to people. Such vibration can be reduced by applying a damping layer [10.6.3, 10.6.12] of viscoelastic polymer of loss modulus E''_d and thickness t_d to the metal plate which has storage modulus E'_p and thickness t_p [10.6.12].

For *axial* vibration of a plate with a damping layer, the effective loss tan δ_{eff} is [10.6.5]

$$\tan\delta_{eff} = \frac{E'_d \tan\delta_d t_d}{E'_p t_p + E'_d t_d}. \qquad (10.6.9)$$

If the damping layer is much more compliant than the plate and it is not too thick,

$$\tan\delta_{eff} \approx \frac{E'_d \tan\delta_d}{E'_p} \frac{t_d}{t_p}. \qquad (10.6.10)$$

For *bending* vibration of a plate with a damping layer the effective loss tan δ_{eff} is

$$\tan\delta_{eff} \approx E'_d \tan\delta_d \frac{t_p \frac{1}{4}(t_d+t_p)^2}{\frac{1}{12}E'_p t_p^3 + E'_d t_d \frac{1}{4}(t_d+t_p)^2}. \qquad (10.6.11)$$

Again if the damping layer is not too thick,

$$\tan \delta_{eff} \approx 3 \frac{E'_d \tan \delta_d}{E'_p} \frac{t_d}{t_p},$$ (10.6.12)

in which t_p is thickness of the plate. The effective damping for a plate with two damping layers was derived via the correspondence principle in Example 5.1.

Consequently, the loss modulus $E'' = E'_d \tan \delta_d$ of the unconstrained damping layer should be maximized for both axial and bending vibration. Axial vibration is the most difficult to damp; in bending there is the advantage that the softer damping layer is some distance from the neutral axis, and hence more effective from a strain energy perspective.

If a thin layer of metal foil is cemented to the polymer layer, it is called a constrained layer [10.6.4]. Substantial shear strains can be induced in the polymer material during the bending of the plate, increasing the damping of the structure.

$$\tan \delta_{eff} = \tan \delta_d \frac{12 \frac{t_d^2 E'_{con}}{t_p^2 E'_p} f(g)}{1 + \left(12 \frac{t_d^2}{t_p^2} + 2 \right) \frac{E'_{con}}{E'_p} f(g)},$$ (10.6.13)

in which E'_{con} is Young's modulus (stiffness) of the constraining layer, E'_d is Young's modulus of the damping layer, E'_p is Young's modulus of the plate, and $f(g)$ is a function of a shear parameter g, given by the following [10.6.4, 10.6.5]

$$g = \frac{G_d}{E'_{con} t_{con} t_d q^2} = \frac{G_d \lambda^2}{E'_{con} t_{con} t_d 4\pi^2},$$ (10.6.14)

with G_d as the shear modulus of the damping layer, q as the wave number associated with vibration, and λ as the wavelength of bending waves. Damping in the constrained layer method is strongly frequency dependent as a result of the wavelength λ in the shear parameter. Even if the damping $\tan \delta_d$ of the polymer layer is constant, the overall damping $\tan \delta_{eff}$ exhibits a peak, less than two decades wide, in frequency. The frequency ν_{peak} of that peak is given by

$$\nu_{peak} = \frac{1}{22} \frac{c_L t_p}{t_{con}} G_d \sqrt{1 + \tan^2 \delta_d},$$ (10.6.15)

in which c_L is the speed of bending waves in the plate. So, as temperature changes, the shear modulus G_d changes substantially, and the frequency of the peak shifts too. The peak in the damping vs. frequency can be significantly broadened on the low-frequency end by segmenting the constraining layer [10.6.3]. Other variants of the damping-layer method include multiple layers with different properties, or use of a filler in the polymer to provide an internal constraint. Enhanced overall damping may also be achieved by mounting the damping layer upon standoff spacers [10.6.13]. The rationale is to move the damping layer as far from the neutral axis as possible to maximize strain levels in the layer.

Since Eq. 10.6.13 is complicated, we see that design of a constrained layer damping system is more involved than for free layers. Tapes, in this case containing both a polymer and a metal layer, are available for damping applications. Damping with the constrained layer method increases with the damping, $\tan \delta_d$, of the damping layer and the thickness of the constraining layer [10.6.4]. Tan δ_{eff} reaches a peak for g in the range 0.1 to 1. If one uses a compliant damping material with a small value of G_d, then it must have a greater thickness t_d to maintain the value of g. If a minimum weight design is desired, the product E tan δ enters the design process indirectly. Methods involving polymer layers are well known [1.6.8, 10.6.14, 10.6.15]. These layers offer acceptable performance in suppressing bending vibration of thin plates. They have the disadvantages of temperature sensitivity, flammability, narrow effective range of temperature and frequency, and comparative inability to control vibration in deformation modes other than bending. For materials of the highest damping (tan δ > 1), the full width at half maximum of the damping peak at constant frequency may be only about 18°C [10.6.16] although the peak may be broadened [10.6.17, 10.6.18] somewhat by non-stoichiometric compositions. This breadth is adequate for machinery which operates near room temperature, or at another nearly constant temperature, but is insufficient for aircraft in which the skin temperature may vary over a considerable range.

In some applications it is impractical to use polymer laminates. Several high-damping metals such as CuMn are available, as described in §7.4.3, and some of their applications are described in §10.6.8.

§10.6.5 Piezoelectric transducers

Piezoelectric materials exhibit coupling between the electric polarization and mechanical deformation. They are used in vibration as frequency standards, as spark generators in stoves, and in transducers. For materials used in ultrasonic transducers, a high damping is often desirable for the following reasons. In applications such as ultrasonic flaw detection and in medical diagnostic ultrasound, a short pulse is transmitted which contains only a few oscillations at the transducer's resonant frequency. The reason is that a short pulse permits greater resolution of defects. The Fourier transform of a short pulse in the time domain contains a wide distribution of frequency.

Recall from §3.5 that $\Delta\omega/\omega_0 \approx [\sqrt{3}]\tan\delta$. Therefore, a large loss tangent is desirable to obtain response over a relatively wide band of frequency. Quartz is low loss and is unsuitable for this type of application. Some piezoelectric ceramics, such as lead metaniobate, exhibit a high loss and are used in ultrasonic transducers. Damping can also be introduced by means of a backing layer applied to the transducer; this layer may contain lead particles in a polymer matrix. In addition to transducers [10.6.19, 10.6.20], there are many other applications of piezoelectric materials [10.6.21].

§10.6.6   Aircraft

Engines and associated machinery used in aircraft can produce considerable noise and vibration. Several examples follow.

§10.6.6.1   Case study: vibration fatigue of jet engine inlet

In the TF-30-P100 jet engine used in military aircraft, the entering air first encounters stationary inlet guide vanes which guide air into the first stage of the engine's turbine [10.6.4, 10.6.22]. Cracks were observed in the inlet guide vanes. Flight test measurements disclosed high vibratory stresses in the vanes, leading to high cycle fatigue. Vibration measurements disclosed high stresses corresponding to resonant torsion and bending modes from 3 to 4 kHz. A viscoelastic damper was designed to reduce the vibration. The design was driven by the fact that the inlet air varies in temperature from 0 to 125°F under most conditions, and that on occasion deicing procedures generated transient temperatures exceeding 400°F. A constrained-layer damping approach was chosen. Laboratory tests compared the effectiveness of several compositions of layer material. The final damping wrap consisted of a constraining layer of aluminum 0.005 in. (0.13 mm) thick and two viscoelastic layers, 0.002 in. (0.05 mm) thick, of ISD-830 material (3M Corp.) for the concave surface and ISD-112 for the convex surface. An additional aluminum layer, 0.002 in. (0.05 mm) thick under the viscoelastic layer was incorporated to avoid air entrapment between layers; this was bonded to the titanium surface of the vane with structural epoxy, type AF 126, nominally 0.005 in. (0.13 mm) thick. The damping wraps were installed on engines, and no further cracks were observed in the vanes.

§10.6.6.2   Case study: helicopter noise

A rescue helicopter, designated HH-53C and manufactured by Sikorsky Aircraft Co., was noisy: sound levels approaching 120 dB (with respect to 0.00002 µbar) were measured in the cabin. The noise came from the turbine and from gears, but was exacerbated by resonance of the fuselage skin. Noise can be reduced by placing standard acoustic blankets consisting of a layer of fiberglass sandwiched between layers of vinyl cloth inside the fuselage. This approach was not used because it would interfere with maintenance, inspection, and repair of the helicopter. A damping layer approach was therefore explored [10.6.4]. The vibration and temperature environment was first characterized via accelerometers and thermocouples applied to the fuselage. A constrained layer approach was chosen. Although only a small portion of the structure was covered with damping treatments, fuselage skin accelerations were reduced by up to 12 dB and high-frequency noise was reduced by 5 to 11 dB in the cabin.

§10.6.6.3 Case study: helicopter exhaust stack

Some helicopters (type CH-54) have experienced problems of vibration-induced fatigue of exhaust stacks [10.6.4, 10.6.23]. Attempts were made to reduce vibration amplitudes by increasing structural stiffness but these failed since the noise spectrum from the engines was broadband. The exhaust stacks exhibited many lightly damped ($\delta = 0.001$ to 0.1) resonances over the audio range. Polymer damping layers were not practical in view of the high temperatures involved, up to 800°F. The final damping material chosen was a vitreous enamel (Corning Glass No. 8363) which survives high temperature and exhibits a damping peak of about 0.5 at 100 Hz, near 800°F. Testing included trials at room temperature with polymer dampers, tests of enamel on a steel cantilever at high temperature, and finally tests on the actual helicopter exhaust stack. A free layer design 0.25 mm thick was chosen. The free layer approach was facilitated by the relatively high dynamic Young's modulus of the glass, about 17 GPa, at the damping peak. The composite peak loss factor was about 0.15 at 100 Hz and 800°F. This was sufficient to prevent the metal fatigue of the pipe.

§10.6.7 Solid fuel rockets

Energy released during the burning of solid rocket propellant has the side effect of amplifying acoustic waves in the rocket chamber [10.6.24]. This amplification can cause oscillations of sufficient amplitude to crack or break up the solid propellant. The solid propellant has, by virtue of its binder, viscoelastic properties [10.6.25, 10.6.26]; therefore, it attenuates stress waves. Higher values of tan δ have the effect of stabilizing combustion in the rocket and preventing fracture of the propellant. One propellant exhibited E = 1.1 GPa and a large tan δ of 0.18 at acoustic frequencies [10.6.25].

§10.6.8 Uses of high-loss metals

High-loss metals are used in situations in which polymer layers are inappropriate. A classic example is the use of copper–manganese alloy to reduce vibration and radiated noise from naval ship propellers [10.6.27]. This alloy has been used to replace a low-damping metal for valves, which had previously failed by fatigue, in a pump [10.6.28]. It has also been used to reduce noise in a pneumatic rock crusher. The damping capacity of cast iron makes it preferable to steel for large machine tool frames. Zinc–aluminum alloys have been used to reduce vibration in handheld pneumatic hammers, engine mounts, and camshaft drive pulleys in engines [10.6.29].

§10.6.9   High-loss composites

Damping layers of improved performance have been made of composite materials. Stiff inclusions [10.6.30–10.6.32] embedded in a high-damping polymer matrix can give rise to enhanced performance of the damping layer. Such composites are in harmony with the principles articulated in §9.3. Compliant, high-loss layers have been embedded in composites as constrained layers subject to shear [10.6.33].

§10.6.10   Sports equipment

Damping layers have been used to reduce vibration in sports equipment such as golf clubs [10.6.34, 10.6.35]. Piezoelectric damping elements have been used commercially in skis [10.6.36] to damp vibration which can cause the skier to lose control at high speeds.

§10.6.11   Seat cushions, automobiles: protection of people

§10.6.11.1   Effects of vibration on the body

People in moving vehicles can suffer a variety of detrimental effects from dynamic accelerations as shown in Fig. 10.6. Whole body vibration can adversely influence the ability of people to perform various tasks [10.6.37]. For example, vision can be degraded by vibration of the observed object, the eyes, or both. The eye can track the motion of objects at low speeds or low frequencies, but a rapidly vibrating object will appear blurred. People who perceive vibration may be annoyed or distracted. The body perceives acceleration most easily near 7 Hz at which the perception threshold is less than 0.01 m/s^2, but at lower or higher frequency it is more difficult to perceive acceleration. The threshold approaches 0.1 m/sec^2 at 0.1 Hz and at 100 Hz. People can experience motion sickness and may vomit if oscillated at low frequency. The range 0.1 to 0.5 Hz is most provocative of motion sickness. It is desirable to reduce vibration and noise within vehicles to reduce suffering in those people who must travel. To that end, viscoelastic materials can be of use.

The low-frequency oscillations which give rise to motion sickness are at too low a frequency to be attenuated by seat cushions or suspension systems. Motions sickness may be prevented or reduced by location of ship passengers or crew where translational movement is minimum and by design of the person's visual environment.

§10.6.11.2   Reduction of vibration

Seat cushions can play a role in reducing the body's exposure to vibration. Seats in vehicles have compliant cushions for several reasons [10.6.37]. Cushions distribute pressure around the bony prominences (ischial tuberosities) in the pelvis; they are soft to drop into; they usually have a low heat capacity so extreme temperatures are not so objectionable to the user; they provide friction to prevent sliding; and they attenuate vibration. These benefits are not necessarily achieved with the same physical characteristics. Moreover, the optimal dynamic behavior of a seat depends on the spectrum of vibration frequencies and on the desired criterion, e.g., comfort or preservation of health. To analyze the isolation of the human body from vibration, one cannot simply use a single degree of freedom model such as the one presented above, since the body itself is not a rigid mass. The body cannot be regarded as rigid above about 2 Hz. Transmissibility coefficients have been determined for pairs of points on the human body, such as the seat and the head of a seated person. The body is heavily damped, so that the transmissibility attains a peak of no more than 1.5 at about 8 Hz. As one might expect, the curves of transmissibility vs. frequency depend on posture. Vibrations in vehicles contain a spectrum of frequencies; in general there will be some stimulus at the resonance frequency (typically near 4 Hz) associated with the compliance of the seat cushion and the mass of the body. Unless adequate damping is present, the cushion will actually make matters worse by amplifying the amplitude of vibration.

As for low-frequency isolation of the automobile itself, the suspension system of automobiles contains springs and "shock absorbers" which are viscous dampers or dashpots. The resonance of such a system is at a relatively low frequency and is heavily damped, as can be demonstrated by abruptly pressing down on a fender of a parked car, and observing the free decay of vibration. The natural frequency of the car's suspension is well below the principal frequency content contained in excitation by bumps. Viscoelastic damping in the tires also contributes to the isolation of the auto body from bumps.

Reduction in automobile interior noise has been achieved by laminates consisting of layers of polymer resin and steel for automobile structural parts [10.6.38, 10.6.39]. Interior noise reduction of 5 dB has been achieved by replacing a conventional dash panel with such laminates. The resin layer consists of a thermosetting polyester resin blended with spherical nickel particles as a filler. The filler permits the laminate to be spot welded by forming electrically conductive paths between the steel layers.

§10.6.12 *Active vibration control: "smart" structures*

Active methods have been used in some buildings to reduce sway due to wind or earthquakes. For example, as discussed in §10.6.3, a tuned damper [10.6.10] in a building is activated by pumps triggered by accelerometers which sense a threshold acceleration. Hybrid systems for buildings [10.6.40] incorporate both tuned mass dampers and weights, driven by computer-controlled actuators, in response to measured accelerations. Active methods have also been used in the design of truss structures to be used in space stations and other spacecraft. Piezoelectric and magnetostrictive materials have been used as actuators.

In bridges, vibration of cables or other structural components can be a problem, particularly in longer bridges which are more vulnerable to instability on windy days. Such vibration can be damped passively [10.6.41] or actively [10.6.42, 10.6.43]. The active control system consists of a sensor for deformation, velocity, or load; a feedback amplifier; and an actuator to generate an opposing force. Analysis of active damping is simplified if one uses collocated actuator and sensor pairs [10.6.44, 10.6.45]. Since the actuators in an active damping system are driven via an external energy source, it is possible that such a system can become unstable and go into oscillation. The development of control modalities which guarantee absorption of vibration energy is therefore attractive, and aided by use of collocated actuator and sensor pairs.

§10.7   *Rolling friction*

Rolling friction arises principally from viscoelasticity in the materials which are in rolling contact [10.7.1]. This is physically reasonable since rolling results in periodic variation in stress; in a viscoelastic material, dissipation of energy occurs, and that dissipation is manifested as rolling friction. The physical origin of rolling friction was not immediately obvious to early scientists [10.7.2]. Coulomb [10.7.3] believed that rolling friction was due to bumps or asperities in one of the surfaces. Experiments later showed that, contrary to such a notion, rolling friction depends on load, speed, and other variables. Microscopic slip was then considered as a possible cause, but experiments by Bowden and Tabor [10.7.4] demonstrated that such slip has a minimal effect on rolling friction. Specifically, they drilled a hole in a roller which was painted to leave an imprint on a rubber substrate. The imprint was observed to be elliptical, indicating deformation of the rubber, and sharp, indicating minimal slip.

The problem of determining the coefficient of rolling friction from viscoelastic properties is not amenable to the correspondence principle, and so is difficult to solve in exact form. Even so, some scaling arguments are sufficient to elucidate several aspects of rolling resistance. The coefficient of rolling friction λ is defined as the ratio of the component of tangential force, which decelerates the object, to normal force, which supports its weight. Since λ is dimensionless, and since it must be zero in the absence of dissipative processes, one might naively suppose it to be proportional to tan δ. However, λ cannot be arbitrarily large, for the following reason. The net force vector must arise at some point within the contact region of radius a assuming nonsticky contact. Therefore, for a ball of radius R,

$$\lambda \le \frac{a}{R}.$$

(10.7.1)

If the size of the contact region is known from observation, an upper bound on the rolling resistance may be calculated [10.7.5]. The size of the contact region may also be determined via elasticity theory. For a rigid ball in contact with a flat deformable surface of Young's modulus E and Poisson's ratio v, by force P, the contact region radius a is given by elasticity theory [10.7.6] as

$$a = \left\{ \frac{3}{4} P \frac{1-v^2}{E} R \right\}^{1/3} . \qquad (10.7.2)$$

This is a special case of contact of two deformable spheres. If the contact region in the viscoelastic case has a similar size, the coefficient of rolling friction λ is bounded by

$$\lambda \le \left\{ \frac{3}{4} (1-v^2) \right\}^{1/3} P^{1/3} \frac{1}{E^{1/3}} \frac{1}{R^{2/3}} . \qquad (10.7.3)$$

Here the stiffness E is interpreted as a dynamic modulus at the frequency of rolling.

Flom and Bueche [10.7.7] presented a solution for the rolling resistance of a sphere on a viscoelastic substrate. It was developed assuming a material describable by the Maxwell model, and is of cumbersome form. The rolling resistance arises due to an asymmetric pressure distribution which develops in viscoelastic substrate materials. Some particular cases are of interest. Here ϕ is defined as the ratio of contact region radius to ball radius. G is the shear modulus.

$$\lambda = \tan \delta, \text{ for } \tan \delta << \phi = \frac{a}{R}. \qquad (10.7.4)$$

$$\lambda = 0.434\phi = 0.243P^{1/3} \frac{1}{G^{1/3}} \frac{1}{R^{2/3}}, \text{ for } \tan \delta = \phi = \frac{a}{R}. \qquad (10.7.5)$$

$$\lambda = 0.510\phi = 0.248P^{1/3} \frac{1}{G^{1/3}} \frac{1}{R^{2/3}}, \text{ for } \tan \delta = 2\phi = \frac{2a}{R}. \qquad (10.7.6)$$

For the particular case of large tan δ, λ ≈ 0.590ϕ so the inequality in Eq. 10.7.1 is satisfied in all cases. Moreover, for large tan δ, the dependence of rolling resistance on ball radius, vertical load, and substrate stiffness is the same as anticipated above in Eq. 10.7.3. Analysis of rolling of rigid cylinders on a viscoelastic substrate [10.7.8] disclosed results similar to the above. Rolling resistance attains a peak value as a function of rolling speed. Experiments on balls of various kinds of rubber [10.7.9] confirmed the general features of the viscoelastic theory of rolling resistance.

Rolling friction is important in tires, since such friction influences fuel consumption. Representative values [10.7.2] for the coefficient of rolling friction are 0.004 for bicycle tires, 0.008 for truck tires, 0.01 for radial automobile tires, and 0.015 for bias ply automobile tires. Sliding friction in polymers is also linked to viscoelastic properties [1.6.18] in that the making and breaking of contacts is governed by the same kinds of molecular processes as are responsible for viscoelasticity. In tires, it is desirable to minimize rolling friction to reduce fuel consumption. For that purpose a low-loss rubber is best [10.7.10]. It is also desirable to maximize sliding friction to prevent skidding (and to prevent suffering due to possible injury to the driver or to others) during braking, turning, and acceleration. For that purpose a high-loss rubber is best. Tires made of high-loss rubber are also relatively quiet and smooth riding since vibrations are heavily damped. Since these requirements are contradictory, different rubber compositions are used for racing and for normal driving. A degree of compromise can be achieved by using a low-loss rubber in highly strained regions of the tire, and a higher loss rubber for the portion in contact with the road. Rolling friction also depends on the air pressure within the tire since pressure alters the cyclic deformation during rolling. Specifically, high pressure reduces the size of the contact patch during rolling, and hence the rolling friction. Conversely, an increase in the load upon the tire increases the contact patch size, and hence the rolling friction.

Tires undergo two principal types of deformation [10.7.11]. The first consists of the flattening of the bottom of the tire tread against the road. During highway driving at 100 km/h (62 mph) the tire rotates about 20 times per second, so the fundamental frequency of oscillatory strain in the tire is about 20 Hz. The second deformation consists of small-scale indentations of the tire by irregularities in the road surface. Tires grip the road with the aid of these microscopic conformations. Frequencies associated with this deformation are high, up to 1 MHz in a skidding tire. Therefore, rolling resistance could be reduced without compromising sliding friction by reducing tan δ at low frequency but maintaining tan δ at high frequency. Use of silica (SiO_2) rather than carbon as particulate inclusions in the rubber has such an effect since, unlike carbon, silica forms chemical bonds with the polymer chains in the rubber, reducing the damping at the lower frequencies.

§10.8   Uses of low-loss materials

§10.8.1   Timepieces

Watches and clocks use vibrating objects to generate periodic signals which may be divided into intervals for timing purposes. In earlier mechanical designs, torsional oscillation is set up in a pivoted wheel held by a torsional spring. The ticking sound from such timepieces comes from the escapement mechanism used to convert oscillatory motion into the uniform circular motion used to display the time. In quartz timepieces, vibration at ultrasonic frequency is set up in a crystal of quartz (SiO_2). Quartz is a piezoelectric material: it deforms in response to an electric signal, and when stressed it generates an electric polarization. The piezoelectric property makes quartz useful as a tuning element to control the frequency of an electronic circuit. There are other piezoelectric materials, notably the ceramics, which are less costly than quartz and offer a stronger piezoelectric effect. Quartz is used for timepieces and other frequency standards because its loss tangent is extremely small in comparison with that of other available piezoelectrics (see Sections 7.2 and 7.9). The peak in the resonance curve of compliance vs. frequency is therefore sharp and the resonance frequency is well defined. Moreover, little power is required to maintain vibration at the resonant frequency, which for watches is typically 32.768 kHz [10.8.1]. As a result of the small loss tangent of quartz, timepieces can be made which are precise and accurate and require little power. However, the quality factor Q (recall that $\tan \delta \approx Q^{-1}$) for practical resonators is 50,000 to 120,000 minimum [10.8.2], lower than expected (10^6 to 10^7) from the intrinsic damping of the quartz. The difference is attributed to the resonator mounting which allows vibratory energy to leak away in the supporting structure. That mount must be more robust in a watch, which is subject to jolts, than in an experimental apparatus. In addition, the stability of quartz in relation to temperature, time, and environmental changes is essential in this application [10.8.1].

§10.8.2 *Frequency stabilization and control*

Vibrating piezoelectric crystals are also used to control the frequency of radio transmitters and receivers. As in timepieces, it is advantageous to use a material with as small a loss tangent as possible [10.8.2]. Other desirable characteristics of materials include minimal aging (change of properties with time) and minimal change of properties with temperature.

Tuning forks are used as frequency standards in music, in testing of the ear, and in physics [10.8.3]. To a first approximation the fork can be considered as two cantilever bars joined with an offset at the base. Tuning forks are commonly made of aluminum or aluminum alloy. The low loss tangent of aluminum allows the fork to continue vibrating for a long period after it is struck. Some energy is lost through radiation of sound, in addition to the loss due to viscoelasticity in the material.

§10.8.3 *Gravitational measurements*

The detection of gravitational waves is a problem of great interest in modern physics. Gravitational waves are undulatory deformations in the structure of space–time. They are caused by cosmic events such as the collapse of massive stars. One method which has been used in attempts to detect gravitational waves is to compare the vibration of resonant bar detectors placed in different geographical locations. If both detectors register a signal, it must be from a remote source, possibly cosmological, not a local one, such as passing trucks or earthquakes. The sensitivity of such detectors depends inversely on the loss tangent of the resonant bar [10.8.2]; therefore, low-loss materials have been used to make them.

Measurement of the Newtonian gravitational constant G are performed with a Cavendish balance in which a dumbbell-shaped pair of masses is suspended from a slender wire or fiber. Fixed masses are then placed at different locations near the dumbbell. The period of oscillation is measured for two positions of the attracting masses, and G is inferred. A key assumption in the interpretation of results is that the spring constant of the torsion fiber is independent of frequency. Individual measurements have estimated errors of about 0.1% but different measurements yield values differing by several parts per thousand. Systematic errors in this measurement have recently been attributed to viscoelasticity at low frequency in the suspension wire of the Cavendish torsion balance [10.8.4–10.8.6]. This viscoelasticity, even in supposedly elastic materials such as fused silica, tungsten, or copper–beryllium, gives rise to a spring stiffness which depends on frequency contrary to the usual assumptions.

§10.9   Impact absorption

§10.9.1   Analysis

Viscoelastic materials can be used to reduce the force of impact [10.9.1]. In the analysis of the impact, a compliant elastic layer is first placed between the impactor and the surface to be protected, and its effect is analyzed. Then the layer is assumed to be viscoelastic, and its optimal viscoelastic properties are determined, with the aim of minimizing the impact force.

Consider first an elastic impact buffer considered as a one-dimensional mass–spring system. The massless linearly elastic buffer in one dimension is mechanically equivalent to a spring. The displacement U in free vibration of the mass m is

$$U(t) = A \sin(\omega_0 t + \theta), \tag{10.9.1}$$

in which the spring is a massless axial spring, the natural angular frequency is $\omega_0 = \sqrt{Ebc/mh}$, E is the axial modulus of the elastic block, b and c are its cross-section dimensions, h is its axial length, and A and θ are amplitude and phase constants, respectively (depending on the initial displacement and velocity of m). In the case of impact of a moving mass m of velocity V on a spring, Eq. 10.9.1 is used to describe the motion of m. The contact of m and the spring occurs at t = 0. The displacement U of m is therefore

$$U(t) = \frac{V}{\omega_0} \sin(\omega_0 t).$$

The impact force F is obtained as

$$F(t) = m\omega_0 V \sin(\omega_0 t).$$

The impact force is proportional to the spring deflection, or the displacement of m. The maximum impact force F_{max} is $V\sqrt{mEbc/h}$ with respect to the maximum spring deflection U_{max} of $V\sqrt{mh/Ebc}$ at $t = \dfrac{\pi}{2}\sqrt{mh/Ebc}$. F_{max} is related to U_{max} by

$$F_{max} = Ebc\frac{U_{max}}{h},$$

or $F_{max} = Ebc\,\varepsilon$ in which ε is the maximum compressive engineering strain of the spring. Decreasing the value of E results in a decrease of the maximum impact force. However, if the buffer is too soft, it will bottom out, corresponding to $\varepsilon = 1$.

A value of

$$E = \frac{mV^2}{bch\varepsilon^2} \qquad (10.9.2)$$

is the optimal stiffness to minimize F_{max} if the buffer layer is elastic [10.9.1].

Now consider [10.9.1] a one-dimensional viscoelastic buffer upon a substrate impacted by a moving mass. To investigate the impact behavior in terms of the viscoelastic properties of the buffer directly, we approximate the impact as one half cycle of free-decay oscillation. The governing equation for the one-dimensional forced oscillation of a massive viscoelastic block with an attached mass m is similar to that given by Christensen [1.6.1] for the torsional case, and is as follows:

$$F_{ext}(t) - U(t)bc\rho\omega^2 h\frac{ctn(\Omega^*)}{\Omega^*} = m\frac{\partial^2 U(t)}{\partial t^2}, \qquad (10.9.3)$$

in which $F_{ext}(t)$ is the external force, $F_{ext}(t) = F_0 \sin(\omega t)$, $U(t)$ is the displacement of m, b and c are the cross-section dimensions of the viscoelastic buffer, h is the axial length, ρ is the mass density, ω is the angular frequency of harmonic oscillation, $E^*(i\omega)$ or $E'(i\omega)(1 + i\tan\delta)$ is the uniaxial complex modulus, $E'(i\omega)$ is the storage modulus, $\tan\delta$ is the loss tangent, and

$$\Omega^* = \sqrt{\frac{\rho\omega^2 h^2}{E^*(i\omega)}}.$$

The lateral restrictions at the ends of the buffer are neglected. This is considered appropriate since the analysis is on the basis of a one-dimensional problem. Now suppose the buffer mass is much less than that of the impactor as a result of a small density ρ or axial length h. Then we can approximate $\Omega^* \to 0$, and therefore $ctn(\Omega^*) \to 1/\Omega^*$. The buffer is therefore a massless viscoelastic "spring" and the induced deflection and stress are uniform in the axial direction. Then

$$F_{ext}(t) - U(t)E^*(i\omega)\frac{bc}{h} = m\frac{\partial^2 U(t)}{\partial t^2}. \tag{10.9.4}$$

The external force F(t) vanishes in the situation when the mass has struck the buffer and is vibrating freely upon it. The impact, from contact to peak force, is one quarter of one cycle. The solution is then

$$U(t) = Ae^{\omega_0(-\alpha+i\beta)t+\theta i}, \tag{10.9.5}$$

in which

$$\omega_0 = \sqrt{\frac{E'(i\omega)bc}{mh}},$$

$$\alpha = \left(1 + \tan^2\delta\right)^{1/4}\sin\frac{\delta}{2},$$

$$\beta = \left(1 + \tan^2\delta\right)^{1/4}\cos\frac{\delta}{2},$$

and A and θ are constants depending on the initial displacement and velocity of m. This solution for free vibration of the mass–viscoelastic buffer system is identical to that for one-dimensional impact of mass m on a viscoelastic buffer. Based on the initial condition the displacement of m can be written

$$U(t) = \frac{V}{\beta\omega_0}\sin(\beta\omega_0 t)e^{-\alpha\omega_0 t}, \tag{10.9.6}$$

in which V is the initial velocity of moving mass m, and the contact of m and the buffer occurs at t = 0 again. Accordingly, the force induced on m is

$$F(t) = m\frac{d^2 U(t)}{dt^2},$$

$$F(t) = \frac{mV}{\beta}\omega_0\left\{\beta^2 \sin(\beta\omega_0 t) + 2\alpha\beta\cos(\beta\omega_0 t) - \alpha^2\sin(\beta\omega_0 t)\right\}e^{-\alpha\omega_0 t}.$$

$$(10.9.7)$$

By equating $dU(t)/dt$ and $dF(t)/dt$ to zeros, respectively, the maximum displacement U_{max} and the maximum impact force F_{max} are derived in terms of the design parameters as

$$U_{max} = V\sqrt{\frac{mh}{E'(i\omega)bc}}\frac{1}{(1+\tan^2\delta)^{1/4}}e^{-\tan(\delta/2)(\pi/2-\delta/2)}, \qquad (10.9.8)$$

at

$$t = \frac{1}{\omega_0}\frac{\pi-\delta}{2(1+\tan^2\delta)^{1/4}\cos\left(\dfrac{\delta}{2}\right)},$$

and

$$F_{max} = V\sqrt{\frac{mE'(i\omega)bc}{h}}(1+\tan^2\delta)^{1/4}e^{-\tan(\delta/2)(\pi/2-3\delta/2)}, \qquad (10.9.9)$$

at time

$$t = \frac{1}{\omega_0}\frac{\pi-3\delta}{2(1+\tan^2\delta)^{1/4}\cos\left(\dfrac{\delta}{2}\right)},$$

for $\tan\delta \le 1.73$ or

$$F_{max} = 2V\sqrt{\frac{mE'(i\omega)bc}{h}}(1+\tan^2\delta)^{1/4}\sin\left(\frac{\delta}{2}\right), \qquad (10.9.10)$$

at $t = 0$, for $\tan\delta \ge 1.73$.

From Eq. 10.9.8, the optimal stiffness is

$$E'(i\omega) = \frac{mV^2h}{bc}\left(\frac{1}{U_{max}}\right)^2\frac{1}{(1+\tan^2\delta)^{1/2}}e^{-\tan(\delta/2)(\pi-\delta)} \qquad (10.9.11)$$

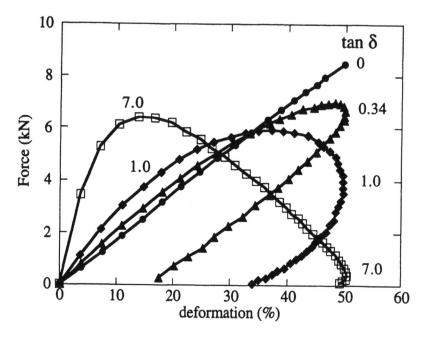

Figure 10.7   Effect of tan δ on impact force, including the transient term, with the storage modulus held constant. (Adapted from Chen and Lakes [10.9.1].)

if there is a limit U_{max} on the maximal deflection. By substituting, F_{max} becomes

$$F_{max} = \frac{mV^2}{U_{max}} e^{-\tan(\delta/2)(\pi - 2\delta)}$$

(10.9.12)

for tan δ ≤ 1.73, or

$$F_{max} = 2\frac{mV^2}{U_{max}} \sin\left(\frac{\delta}{2}\right) e^{-\tan(\delta/2)(\pi/2 - \delta/2)}$$

(10.9.13)

for tan δ ≥ 1.73.

   Curves of force vs. deformation for different values of tan δ are shown in Fig. 10.7.

   Differentiating with respect to δ, one finds the optimal loss tangent to minimize the impact force to be

$$\tan \delta = 0.4$$

if the stiffness is held constant, and

$$\tan \delta = 1.1$$

if the stiffness is treated as a design variable. For $\tan \delta = 1.1$ the optimal value of $E'(i\omega)$ is 24% of the value for the optimized buffer, based on Eq. 10.9.11. The resulting impact force is 52% of the value found in the optimized elastic buffer, based on Eq. 10.9.2. So the effect of buffer viscoelasticity is to substantially reduce the peak impact force in comparison with an elastic buffer of the same thickness.

More sophisticated analysis [10.9.1] including the transient term does not change the numerical results much, and it does not change the conclusion. Specifically, the angular frequency at which one refers to $\tan \delta$ is ω_0, the natural angular frequency of the mass–buffer system. The impact actually contains a distribution of frequencies, so a more complete analysis must include assumptions or input data regarding the frequency dependence of the damping. Such a requirement gives rise to considerable complication, requiring numerical solution. If the indenter is spherical rather than one dimensional, the optimal loss tangent is greater than in the above case, since the force–deformation characteristic for this case is a nonlinear function, concave up. Again, the peak impact force can be reduced by almost a factor of two by using a viscoelastic material rather than an elastic material for the buffer.

The conclusion is that force associated with impact from a moving object can be minimized by an elastic buffer of appropriate stiffness. The force can be further reduced by an additional factor of about two if the buffer is made viscoelastic. The optimal loss tangent for this application is large; it can be achieved in viscoelastic elastomers.

§10.9.2   Bumpers and pads

As for specific practical applications of viscoelastic materials in impact absorption, viscoelastic elastomers are used in automobile bumpers [10.9.2] and as bumpers and crash stops to isolate computer disk drives from mechanical shock during power failures and during shipment, as well as to protect equipment which could be damaged by being dropped. Car bumpers may contain lumped element damping units [10.9.3]. Moving parts in computer disk drives are also damped [10.9.4] to improve tracking and response speed. Viscoelastic materials are also used as bushings in printing presses to minimize shock forces. Viscoelastic foams are used for seats for airplanes, spacecraft, bicycles, and trucks to reduce the user's discomfort. Viscoelastic foams are also used in padding for athletic equipment, e.g., helmets, ski boots, wrestling mats, and weight-training equipment.

§10.9.3   Shoe insoles, athletic tracks, and glove liners

Viscoelastic elastomers are used in shoe insoles to reduce impacts transmitted to a person's skeleton. Some measurements have been reported of the effect of shoe mechanical properties on forces on or vibration in the human body. Accelerations of the tibia and skull during walking show peaks [10.9.5] of about 5 and 0.5 g, respectively, when hard heels are worn (g is the acceleration due to gravity). Peak accelerations are reduced by about a factor of two when compliant heels are worn. Viscoelastic inserts in shoes reduced leg accelerations by up to a factor of 1.8 [10.9.6]. Some investigators report a reduction in back and joint pain and other problems in athletes who use shoes with viscoelastic isolators; moreover, people suffering from pain due to arthritis [10.9.7] or occupational exposure to prolonged standing on hard surfaces [10.9.8] observed significant reduction in pain after using the viscoelastic inserts. People with joint diseases appear to have insufficient shock absorbing capacity in the tissues [10.9.9]. A variety of the available viscoelastic elastomer materials (with trade names Sorbothane, Implus, and Noene) have been compared [10.9.10] for damping and stiffness.

Compliant polymers have also been used for the surfaces of running tracks. There is considered to be an optimal compliance which results in the fastest speed of runners in racing events [10.9.11]. The improvement in speed for distance events on the "tuned" track was about 3% compared with a hard surface. Compliant track surfaces also have been reported to be more comfortable for runners and to result in fewer injuries to competitive athletes. A running surface which is too compliant, as one might expect, slows down the runners. Analyses of tracks thus far have incorporated surface compliance but not damping.

A related application of viscoelastic elastomers is in liners of work gloves used by operators of tools which generate severe vibration, such as jackhammers and chipping tools. Vibration can cause circulatory disturbances in the hands and arms. Long exposure to severe vibration can result in numbness and blanching of the fingers [10.9.12–10.9.14], a condition known as Raynaud's syndrome, or "vibration white finger". The vibration suppresses blood circulation to the extent that fingers turn white and numb. Use of high-damping metals in the tool itself has also been tried to reduce suffering due to this problem.

§10.10   Rebound of a ball

§10.10.1   Analysis

A ball dropped on a viscoelastic substrate rebounds to a height lower than the height from which it was dropped. In this section, the height of rebound is related to the loss tangent of the material [10.9.1, 10.10.1].

Tan δ of viscoelastic materials can be determined from the degree of rebound in a one-dimensional rebound test via the solutions [10.9.1] of $U(t)$ and $F(t)$ given above. A mass m dropped from height H_0 on a massive block of material rebounds back with velocity V_1 following an impact time t_{imp} and back to height H_1. The ratio of velocities is defined as the *coefficient of restitution*, e. Specifically,

$$e = \frac{\text{velocity of separation}}{\text{velocity of approach}}. \tag{10.10.1}$$

Since the potential energy is mass x gravity x height (mgH, with g as the acceleration due to gravity) and the kinetic energy is $\frac{1}{2} mv^2$, the height ratio is

$$\frac{H_1}{H_0} = e^2 \equiv f. \tag{10.10.2}$$

The coefficient of restitution depends on the properties of both the ball and the material upon which it bounces. If one material is much stiffer than the other, strain energy in the stiff material is negligible, and the damping properties of the more compliant material dominate the behavior.

The impact time t_{imp} is obtained by equating F(t) to zero [10.9.1] in the above analysis, Eq. 10.9.7, so that

$$t_{imp} = \frac{1}{\omega_0} \frac{\pi - \delta}{\left(1 + \tan^2 \delta\right)^{1/4} \cos\left(\dfrac{\delta}{2}\right)}, \tag{10.10.3}$$

with

$$\omega_0 = \sqrt{\frac{Ebc}{mh}}.$$

V_1 is obtained by substituting t_{imp} in the time derivative of U(t). Tan δ is therefore related to the rebound height ratio H_1 / H_0 by

$$\frac{H_1}{H_0} = \left(\cos\delta + \tan\frac{\delta}{2}\sin\delta\right) e^{-2\tan(\delta/2)(\pi-\delta)}. \tag{10.10.4}$$

Again, more sophisticated analysis is possible [10.9.1] but the results are similar (Fig. 10.8). If tan δ is small, the relation above can be approximated as follows:

$$\tan\delta \cong \frac{1}{\pi}\ln\left(\frac{H_0}{H_1}\right). \tag{10.10.5}$$

This is the form obtained from a study of the approximate solution for free decay of vibration as done in Example 3.3. To conclude, a rigid ball dropped on a viscoelastic substrate bounces high if the loss tangent is small, and does

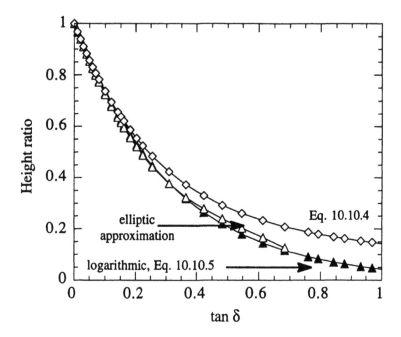

Figure 10.8   Diagram of several approximations for the relation between height ratio and loss tangent for ball rebound. The elliptical approximation, which neglects the initial transient, is not shown over the range in which the approximation radically breaks down.

not bounce much if the loss tangent is large. Similar results can be expected for a viscoelastic ball on a rigid substrate.

§10.10.2   *Applications in sports*

In many sports, the rebound of a ball following an impact is highly relevant to the character of the sport. For example, in baseball [10.10.2, 10.10.3], the batter likes to hit the ball as far as possible. The baseball itself consists of a winding of wool and cotton yarn around a core of cork, and covered by cowhide [10.10.2]. The ball is rather inelastic: it exhibits a height ratio of 0.3 in a drop test. This corresponds to a coefficient of restitution of only about 0.55 [10.10.3]. The coefficient of restitution of a major league baseball is required to be e = 0.546 ± 0.032 [10.10.4] based on a standard impact with a block of ash wood. Even such a small allowable variation can result in as much as a 15-ft (4.6-m) difference in the range of a well-hit baseball. It is not really meaningful to speak of a loss tangent for a baseball, since it is actually a heterogeneous structure. Since the ball's constituents are natural polymers, the viscoelasticity depends on temperature and humidity. Adair [10.10.2] adduces several apocryphal stories of efforts by home teams to gain advantage by cooling some or all the baseballs prior to the game. The rules have been changed so that all balls must be provided to the umpires 2 h before game time. The rebound velocity of the ball depends on the properties of both the bat and ball. Since the bat is stiffer than the ball, its viscoelastic properties are less important. Even so, since aluminum bats dissipate less mechanical energy than wood ones, it is possible to hit the ball further with an aluminum bat. The mechanical damping of the bat also manifests itself in the sound of the impact: a "ping" sound in the case of an aluminum bat which has low damping.

In a hollow ball, the rebound depends on the pressure of the gas within the ball. The reason is that the compressed gas acts as an elastic spring element. Rebound resilience of a hollow ball depends on the internal pressure by virtue of the contact region size which depends on pressure. Tennis balls, for example, are pressurized to increase rebound resilience. The rebound of a tennis ball decreases over a period of weeks due to diffusion of gas through the rubber. The balls are stored in pressurized cans to prevent any loss of internal pressure prior to use. Rebound of tennis balls used in tournament play is restricted: the height ratio must be from 0.53 to 0.58 [10.10.5, 10.10.6]. Footballs and basketballs are large enough to accommodate a valve so that they can be inflated with air to control rebound characteristics.

Golf balls [10.10.6] are typically made of rubber thread wrapped under tension around a core. The coefficient of restitution e for a drive in golf is about 0.7. Since the golf club has been observed to exhibit no appreciable deformation during impact, the value of e depends almost entirely on the ball. Over a temperature range of 0 to 27°C, the value of e for the ball changes from 0.64 to 0.75. For this reason a golf shot does not carry as far on a cold day as on a hot day. Some golfers keep golf balls warm in a pocket, or even in a battery-powered electric heater, on cold days. Rubber has such a low thermal conductivity that the thermal time constant for a golf ball is about an hour. Therefore, it is insufficient to prewarm the ball briefly. As in the case of baseball and tennis, there is a restriction on the coefficient of restitution of golf balls considered legal for sporting competition.

High rebound balls include Ping-Pong balls, handballs, and Superballs. A Ping-Pong ball dropped on a concrete floor exhibited a coefficient of restitution e of 0.85, and for a handball, 0.82 [10.10.7]. A Superball is even more responsive; it has e = 0.94 corresponding to a low loss tangent (from Eqs. 10.10.2 and 10.10.5, tan $\delta \approx$ 0.04). It is an interesting toy. The Superball can bounce in unexpected directions due to coupling between translational and rotational degrees of freedom [10.10.8, 10.10.9].

The game of squash, by contrast, requires use of a viscoelastic ball with a low degree of rebound. The squash ball is composed of a proprietary blend of rubber containing as many as 15 ingredients [10.10.10]. A "slow" (yellow dot) squash ball, in a drop test [10.10.11] from 1 m, exhibited a height ratio of 0.2, corresponding to a high effective loss tangent. By Eq. 10.10.4, tan $\delta \approx$ 0.7. The actual tan δ of the rubber in the squash ball may actually be higher than this since the ball is hollow; and the air inside, though not pressurized, offers some elastic response. The World Squash Federation specifies a rebound height ratio (given as a percentage) of 12% minimum at 23°C, and 26 to 33% at 45°C assuming a drop height of 2.54 m (100 in.). A specification is given at 45°C since such a temperature has been measured in balls during vigorous play [10.10.10]. Most of the energy of impact between the ball and racket or ball and playing surface is converted into heat as a result of viscoelasticity in the ball.

§10.11   Winding of tape

Materials stored in the form of rolls include paper, plastic sheets, and magnetic tape for computers and for audio and video recording. As for tapes, the requirement for high-quality reproduction of stored information implies that the dimensions of the tape be essentially unchanged during and between read and write operations [10.11.1]. When wound, the sheet or tape is under tension and is applied one layer at a time over a core. The materials involved are viscoelastic. Therefore, the final stress state in the layers will depend on the speed of the winding, time of storage, and any starts or stops in the process. A fairly complicated analysis of this problem has been presented [10.11.1].

§10.12   Viscoelastic gels in surgery

Viscoelastic pastes and gels are used to facilitate various kinds of surgery. Methylcellulose has a long history of use in ophthalmic procedures. Sodium hyaluronate is a large polysaccharide molecule which forms a viscoelastic gel with water. It occurs in the connective tissues of vertebrates [10.12.1] and in the vitreous humor of the eye. Viscoelastic gels based on methylcellulose or purified sodium hyaluronate have been used in surgery, for cell protection, in maintenance of tissue spaces, in tissue lubrication, and in tissue manipulation. Use of such viscoelastic pastes has been of particular use in ophthalmic surgery [10.12.2] to maintain tissue spaces, to replace the vitreous humor [10.12.3], and to manipulate and protect the delicate tissues of the eye. Manipulation of tissue in a space filled with viscoelastic paste is facilitated by the damping of disturbances. Moreover, bleeding is minimized by the viscous resistance of the paste. Purified sodium hyaluronate has been used in tendon repair [10.12.4], and a gel of carboxymethylcellulose and polyethylene oxide has been used for spine surgery and neurosurgery [10.12.5]. The gel is used to prevent scar tissue formation after surgery.

§10.13   Food products

Mechanical loads are applied to food products during harvesting, transportation, and storage. Viscoelastic behavior of various fruits and vegetables have been studied with the aim of achieving better quality of food products in the marketplace. To that end, viscoelastic properties have been measured [10.13.1, 10.13.2] and some results are discussed in §7.10. Such data can aid in the design of food handling equipment and procedures.

Viscoelasticity is the basis of informal tests of the ripeness of melons. Many people tap melons before buying them, to judge from the sound whether they are ripe. This is an example of free decay of vibration following an impulse.

Perceived freshness of some foods depends on the glass transition temperature T_g of constituents. For example, a fresh cookie [10.13.3] had a T_g of 60°C. It would be consumed near room temperature, in the glassy state, perceived as crunchy. Exposure to the atmosphere resulted in a weight gain of 5% and a reduction of the T_g to 10°C, most likely due to the plasticizing effect of the adsorbed moisture. The texture would then be perceived as leathery or rubbery.

§10.14   Hand strength exerciser

A viscoelastic putty [10.14.1] is used for exercising the gripping strength of the hand, either for rehabilitation after injury, or for strength development for sports such as rock climbing. The putty exhibits substantial creep and at long times is a viscoelastic liquid. It therefore conforms well to the shape of the hand when squeezed slowly, but is stiff when squeezed rapidly. The response of the material also encourages the user to squeeze it.

§10.15   Viscoelastic toys

Silly Putty® is a highly viscoelastic silicone polymer [10.15.1] sold as a toy. It exhibits extreme rate dependence. When dropped on a hard surface it bounces back to a substantial fraction of the original height. When stretched slowly, it draws out like taffy. The material flows like a liquid when left alone for some time.

Gelatinous materials have also been used as toys. For example, a transparent gelatinous material has been fabricated as a melt blend admixture of poly(styrene–ethylene–butylene–styrene) triblock copolymer with high levels of a plasticizing oil [10.15.2, 10.15.3]. This material has been marketed under the name GlueSlug®; it exhibits a large maximum elongation combined with low stiffness. Other applications, such as flexible lenses, light pipes, hand exercise grips, and acoustical isolators have been suggested in the patents.

§10.16   Tissue viscoelasticity in medical diagnosis

Physicians are able to obtain important diagnostic information by pressing the patient's body with one finger and tapping (percussing) with another finger. The sounds produced and the tactile sensations felt reveal to the

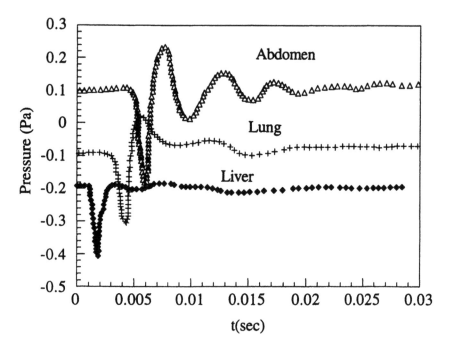

Figure 10.9   Waveforms associated with response to diagnostic percussion impuls-
es upon different parts of the human body. (Adapted from Murray and Neilson
[10.16.1].)

examiner the condition of the underlying tissue [10.16.1, 10.16.2]. Percussion
as a diagnostic tool dates from a publication in 1761 by Auenbrugger
[10.16.3].

Oscilloscope tracings of percussion sounds [10.16.1] disclose damped
oscillations similar to the curves for free decay of vibration described in §3.6.
Different tissues, and comparison of diseased and healthy states of a single
tissue, reveal waveforms of considerable difference in the rate of decay of
vibration. The percussion sound [10.16.1] described by a physician as "dull"
corresponded to a single pressure impulse of less than 3 msec; "resonant"
sound consisted of two to three oscillations with a duration of 15 msec; while
"tympany" consisted of a damped sinusoid with center frequency 200 to 600
Hz, and lasting more than 40 msec, as shown in Fig. 10.9. Lung lesions up
to 50 mm deep and 20 to 30 mm in diameter can be detected by diagnostic
percussion [10.16.4]. Percussion has also been used to evaluate the degree
of bone fracture healing [10.16.5].

Diagnostic percussion is an informal test of structural resonance in the
human body. As we have seen in §3.6, the free decay of vibration in a
structure depends on the loss tangent of the material. The waveforms
described above indicate that structures in the body are heavily damped.

§10.17   No-slip flooring, mats, and shoe soles

Cork is at times used for floor surfaces which are claimed not to become slippery even if wet or polished [10.17.1]. Viscoelastic rubbery materials are used as mats in bathtubs to prevent falls on wet and soapy surfaces. The friction between a shoe or foot and a floor surface arises from two sources: adhesion due to atomic bonds between the surfaces, and resistance due to irreversible loss of mechanical energy due to viscoelastic deformation in one or both of the surfaces. Friction due to adhesion is abolished by polishing or by lubrication by water or oil; therefore, polished or lubricated hard surfaces are slippery. If one surface is compliant and viscoelastic, the energy lost in deforming it will provide sliding friction, even if the surfaces are polished or lubricated. Use of viscoelastic materials in these applications can reduce suffering due to injury from falls.

§10.18   Applications involving thermoviscoelasticity

When one cuts or machines stiff polymers, the cutting process causes a rise in temperature, which reduces the stiffness and increases the loss tangent at frequencies associated with the cutting process. Continued deformation of the polymer gives rise to a further temperature rise as mechanical energy is converted into thermal energy by viscoelastic loss. If the heat cannot escape rapidly enough, a thermal runaway process can occur in which the polymer part to be machined overheats and melts. The machinist soon learns to cut slowly enough to prevent this from occurring.

In the making of tempered glass, a slab of red-hot glass is extruded between rollers. Jets of air are directed at the free surfaces to cool the glass. The effect of thermal contraction, heat conduction, and thermoviscoelasticity is that the cooled glass acquires a distribution of residual stress in which the free surfaces are under compression. Such a distribution is beneficial in a brittle material such as glass since fracture initiates at cracks or defects under tension. Since most of the defects are at the surface as a result of scratches or other damage, a superposed compressive stress improves the overall strength of the glass plate when it is bent.

§10.19   Satellite dynamics

Early satellites were given a spinning motion to stabilize them. The spin was about an axis of minimal inertia, similar to the spin of a rifle bullet. For example, the Explorer 1 satellite was given a spin of this type. The satellite had a stable orientation for a few minutes after which it gradually tumbled out of alignment. The cause is the viscoelasticity of the materials in the satellite as analyzed below. Specifically, the stability of a spinning object such as a satellite is examined in view of the Euler angles θ, ϕ, and ψ, [10.19.1] based on the analysis of Thomson and Reiter [10.19.2].

In a freely spinning body of revolution with principal moments of inertia A, A, and C, precession occurs if the spin axis is not aligned with a principal axis of inertia. The spin angular velocity is defined as $\dfrac{d\phi}{dt}$; the precession angular velocity, as $\dfrac{d\psi}{dt}$; and θ, as the angle between the angular momentum L and the 3 axis.

$$\frac{d\psi}{dt} = \frac{C}{(A-C)\cos\theta}\frac{d\phi}{dt}.$$ (10.19.1)

The angular velocity about the z (or 3) axis is, from the definition of the Euler angles,

$$\omega_3 = \frac{d\phi}{dt} + \frac{d\psi}{dt}\cos\theta.$$ (10.19.2)

The angular momentum L is constant since it is assumed that there are no external torques.

Consider first the question of stability. For small misalignment θ,

$$L = C\omega_0,$$ (10.19.3)

$$\omega_3 = \omega_0 \cos\theta,$$
$$\frac{d\psi}{dt} = \frac{C}{A}\omega_0.$$ (10.19.4)

The kinetic energy T of rotation is

$$T = \frac{1}{2}A(\omega_1^2 + \omega_2^2) + \frac{1}{2}C\omega_3^2,$$ (10.19.5)

but for small misalignment θ,

$$T = \frac{1}{2}C\omega_0^2\left[1 + \left(\frac{C}{A}-1\right)\sin^2\theta\right].$$ (10.19.6)

If the kinetic energy T is dissipated due to viscoelasticity in the satellite, there must be a change in θ. By differentiating, the rate of change is

$$\frac{dT}{dt} = C\omega_0^2\left(\frac{C}{A}-1\right)\sin\theta\cos\theta\frac{d\theta}{dt}. \qquad (10.19.7)$$

Since T decreases with time in a dissipative material,

$$\frac{d\theta}{dt} < 0 \text{ if } \frac{C}{A} > 1, \text{ but } \frac{d\theta}{dt} > 0 \text{ for } \frac{C}{A} < 1. \qquad (10.19.8)$$

So spin about the principal axis of minimum moment of inertia ($C/A < 1$) is unstable in the presence of dissipation.

Now consider the rate at which the angle θ changes as a result of damping in the satellite. The precession generates a time varying stress in the object with a frequency $2\pi\left\{\frac{d\phi}{dt}\right\}^{-1}$. To calculate that stress, consider a simple specific model of a spinning satellite, consisting of two identical parallel rigid disks of mass m connected at their centers by a viscoelastic tube of length L, dynamic Young's modulus E, and loss tangent tan δ. The mass moment of inertia of each disk is C_1 about its polar axis and A_1 about its diametric axis.

The gyroscopic moment about each disk is

$$M_G = C_1\left[\frac{d\phi}{dt}+\frac{d\psi}{dt}\cos\theta\right]\frac{d\psi}{dt}\sin\theta - A_1\left(\frac{d\psi}{dt}\right)^2\sin\theta\cos\theta, \qquad (10.19.9)$$

in which the moments of inertia about the center of mass are

$$C = 2C_1,$$

$$A \approx 2\left(A_1 + mL^2\right),$$

so

$$M_G = \frac{1}{2}\left\{C\left(\frac{d\phi}{dt}+\frac{d\psi}{dt}\cos\theta\right)\frac{d\psi}{dt}\sin\theta - A\left(\frac{d\psi}{dt}\right)\sin\theta\cos\theta\right\}$$
$$+ mL^2\left(\frac{d\psi}{dt}\right)^2\sin\theta\cos\theta, \qquad (10.19.10)$$

but the term in the { } braces is the moment about the center of mass; that moment is zero for a freely spinning object. So the gyroscopic moment is

$$M_G = mL^2 \left(\frac{d\psi}{dt} \right)^2 \sin\theta \cos\theta = FL\cos\theta, \qquad (10.19.11)$$

with F as the centripetal force of the precessing disk.

The bending moment on the connecting tube is

$$M_z = M_G \frac{z}{L}, \qquad (10.19.12)$$

and the maximum stress is

$$\sigma = \frac{M_z y}{I} = mL^2 \left(\frac{d\psi}{dt} \right)^2 \sin\theta \cos\theta \frac{zy}{LI}. \qquad (10.19.13)$$

Since

$$\frac{d\psi}{dt} = \frac{C}{A} \omega_0, \qquad (10.19.14)$$

$$\sigma = \frac{1}{2} mL^2 \left(\frac{C}{A} \right)^2 \omega_0^2 \frac{zy}{LI} \sin\theta \cos\theta.$$

The energy dissipation per unit volume per cycle is

$$\Gamma = \frac{1}{2} \tan\delta \frac{\sigma^2}{E}, \qquad (10.19.15)$$

so

$$\Gamma = \frac{\tan\delta}{48\pi E} \left(\frac{mL^2 y}{I} \right)^2 \cdot V \left(\frac{C}{A} \right)^4 \left(\frac{C}{A} - 1 \right) \omega_0^5 \frac{zy}{LI} \sin^2\theta \cos^3\theta, \quad (10.19.16)$$

with V as the volume of material.

Since $d\theta/dt$ is proportional to this Γ via Eq. 10.19.7, tumbling is initiated by a small initial misalignment, increases to a peak value, and then decreases as θ approaches 90° [10.19.2, 10.19.3].

An instability of this type was observed in the early Explorer I satellite, which was given a spin about its longitudinal axis of minimum moment of inertia. In about 90 min, θ went from near 0° to about 60°. Later satellites were stabilized by spinning about an axis of maximal moment of inertia; hence, they tend to be oblate in shape rather than prolate.

§10.20 Structural damping

§10.20.1 Rationale

High-loss, flexible polymer layers may be used, as described in §10.6.4, to reduce flexural vibration of plates. It is more difficult to reduce axial vibration, since the attached layer would require both high stiffness and high damping. Moreover, some environments are hostile to polymers and in some designs, the extra space occupied by a polymer layer would be problematic. Figure 7.1 shows that high stiffness combined with high damping does not occur among common materials. There are some alloys, such as CuMn, which offer high stiffness with moderate damping, as discussed in §7.4.3. These materials have the disadvantage of being nonlinear: the damping is much less at small strain than at large strain. A linear material with high stiffness and high loss would be useful in many applications involving vibration reduction, particularly in the aerospace field [10.20.1].

§10.20.2 New materials

The viscoelastic mechanisms discussed in Chapter 8 have been explored primarily for the purpose of understanding existing materials; however, they may also be used to guide the synthesis of new materials. For example, composite materials may be designed to achieve a measure of thermo-elastic damping [10.20.2], by proper choice of constituents. Damping achieved in this way is advantageous in that it is hardly dependent on temperature. However, it is difficult to achieve much thermoelastic damping (tan δ greater than 0.01) using common materials as is evident from Table 8.1. Piezoelectric materials are more promising in that the coupling between the stress and electric field can be much stronger than the coupling between the stress and temperature field for available materials as discussed in §8.10. A resistive circuit element can be attached to a piezoelectric inclusion to achieve substantial damping [10.20.3–10.20.5]. A current challenge is how to make a fine-grained composite with resistively damped piezoelectric elements. A third possibility is the development of composite materials with optimal architecture for high damping, as discussed in §9.3 and §9.4. The phase which provides the high damping need not be stiff or strong provided that the composite geometry permits the other phase to contribute stiffness and strength. Composites with $E = 161$ GPa (more than twice as stiff as aluminum) and tan $\delta = 0.1$ (comparable to a glassy polymer) have been made [10.20.6].

§10.21 Viscoelasticity in scientific investigations

Determination of viscoelastic properties can be used as a probe into various microphysical phenomena in materials. For example, the true and effective diffusion coefficients can differ as a result of differences in the thermodynamic activity of the diffusing substance [10.21.1]; calculations based on measurement of concentration vs. distance give the effective diffusion coefficient. True diffusion coefficients in alloys have been determined indirectly by evaluation of relaxation times via measurement of damping over a wide range of temperature. As another example, it is possible to achieve a qualitative mapping of viscoelastic properties using the atomic force microscope [10.21.2]. Viscoelasticity, as evaluated via the out-of-phase component of an electrical signal at 2 kHz, appears qualitatively as image contrast.

In geophysics, one can use seismic waves to infer cross-sectional maps of properties of matter on Earth: seismic tomography [10.21.3]. In the upper mantle of Earth, $Q \approx 100$ (tan $\delta \approx 0.01$), and in the lower mantle, $Q \approx 300$ (tan $\delta \approx 0.003$) for shear waves. Temperature increases with depth; therefore, the derivative of wave speed with temperature is important in interpretation. By virtue of the Kramers–Kronig relations developed in §3.3, damping gives rise to frequency dependence of the wave speed; moreover, since the damping processes are thermally activated, changes in temperature alter damping as well as velocity. In another geophysical example, creep of rock structures on Earth is measured [10.21.4] and related to changes in stress associated with earthquakes. Ultimately, such measurements may be of use in earthquake prediction. In yet another example, viscoelasticity of Moon rocks, inferred from seismic studies, was used to better understand the nature of those rocks as discussed in §7.5.

Many scientific instruments are vulnerable to vibration [10.21.5]. For example, scanning probe microscopes such as the scanning tunneling microscope and atomic force microscope are capable of resolving individual atoms on a size scale below 0.2 nm. An atomically sharp tip supported by a microscopic cantilever is scanned across the specimen. Tip deflection is measured via a reflected laser beam and held constant by a feedback circuit driving a piezoelectric actuator. In view of the high resolution desired, vibration must be minimized. Structural beam elements within the instrument may be made of magnesium, which has a low density ρ and comparatively high damping. Stresses are low in such an instrument, so the limited strength of magnesium is not a problem. Moreover, magnesium has a high value of E/ρ^2, which is favorable for bending stiffness per unit weight. As another example consider the optical tables used to support laser equipment used in holography, interferometry, and other optical procedures which are sensitive to vibration. The table is supported upon pneumatic legs in which pressurized air supports the table via force upon a piston. The legs provide compliant support so that the table-leg system has a low natural frequency v_0 near 1 Hz. Following Eq. 10.6.7, the transmissibility for vibration goes as $(v_0/v)^2$ for frequencies v well above the natural frequency. Therefore, little vibration above 1 Hz gets through to the table. The table is made using a honeycomb sandwich construction which confers a high rigidity per unit mass, and hence a high fundamental resonance frequency. That resonance is damped by a tuned damper [10.21.6] consisting of a mass supported by flexure springs and immersed in a viscous fluid.

§10.22 Relaxation in musical instruments

Players of stringed instruments are aware that nylon strings rapidly go out of tune [10.22.1]; in comparison, metal strings retain their tuning. The pitch of the note generated by the string is calculated as follows. The speed of sound c in a string of mass m and length L, under tensile force T is

$$c = \sqrt{\frac{T}{m/L}}.$$

This may be expressed in terms of the stress σ and density ρ.

$$c = \sqrt{\frac{\sigma}{\rho}}.$$

The frequency f of the lowest (fundamental) natural frequency is

$$f = \frac{c}{\lambda} = \frac{1}{\lambda}\sqrt{\frac{\sigma}{\rho}}.$$

This frequency governs the pitch of the tone. The length of the string corresponds to half a wavelength λ, since the string is fixed at each end. The musician tunes the instrument by adjusting the extensional displacement of the string by means of a screw. After tuning, the string is under essentially constant axial strain, and it undergoes relaxation of stress, which causes the pitch of tones derived from the string to decrease with time. Therefore, the musician finds it necessary to again tune strings made of a polymer such as nylon. Strings made of catgut contain natural polymers which also undergo relaxation. Wood in the instrument is also viscoelastic, but provided the instrument is much stiffer than the strings, displacements due to creep in the wood are small. Strings also can go out of tune as a result of temperature changes. As for the free decay of vibration in a plucked string, energy is radiated as sound, and this radiation can be the predominant cause of the free decay of the vibration in a string in the instrument.

§10.23 Ultrasonic testing

Ultrasonic waves are used for materials characterization as discussed in §6.8, as a method of nondestructive evaluation to detect flaws in machine parts, and in medical diagnosis. Flaws or diseased tissues are detected via the reflections caused by mismatch in acoustic impedance. Waves at higher frequencies offer superior resolution since the corresponding short wavelength allows small features to be visualized. However, wave attenuation due to tan δ tends to increase with frequency, and this attenuation, combined with the depth of the flaw, limits the highest frequency which can be practically used [10.23.1]. For that reason higher frequencies are used in the diagnosis of shallow organs such as the eye, and lower frequencies are used for larger organs such as the liver. In some tests, such as evaluation of fatigue crack damage in machine parts [10.23.2, 10.23.3], damping or attenuation may be associated with the structural feature under study.

§10.24 *Summary*

We have considered the effect of viscoelastic behavior in the performance of materials in particular applications, as well as the implications in design. The influence of viscoelasticity can be beneficial, as in cases when it is used explicitly in design to achieve design goals. Cases have been presented in which high or low mechanical damping were used to achieve specific objectives. If viscoelasticity is ignored in the design process, the results of design calculations may not correspond to physical reality, leading to an unsuccessful design. Proper use of viscoelastic materials can aid in the design of devices which serve to reduce suffering.

Examples

Example 10.1

Consider the use of piezoelectric materials as stiff damping elements. A piezoelectric ceramic is assumed to be connected to an external electric circuit. Determine the maximal value of tan δ attainable with available piezoelectric materials connected to a resistive circuit, and discuss. A one-dimensional treatment will suffice for this example.

Solution

In Example 8.6 the energy concepts used in the definition of coupling coefficient K were related to the limiting stiffnesses considered in the definition of relaxation strength to obtain for a Debye peak,

$$\tan \delta_{max} = \frac{1}{2} \frac{K^2}{\sqrt{1-K^2}}. \tag{E10.1.1}$$

Piezoelectric elements have been studied and used as damping elements. A form identical to Eq. E10.1.1 was obtained by a different approach for a piezoelectric element loaded with a resistor [10.20.5].

If the piezoelectric element is loaded by a short circuit, it is at constant (zero) electric field; if it is electrically free, it is at constant electric displacement. At low frequency the piezoelectric element is considered equivalent to a capacitor of value C. If an external resistor is connected, any transient charge on that capacitor decays with a time constant τ = RC. The largest available K in common piezoelectric materials is about 0.75 in lead titanate zirconate ceramics, for load in the 3 direction and polarization in the 3 direction. This gives tan δ_{max} = 0.43 for a Debye peak. This is a large loss tangent in view of the relatively high stiffness (about 70 GPa) of piezoelectric ceramics. Overall damping would not be so high since only a small part of a structure would consist of piezoelectric damping elements. Moreover, in some applications the designer desires a high tan δ over a wider range of frequency than a Debye peak. Even so, the concept is attractive for structural damping.

Example 10.2

How fast does a golf ball leave the club during a drive shot? How much energy is dissipated? How much does the golf ball warm up as a result of the energy dissipation? What is the effective tan δ of the golf ball? Assume [10.10.6] the club head has a mass of 0.2 kg and is moving at 50 m/sec, that the ball is initially stationary and has a mass of 0.046 kg, that the heat capacity of rubber is 1700 J/kg °C, and that the coefficient of restitution is e = 0.7.

Solution

Consider [10.10.6] conservation of momentum: the total momentum of ball and club must be the same before and after the impact (denoted by underlined symbols).

$$m_{ball}v_{ball} + m_{club}v_{club} = m_{ball}\underline{v}_{ball} + m_{club}\underline{v}_{club}.$$

Writing the definition of coefficient of restitution in symbolic form gives ·

$$e = \frac{\underline{v}_{ball} - \underline{v}_{club}}{v_{club} - v_{ball}}.$$

By solving for the velocities,

$$\underline{v}_{ball} = \frac{m_{club}\left\{v_{club}(1+e) - ev_{ball}\right\} + m_{ball}v_{ball}}{m_{ball} + m_{club}},$$

$$\underline{v}_{club} = \frac{m_{ball}\left\{v_{ball}(1+v_{club}e) - e\right\} + m_{club}v_{club}}{m_{ball} + m_{club}}.$$

By incorporating the assumption that the ball is initially stationary,

$$\underline{V}_{ball} = \frac{m_{club} V_{club}(1+e)}{m_{ball} + m_{club}},$$

$$\underline{V}_{club} = \frac{V_{club}(m_{club} - em_{ball})}{m_{ball} + m_{club}}.$$

By substituting the assumed numbers,

$$\underline{V}_{ball} = 70 \, \text{m/sec}, \quad \underline{V}_{club} = 34 \, \text{m/sec}.$$

The loss in energy is

$$\Delta U = \frac{1}{2} m_{club} V_{club}^2 - \frac{1}{2} m_{club} \underline{V}_{club}^2 - \frac{1}{2} m_{ball} \underline{V}_{ball}^2.$$

By substituting,

$$\Delta U = \frac{m_{ball} m_{club} V_{club}^2 (1 - e^2)}{2(m_{ball} + m_{club})} = 23 \, \text{J}.$$

By considering the mass of the ball and the given heat capacity, the temperature rise is $\Delta T = 0.3 \, °C$, which is unimportant.

The effective $\tan \delta$ of the golf ball is given as a function of the height ratio (Eq. 10.10.5)

$$\tan \delta \cong \frac{1}{\pi} \ln\left(\frac{H_0}{H_1}\right),$$

but the height ratio is $e^2 = 0.49$ from the given value $e = 0.7$, so

$$\tan \delta \cong 0.24.$$

Actually, the coefficient of restitution was given for a golf club impact, which involves large deformation of the ball under which materials behave non-linearly. Moreover, the ball is not a homogeneous material. Therefore, one speaks of an "effective" loss tangent.

Example 10.3

Traditionally, applicants for police training must be above a threshold height. If a person is a half inch or one inch too short to qualify, is there any hope for acceptance?

Solution

Creep in spinal disks results in a considerable variation in a person's height during the day. People are taller in the morning than the evening since the spinal disks are substantially unloaded during sleep. The applicant may try to take the height test in the morning. If more height is needed, a few days of bed rest will allow further recovery of creep strains in the spine.

Example 10.4

How does three-dimensional deformation influence the use of viscoelastic rubber in such applications as shoe insoles to reduce impact force in running, or wrestling mats to reduce impact force in falls?

Solution

Refer to Example 5.9, in which deformation under transverse constraint is analyzed. Rubbery materials are much stiffer when compressed in a thin-layer geometry than they are in shear or in simple tension; they are too stiff to perform the function of reducing impact. Compliant layers can be formed by corrugating the rubber to provide room for lateral expansion or by using an elastomeric foam, which typically has a Poisson's ratio near 0.3. Corrugated rubber is used in shoe insoles and in vibration isolators for machinery. Foam is used in shoes and in wrestling mats.

Problems

10.1 Calculate the rolling resistance of a tire. Specifically, infer the effective horizontal force of resistance by determining the power dissipated in the tire due to its viscoelasticity. The coefficient of rolling friction is defined as the ratio of the horizontal resistance force to the vertical force. Assume that E^* of the rubber is known. This problem may be considerably simplified by treating the tire as a thin-walled cylindrical pressure vessel with a flat region corresponding to contact with the road. How does the rolling resistance depend on the air pressure in the tire? How does it depend on the speed of the vehicle?

10.2 Calculate the effective tension–compression damping of a steel plate with a single layer of viscoelastic polymer glued to it. Assume all thicknesses and material properties are known.

10.3 Calculate the effective flexural damping of a steel plate with a single layer of viscoelastic polymer glued to it. Assume all thicknesses and material properties are known.

10.4 Obtain an approximate value for the optimal stiffness for a knee pad intended to minimize impact force upon the knee in sports. Assume a reasonable value for the thickness of the pad. Is it more beneficial to make the pad of a viscoelastic material or to incorporate a plateau region in the stress–strain curve (such as by elastic collapse of a foam)?

10.5 Suppose a material were developed with tan δ < 0. What causal mechanisms might be used to achieve such a result? What might such a material be used for? Design a sport which could take advantage of a ball with such behavior.

10.6 If a squash ball is available, infer the tan δ of the rubber in the ball via a rebound test. How does rebound depend on temperature? Are your results consistent with the expected behavior of a polymer in the rubbery and leathery regions. If you have access to test equipment, measure the tan δ of the rubber. Compare with the results of a rebound test.

10.7 If you have a guitar or other stringed instrument which can use nylon or catgut strings, a microphone, and a method to determine frequency, then perform the following evaluation of stress relaxation. Install and tune a new string, and measure the frequency as a function of time after tuning [10.22.1]. Compare with a metal string. Interpret your results in terms of what you now know about viscoelastic behavior.

10.8 How does the rebound resilience of a hollow ball differ from that of a solid one?

References

10.2.1   EAR, Division of Cabot Corp., 7911 Zionsville Rd., Indianapolis, IN.

10.2.2   Gardner, R., Jr., "Earplugs", US Patents 3,811,437 (1974); RE 29,487 (1977).

10.3.1   Bazant, Z., *Mathematical Modeling of Creep and Shrinkage of Concrete*, J. Wiley, NY, 1988.

10.3.2   Gordon, J. E., *Structures*, Penguin, Harmondsworth, Middlesex, UK, 1983.

10.3.3   Backman, D. G. and Williams, J. C., "Advanced materials for aircraft engine applications", *Science* 255, 1082–1087, 1992.

10.3.4   Ashby, M. F. and Jones, D. R. H., *Engineering Materials*, Pergamon, Oxford, 1980.

10.3.5   Willett, F., "The case of the derailed disk drives", *Mech. Eng.* 110, 42–44, 1988.

10.3.6   Eagle, R., "A pain in the back", *New Scientist* 84, 170–172, 1979.

10.3.7   Althoff, I., Brinckmann, P., Frobin, W., Sandover, J., and Burton, K., "An improved method of stature measurement for quantitative determination of spinal loading", *Spine* 17, 682–693, 1992.

10.3.8   Magnusson, M., Hult, E., Lindström, I., Lindell, V., Pope, M., and Hansson, T., "Measurement of time-dependent height loss during sitting", *Clin. Biomech.* 5, 137–142, 1990.

10.3.9   Kasra, M., Shirazi-adl, A., and Drouin, G., "Dynamics of human lumbar intervertebral joints: experimental and finite element investigations", *Spine* 17, 93–102, 1992.

10.3.10 Formby, W. A., Nanda, R., and Currier, G. F., "Longitudinal changes in the adult facial profile", *Am. J. Orthod. Dentofacial Orthod.* 105, 464–476, 1994.

10.3.11 Gunner, C. W., Hutton, W. C., and Burlin, T. E., "An apparatus for measuring the recoil characteristics of human skin *in vivo*", *Med. Biol. Eng. Comput.* 17, 142–144, 1979.

10.3.12 Payne, P. A., "Measurements of properties and function of skin", *Clin. Phys. Physiol. Meas.* 12, 105–129, 1991.

10.3.13 Largo, R. H. and Duc, G., "Head growth and changes in head configuration in healthy preterm and term infants during the first six months of life", *Hel. Paediatr.* 32, 431–442, 1977.

10.3.14 AAP Task Force on Infant Positioning and SIDS: positioning and SIDS, *Pediatrics* 89, 1120–1126, 1992.

10.3.15 Clarren, S., Smith, D., and Hanson, J., "Helmet treatment for plagiocephaly and congenital muscular torticollis", *J. Pediatr.* 94, 443, 1979.

10.3.16 Pomatto, J. K., Littlefield, T. R., Manwaring, K., and Beals, S. P., "Etiology of positional plagiocephaly in triplets and treatment using a dynamic orthotic cranioplasty device", *Neurosurg. Focus* 2, 1–4, 1997.

10.3.17 Das, B. M., *Principles of Geotechnical Engineering*, 2nd ed., PWS, Kent and Boston, 1990.

10.3.18 Terzaghi, K. and Peck, R. B., *Soil Mechanics in Engineering Practice*, 2nd ed., J. Wiley, NY, 1967.

10.3.19 Hwang, J. S., "Soldering and solder paste technology", ed. C. Harper, *Electronics Packaging and Interconnection Handbook*, McGraw-Hill, NY, 1991.

10.3.20 Hwang, J. S. and Vargas, R. M., "Solder joint reliability—can solder creep?", *Soldering and Surface Mount Technol.* 5, 38–45, 1990.

10.3.21 Morris, J. W., Jr., Goldstein, J., and Mei, Z., "Microstructure and mechanical properties of Sn–In and Sn–Bi solders", *JOM* 45, 25–27, 1993.

10.3.22 McCormack, M. and Jin, S., "Progress in the design of new lead-free solder alloys", *JOM* 45, 36–40, 1993.

10.3.23 Bartha, L., Lassner, E., Schubert, W. D., and Lux, B., ed., *The Chemistry of Non-Sag Tungsten*, Elsevier, Barking, Essex, UK, 1995.

10.3.24 Howard, W. H. and Williams, M. L., "Viscoelasticity and flatspotting", *Rubber Chem.Technol.* 40, 1139–1146, 1967.

10.3.25 Setright, L. J. K., *Automobile Tyres*, Chapman and Hall, London, 1972.

10.3.26 Cunningham A., Huygens E., and Leenslag J. W., "MDI comfort cushioning for automotive applications", *Cell. Polym.* 13, 461–472, 1994.

10.3.27 Dementjev, A. G., "Deformation of flexible polyether polyurethane foams with bimodal cell structure", *Cell. Polym.* 15, 155–171, 1996.

10.3.28 Dinsdale, S. M., "Decubitus ulcers: role of pressure and friction in causation", *Arch. Phys. Med. Rehabil.* 55, 147–152, 1974.

10.3.29 Yarkony, G. M., "Pressure ulcers: a review", *Arch. Phys. Med. Rehabil.* 75, 908–917, 1994.

10.3.30 Palmieri, V. R., Haelen, G. T., and Cochran, G. V., "A comparison of sitting pressures on wheelchair cushions as measured by air cell transducers and miniature electronic transducers", *Bull. Prosthetics Res.* 10 (33), 5–8, 1980.

10.3.31 Garber, S. L., "Wheelchair cushions: a historical review", *Am. J. Occup. Ther.* 39, 453–459, 1985.

10.3.32 Rose, R. M., Nusbaum, H. J., Schneider, H., Ries, M., Paul, I., Crugnola, A., Simon, S. R., and Radin, E. L., "On the true wear rate of ultra high molecular weight polyethylene in the total hip prosthesis", *J. Bone Joint Surg.* 62-A, 537–549, 1980.

10.3.33 Von Finger, W., "Elastizitat von Composite-fullungsmaterialein", *Dtsch. Zahnaerztl. Z.* 30, 345–349, 1975.

10.3.34 Ruyter, I. E. and Oysaed, H. "Compressive creep of light cured resin based restorative materials", *Acta Odontol. Scand.* 40, 319–324, 1982.

10.3.35 Papadogianis, Y., Boyer, D. B., and Lakes, R. S., "Creep of conventional and microfilled dental composites", *J. Biomed. Mater. Res.* 18, 15–24, 1984.

10.3.36 Papadogianis, Y., Boyer, D. B., and Lakes, R. S., "Creep of amalgam at low stress", *J. Dental Res.* 66, 1569–1575, 1987.

10.4.1 Bassett, R. H., "Time-dependent strains and creep in rock and soil structures", in *Creep of Engineering Materials*, ed. C. D. Pomeroy, Mechanical Engineering Publications, Ltd., London, 1978.

10.4.2 Zyczkowski, M., "Optimal structural design under creep conditions", *Appl. Mech. Rev.* 49, 433–446, 1996.

10.5.1 Feynman, R. P., *What Do You Care What Other People Think?*, W. W. Norton, NY, 1988.

10.6.1 Snowdon, J. C., *Vibration and Shock in Damped Mechanical Systems*, J. Wiley, NY, 1968.

10.6.2 Frolov, K. V. and Furman, F. A., *Applied Theory of Vibration Isolation Systems*, Hemisphere/Taylor and Francis, NY, 1990.

10.6.3 Kerwin, E. M., Jr. and Ungar, E. E., "Requirements imposed on polymeric materials in structural damping applications", in *Sound and Vibration Damping with Polymers*, ed. R. D. Corsaro and L. H. Sperling, American Chemical Society, Washington, DC, 1990.

10.6.4 Nashif, A. D., Jones, D. I. G., and Henderson, J. P., *Vibration Damping*, J. Wiley, NY, 1985.

10.6.5 Cremer, L., Heckl, M. A., and Ungar, E. E., *Structure Borne Sound*, 2nd ed., Springer Verlag, Berlin, 1988.

10.6.6 Fantel, H., "Turntables rise to the challenge of the CD era" (regarding Ariston equipment), *New York Times*, May 3, 1987.

10.6.7 Miles, J. W., "On the annular damper for a freely precessing gyroscope", II, *J. Appl. Mech.* 30, 189–192, 1963.

10.6.8 Chang, C. O. and Chen, M. P., "Elastomer damper for a freely precessing dual spin seeker", *J. Guidance* 16, 221–224, 1992.

10.6.9 Branson, D., *Deformation of Concrete Structures*, McGraw-Hill, NY, 1977.

10.6.10 Taranath, B. S., *Structural Analysis and Design of Tall Buildings*, McGraw-Hill, NY, 1988.

10.6.11 Schueller, W., *The Vertical Building Structure*, Van Nostrand, NY, 1990.

10.6.12 Wetton, R. E., "Design of elastomers for damping applications", in *Elastomers: Criteria for Engineering Design*, ed. C. Hepburn and R. J. W. Reynolds, Applied Science Publishers, Ltd., London, 1979.

10.6.13 Rogers, L. and Parin, M., "Experimental results for stand off passive vibration damping systems", in *Smart Structures and Materials 1995: Passive Damping*, Proc. SPIE Volume 2445, ed. C. D. Johnson, Society of Photo-Optical Instrumentation Engineers, Bellingham, WA, 1995, pp. 374–383.

10.6.14 Oberst, H., "Über die Dämpfung der Biegeschwingungen dünner bleche durch fest haftende Beläge", *Acustica* 2(4), 181–194, 1952.

10.6.15 Oberst, H., "Über die Dämpfung der Biegeschwingungen dünner bleche durch fest haftende Beläge II", *Acustica* 4(1), 433, 1954.

10.6.16 Capps, R. N. and Beumel, L. L., "Dynamic mechanical testing, application of polymer development to constrained layer damping", in *Sound and Vibration Damping with Polymers*, ed. R. D. Corsaro and L. H. Sperling, American Chemical Society, Washington, DC, 1990.

10.6.17 Hostettler, F., "Energy attenuating polyurethanes", US Patent 4,722,946 (1988).

10.6.18 Tiao, W.Y. and Tiao, C. S., "Methods for the manufacture of energy attenu-ating polyurethanes", US Patent 4,980,386 (1990).

10.6.19 Fukumoto, A., "The application of piezoelectric ceramics in diagnostic ul-trasound transducers", *Ferroelectrics* 40, 217–230, 1982.

10.6.20 Berlincourt, D. A., Curran, D. R., and Jaffe, H., "Piezoelectric and piezomag-netic materials and their function in transducers", in *Physical Acoustics*, Vol. 1A, ed. E. P. Mason, Academic, NY, 1964, pp. 169–270.

10.6.21 Rosen, C. Z., Hiremath, B., and Newnham, R., ed., *Piezoelectricity*, American Institute of Physics, NY, 1982.

10.6.22 Henderson, J. P., "Damping applications in aero-propulsion systems", in *Damping Applications for Vibration Control*, ed. P. J. Torvik, ASME Publication AMD-38, 1980.

10.6.23 DeFelice, J. J. and Nashif, A. D., "Damping of an engine exhaust stack", *Shock Vib. Bull.* 48, 75–84, 1978.

10.6.24 McClure, F. T., Hart, R. W., and Bird, J. F., "Acoustic resonance in solid propellant rockets", *J. Appl. Phys.* 31, 884–896, 1960.

10.6.25 Landel, R. F. and Smith, T. L., "Viscoelastic properties of rubberlike com-posite propellants and filled elastomers", *ARS J.* 31, 599–608, 1961.

10.6.26 Nall, B. N., "Acoustic attenuation of a solid propellant", *AIAA J.* 1, 76–79, 1963.

10.6.27 Ritchie, I. G. and Pan, Z. L., "High damping metals and alloys", *Met. Trans.* 22A, 607–616, 1991.

10.6.28 James, D. J., "High damping metals for engineering applications", *Mater. Sci. Eng.* 4, 1–8, 1969.

10.6.29 Ritchie, I. G., Pan, Z. L., and Goodwin, F. E., "Characterization of the damp-ing properties of die-cast zinc–aluminum alloys", *Met. Trans.* 22A, 617–622, 1991.

10.6.30 Alberts, T. E. and Chen, Y., "Fiber enhancement of viscoelastic damping polymers", US Patent 5,256,223 (1993).

10.6.31 Mifune, N., "Vibration damping material", US Patent 5,300,355 (1994).

10.6.32 Miller, D., "Acoustical damping structure and method of preparation", US Patent 3,894,169 (1975).

10.6.33 Hutin, P., "Golf club head having vibration damping means", US Patent 5,316,298 (1994).

10.6.34 Sattinger, S. S., "Internally damped thin-walled, composite longitudinal member having dedicated internal constraining layers", US Patent 5,108,802 (1992).

10.6.35 Artus, J. P., "Vibration-damping device for an instrument having a shaft and a striking head", US Patent 5,277,423 (1994).

10.6.36 Ashley, S., "Smart skis", *Technol. Rev.* 99, 16, 1996.

10.6.37 Griffin, M. J., *Handbook of Human Vibration*, Academic, NY, 1990.

10.6.38 Staff, "New and improved steel products", *Adv. Mater. Process.* 141, 52–56, 1992.

10.6.39 Suzukawa, Y., Ikeda, K., Morita, J., and Katoh, A., "Application of vibration damping steel sheet for autobody structural parts", SAE Paper 920249, 1992.

10.6.40 Culshaw, B., *Smart Structures and Materials*, Artech, Boston, 1996.

10.6.41 Pacheco, B. M., Fujino, Y., and Sulekh, A., "Estimation curve for modal damping in stay cables with viscous damper", *J. Struct. Eng. ASCE* 119, 1961–1979, 1993.

10.6.42 Yang, J. N. and Giannopoulos, F., "Active control and stability of cable stayed bridge", *J. Eng. Mech. Div. ASCE* 105, 677–694, 1979.

10.6.43 Achkire, Y. and Preumont, A., "Active tendon control of cable-stayed bridges", *Earthquake Eng. Struct. Dynamics* 25, 585–597, 1996.

10.6.44 Cannon, R. H. and Rosenthal, D. E., "Experiment in control of flexible structures with noncollocated sensors and actuators", *AIAA J. Guidance* 7, 546–553, 1984.

10.6.45 Benhabib, R. J., Iwens, R. P., and Jackson, R. L., "Stability of large space structures control systems using positivity concepts", *AIAA J. Guidance* 4, 487–494, 1981.

10.7.1 Tabor, D., "The mechanism of rolling friction", *Philos. Mag.* 43, 1055–1059, 1952.

10.7.2 Schuring, D. J., "The rolling loss of pneumatic tires", *Rubber Chem. Technol.* 53, 600–727, 1980.

10.7.3 Coulomb, C. A., "Théorie des machines simples", in *Acad. R. Sci. Mem. Math. Phys.* 10, 1785.

10.7.4 Bowden, F. P. and Tabor, D., *Friction and Lubrication*, J. Wiley, NY, 1956.

10.7.5 Witters, J. and Duymelinck, D., "Rolling and sliding resistive forces on balls moving on a flat surface", *Am. J. Phys.* 54, 80–83, 1986.

10.7.6 Timoshenko, S. P. and Goodier, J. N., *Theory of Elasticity*, McGraw-Hill, NY, 1982.

10.7.7 Flom, D. G. and Bueche, A. M., "Theory of rolling friction for spheres", *J. Appl. Phys.* 30, 1725–1730, 1959.

10.7.8 Hunter, S. C., "The rolling contact of a rigid cylinder with a viscoelastic half space", *J. Appl. Mech.* 28, 611–617, 1961.

10.7.9 Flom, D. G., "Rolling friction of polymeric materials. I. Elastomers", *J. Appl. Phys.* 31, 306–314, 1960.

10.7.10 Luchini, J. R. ed., Rolling Resistance of Highway Truck Tires, Publication SP-546, Society of Automotive Engineers, Warrendale, PA, 1983.

10.7.11 Mullins, J., "Easy rollers", *New Scientist* 146, 31–33, 1995.

10.8.1 Momosaki, E. and Kogure, S., "The application of piezoelectricity to watches", *Ferroelectrics* 40, 203–216, 1982.

10.8.2 Braginskii, V. B., Mitrofanov, V. P., and Panov, V. I., *Systems with Small Dissipation*, University of Chicago Press, IL, 1985.

10.8.3 Rossing, T. D., Russell, D. A., and Brown, D. E., "On the acoustics of tuning forks", *Am. J. Phys.* 60, 620–626, 1992.

10.8.4 Kuroda, K., "Does the time of swing method give a correct value of the Newtonian gravitational constant?", *Phys. Rev. Lett.* 75, 2796–2798, 1995.

10.8.5 Quinn, J. J., Speake, C. C., and Brown, L. M., "Materials problems in the construction of long-period pendulums", *Philos. Mag A* 65, 261–276, 1992.

10.8.6 Maddox, J., "Systematic errors in 'Big G'", *Nature* 377, 573, 1995.

10.9.1 Chen, C.P. and Lakes, R.S., "Design of viscoelastic impact absorbers: optimal material properties", *Int. J. Solids Struct.* 26, 1313–1328, 1990.

10.9.2 Bauer, W., Haberle, F., and Riechers, D., "Bumper for motor vehicles", US Patent 4,427,225 (1984).

10.9.3 Tuggle, R. E., "Shock absorbing bumper", US Patent 3,715,139 (1973).

10.9.4 Erpelding, A.D. and Ruiz, O.J., "Viscoelastically damped slider suspension system", US Patent 5,606,447 (1997).

10.9.5 Light, L. H., McLellan, G. E., and Klenerman, L., "Skeletal transients on heel strike in normal walking with different footwear", *J. Biomech.* 13, 477–480, 1980.

10.9.6   Voloshin, A. and Wosk, J., "Influence of artificial shock absorbers on human gait", *Clin. Orthopaed.* 160, 52–56, 1981.

10.9.7   Voloshin, A. and Wosk, J., "An *in vivo* study of low back pain and shock absorption in the human locomotor system", *J. Biomech.* 15, 21–27, 1982.

10.9.8   Basford, J. R. and Smith, M. A., "Shoe insoles in the workplace", *Orthopedics* 11, 285–288, 1988.

10.9.9   Voloshin, A. and Wosk, J., "Shock absorption of meniscectomized and painful knees: a comparative study", *J. Biomed. Eng.* 5, 157–160, 1983.

10.9.10 Garcia, A. C., Durá, J. V., Ramiro, J., Hoyos, J. V., and Vera, P., "Dynamic study of insole materials simulating real loads", *Foot Ankle Int.* 15, 311–323, 1994.

10.9.11 McMahon, T. A. and Greene, P. R., "Fast running tracks", *Sci. Am.* 239, 148–163, 1978.

10.9.12 Loriga, G., "Pneumatic tools: occupation and health", in *Encyclopedia of Hygiene, Pathology, and Social Welfare*, Vol. 2, International Labour Office, Geneva, Switzerland, 1934.

10.9.13 Wasserman, D., Reynolds, D., Behrens, V., and Samueloff, S., *Vibration White Finger Disease in U. S. Workers Using Pneumatic Chipping and Grinding Tools. II. Engineering Testing*, National Institute for Occupational Safety and Health, DHEW/ NIOSH Publication 82-101, Cincinnati, 1981, pp. 1–89.

10.9.14 Griffin, M. J., *Handbook of Human Vibration*, Academic, NY, 1990.

10.10.1 Calvit, H. H., "Numerical solution of the problem of impact of a rigid sphere onto a linear viscoelastic half-space and comparison with experiment," *Int. J. Solids Struct.* 3, 951, 1967.

10.10.2 Adair, R. K., *The Physics of Baseball*, Harper Collins, NY, 1994.

10.10.3 Adair, R. K., "The Physics of Baseball", *Phys. Today* 48, 26–30, 1995.

10.10.4 Kagan, D. T., "The effects of coefficient of restitution variations on long fly balls", *Am. J. Phys.* 58, 151–154, 1990.

10.10.5 Brody, H., "The tennis-ball bounce test", *Phys. Teacher* 28, 407–409, 1990.

10.10.6 Daish, C. B., *The Physics of Ball Games*, English Universities Press, London, 1972.

10.10.7 Griffing, D. F., *The Dynamics of Sports*, 3rd ed., Dalog, Oxford, OH, 1987.

10.10.8 Garwin, R. L., "Kinematics of an ultraelastic rough ball", *Am. J. Phys.* 37, 88–92, 1969.

10.10.9 Armenti, A., Jr. ed., *The Physics of Sports*, Ameran Institute of Physics, NY, 1992.

10.10.10 Dunlop, Ltd., Barnsley, South Yorkshire, UK, private communication.

10.10.11 Lakes, R. S., technical report, unpublished, 1996.

10.11.1 Lin, J. Y. and Westmann, R. A., "Viscoelastic winding mechanics", *J. Appl. Mech.* 56, 821–827, 1989.

10.12.1 Comper, W. D. and Laurent, T. C., "Physiologic function of connective tissue polysaccharides", *Physiol. Rev.* 58, 255–315, 1978.

10.12.2 Pape, L. G. and Balazs, E. A., "The use of sodium hyaluronate (Healon®) in human anterior segment surgery", *Ophthalmology* 87, 699–705, 1980.

10.12.3 Pruett, R. C., Stephens, C. L., and Swann, D. A., "Hyaluronic acid vitreous substitute. A six year clinical evaluation", *Arch. Ophthalmol.* 97, 2325–2330, 1979.

10.12.4 St.Onge, R., Weiss, C., Denlinger, J. L., and Balazs, E. A., "A preliminary clinical assessment of Na hyaluronate injection for primary flexor tendon repair in no-man's land", *Clin. Orthopaed.* 146, 269–275, 1980.

10.12.5 Peanell, P. E., Blackmore, J. M., and Allen, M., "Viscoelastic fluid for use in spine and general surgery and other surgery and therapies and method of using same", US Patent 5,156,839 (1992).

10.13.1 Lu, R. and Puri, V. M., "Characterization of nonlinear creep behavior of two food products", *J. Rheol.* 35, 1209–1233, 1991.

10.13.2 Rao, M. and Steffe, J. F., *Viscoelastic Properties of Foods*, Elsevier Applied Sciences, London and NY, 1992.

10.13.3 Roos, Y. H., Karel, M., and Kokini, J. L., "Glass transitions in low moisture and frozen foods: effects on shelf life and quality", *Food Technol.* 50, 95–105, 1996.

10.14.1 Smith and Nephew Rolyan, Inc., N93 W14475 Whittaker Way, Menomonee Falls, WI.

10.15.1 Binney and Smith Co. Inc., Easton, PA.

10.15.2 Chen, J. Y., "Thermoplastic elastomer gelatinous compositions", US Patent 4,369,284 (1983).

10.15.3 Chen, J. Y., "Gelatinous elastomeric optical lens, light pipe, comprising a specific block copolymer and an oil plasticizer", US Patent 4,618,213 (1986).

10.16.1 Murray, A. and Neilson, J. M. M., "Diagnostic percussion sounds. I. A qualitative analysis", *Med. Biol. Eng.* 13, 19–28, 1975.

10.16.2 Murray, A. and Neilson, J. M. M., "Diagnostic percussion sounds. II. Computer-automated parameter measurement for quantitative analysis", *Med. Biol. Eng.* 13, 29–38, 1975.

10.16.3 Auenbrugger, L., "Inventum novum ex percussione thoracis humani ut signo abstrusos interni pectoris morbos detegendi", Reprinted 1966, Dawsons, London (translated by Forbes, 1824).

10.16.4 Delp, M. H. and Manning, R. T., in *Major's Physical Diagnosis*, 7th ed., ed. R. H. Major, W. B. Saunders, Philadelphia. PA, 1968, p. 102.

10.16.5 Lippmann, R. K., "The use of auscultatory percussion for the examination of fractures", *J. Bone Joint Surg.* 14, 118–126, 1932.

10.17.1 Gibson, L. J. and Ashby, M. F., *Cellular Solids*, Pergamon, Oxford, 1988.

10.19.1 Symon, K. R., *Mechanics*, 2nd ed., Addison Wesley, Reading, MA, 1964.

10.19.2 Thomson, W. T. and Reiter, G. S., "Attitude drift of space vehicles", *J. Astronaut. Sci.* 7, 29–34, 1960.

10.19.3 Meirovitch, L., "Attitude stability of an elastic body of revolution in space", *J. Astronaut. Sci.* 8, 110–119, 1961.

10.20.1 Fishman, S. G., "Damping in Metal Matrix Composites, Overview and Research Needs, in *Role of Interfaces on Material Damping*, Proceedings of an International Symposium Held in Conjunction with ASM's Materials Week and TMS/AIME Fall Meeting, Toronto, October 13–17, 1985, ed. B. B. Rath and M. S. Misra, ASM 1985, pp. 33–41.

10.20.2 Bishop, J. E. and Kinra, V. K., "Elastothermodynamic damping in composite materials", *Mech. Compos. Mater. Struct.* 1, 75–93, 1994.

10.20.3 Forward, R. L., "Electronic damping of vibrations in optical structures", *J. Appl. Opt.* 18, 690–697, 1979.

10.20.4 Forward, R. L. and Swigert, C. J., "Electronic damping of orthogonal bending modes in a cylindrical mast—Theory", *J. Spacecraft Rockets* 18, 5–10, 1981.

10.20.5 Hagood, N. W. and von Flotow, A., "Damping of structural vibrations with piezoelectric materials and passive electrical networks, *J. Sound Vib.* 146, 243–268, 1991.

10.20.6 Brodt, M. and Lakes, R. S., "Composite materials which exhibit high stiffness and high mechanical damping", *J. Compos. Mater.* 29, 1823–1833, 1995.

10.21.1 Krishtal, M., "Determination of the true diffusion coefficients and of the thermodynamic activity by the internal friction method", in *Internal Friction in Metals and Alloys*, ed. V. S. Postnikov, F. N. Tavadze, and L. K. Gordienko, Consultants Bureau, NY, 1967.

10.21.2 Nysten, B., Legras, R., and Costa, J. L., "Atomic-force microscopy imaging of viscoelastic properties in toughened polypropylene resins", *J. Appl. Phys.* 78, 5953–5958, 1995.

10.21.3 Karato, S., "Importance of anelasticity in the interpretation of seismic tomography", *Geophys. Res. Lett.* 20, 1623–1626, 1993.

10.21.4 Lienkaemper, J. J., Galehouse, J. S., and Simpson, R. W., "Creep response of the Hayward fault to stress changes caused by the Loma Prieta earthquake", *Science* 276, 2014–2016, 1997.

10.21.5 Cebon, D. and Ashby, M. F., "Materials selection for precision instruments", *Meas. Sci. Technol.* 5, 296–306, 1994.

10.21.6 Staff, "Tuned damping—the key to effective structural damping", Newport Corp., Irvine, CA.

10.22.1 Vilela, P. M., Moscoso, R. A., and Thompson, D., "What every musician knows about viscoelastic behavior", *Am. J. Phys.* 65, 1000–1003, 1997.

10.23.1 Webster, J. G., *Medical Instrumentation*, Houghton Mifflin, Boston, 1978.

10.23.2 DiBenetto, A. T., Gauchel, J. V., Thomas, R. L., and Barlow, J. W., "Nondestructive determination of fatigue crack damage using vibration tests", *J. Mater.* 7, 211–215, 1972.

10.23.3 Schultz, A. B. and Warwick, D. N., "Vibration response: a nondestructive test for fatigue crack damage in filament reinforced composites", *J. Compos. Mater.* 5, 394–404, 1971.

appendix 1

Mathematical preliminaries: functionals and distributions

Introduction

The relationship between two physical quantities is commonly thought of as describable by a function f(x) which assigns a number to each numerical value of some independent variable x. In some physical systems, a mathematical description of greater generality is useful [A1.1]. For example, it is impossible to measure a physical quantity at an exact value of x but only over some small interval of x. As another example, the response of a system to an impulsive input is of interest in many contexts. A *functional* is a more general mathematical description which is useful in such situations. A functional is a rule which assigns a number to each function in a set of "testing functions". A *distribution* is a particular type of functional. Distributions are useful in that a concept such as the Dirac delta, which describes an impulsive stimulus, is correctly described as a distribution but not as a function.

Functionals and distributions

A *functional* is defined as an assignment F(x) of a number to each element x of a vector space [A1.2]. Examples of functionals include the scalar product and the norm of a vector. A functional is *linear* if

$$F(a_1 x_1 + a_2 x_2) = F(a_1 x_1) + F(a_2 x_2).$$

The set of functions x(t) on some interval $a \leq t \leq b$ may be regarded as a "space" L_p. The norm is given by

433

$$\left\|x_p\right\| = \left[\int_a^b |x(t)|^P dt\right]^{1/p} .$$ (A1.1)

Bounded linear functionals F(x) in this space can be constructed as Stieltjes integrals constructed as limits of sums of step functions. One can write a functional as

$$F(x) = \int_a^b x(t)dg(t),$$

for continuous functions x(t).
    A *distribution* is defined [A1.1] as a linear, continuous functional.

Heaviside unit step function

The Heaviside unit step function $\mathcal{H}(x)$ is defined as follows:

$$\mathcal{H}(x) = \begin{pmatrix} 0 \\ 1/2 \\ 1 \end{pmatrix} \text{ if } \begin{pmatrix} x < 0 \\ x = 0 \\ x > 0 \end{pmatrix}$$ (A1.2)

so that

$$\mathcal{H}(x-a) = \begin{pmatrix} 0 \\ 1/2 \\ 1 \end{pmatrix} \text{ if } \begin{pmatrix} x < a \\ x = a \\ x > a \end{pmatrix}.$$

    Observe that the function $\mathcal{H}(x)$ is discontinuous. Therefore, $d\mathcal{H}(x)/dx$ does not exist at $x = 0$. Since that derivative appears in the Boltzmann integral for a creep or relaxation procedure, a method of dealing with such derivatives is of use. Distributions such as the Dirac delta are used for that purpose and others.

Dirac delta

The Dirac delta $\delta(x)$ is defined by

$$\int_{-\infty}^{\infty} \delta(x)dx = 1; \quad \delta(x) = 0 \quad \text{if } x \neq 0.$$

There is no such function with such properties. The Dirac delta, although sometimes called the delta function or impulse function, is actually a distribution. Observe that:

$$\int_{-\xi}^{\xi} \delta(x)dx = 1,$$

with ξ as an arbitrarily small but nonzero number, and the Dirac delta exhibits the sifting property

$$\int_{-\infty}^{\infty} \delta(x)f(x)dx = f(0).$$

The sifting property of $\delta(x - a)$ with a as a real number is demonstrated [A1.3] by a change of variable $x = X + a$.

$$\int_{-\infty}^{\infty} f(x)\delta(x-a)dx = \int_{-\infty}^{\infty} f(X+a)\delta(X)dX$$

$$= f(X+a)\big|_{X=0} = f(a).$$

To develop a more formal understanding of the delta in connection with familiar functions, consider a sequence of functions $\phi_n(x)$. The sequence is called a delta sequence if

$$\lim_{n\to\infty} \int_{-\infty}^{\infty} \phi_n(x)f(x)dx = f(0).$$

Although $\phi_n(x)$ does not converge to any function, we say in the sense of distributions that $\phi_n(x)$ converges to $\delta(x)$. Examples of delta sequences include [A1.1–A1.5]

$$\phi_n(x) = \begin{pmatrix} 0 \\ \dfrac{n}{x_0} \\ 0 \end{pmatrix} \text{ if } \begin{pmatrix} x < 0 \\ 0 < x < \dfrac{x_0}{n} \\ x > \dfrac{x_0}{n} \end{pmatrix} \tag{A1.3}$$

$$\phi_n(x) = \frac{n}{\sqrt{\pi}} e^{-n^2 x^2},$$

$$\phi_n(x) = \frac{n}{2} e^{-n|x|},$$

$$\phi_n(x) = \frac{\sin nx}{\pi x}, \qquad\qquad\qquad\qquad \text{(A1.4)}$$

$$\phi_n(x) = \frac{n}{\pi(1 + x^2 n^2)}.$$

As n becomes large, each function attains a higher peak but becomes narrower in width.

The sifting property of the delta may be demonstrated [A1.4] using the sequence, Eq. A1.3.

$$\int_{-\infty}^{\infty} \phi_n(x)f(x)dx = \frac{n}{x_0} \int_0^{x_0/n} f(x)dx.$$

By the mean value theorem,

$$\int_0^{x_0/n} f(x)dx = \frac{x_0}{n} f\left(Q\frac{x_0}{n}\right), \quad 0 \le Q \le 1,$$

so

$$\int_{-\infty}^{\infty} \phi_n(x)f(x)dx = f\left(Q\frac{x_0}{n}\right), \quad 0 \le Q \le 1.$$

By letting n become large,

$$\int_{-\infty}^{\infty} \delta(x)f(x)dx = f(0).$$

This is the sifting property, valid for all functions f which are continuous in the neighborhood of the origin.

We may also consider "Heaviside sequences" $\mathcal{H}_n(x - a)$ such that

$$\frac{d\,\mathcal{H}_n(x-a)}{dx} = \phi_n(x-a).$$

Due to the properties of the Dirac delta as shown below, $\mathcal{H}_n(x)$ must converge to the Heaviside step function $\mathcal{H}(x)$. In this sense we can regard the derivative of the step function as the delta. By integrating both sides,

$$\int_{-\infty}^{\infty} \frac{d\,\mathcal{H}_n(x-a)}{dx}\,dx = \int_{-\infty}^{\infty} \phi_n(x-a)\,dx,$$

so

$$\mathcal{H}_n(\infty) - \mathcal{H}_n(-\infty) = \int_{-\infty}^{\infty} \phi_n(x-a)\,dx.$$

In the limit,

$$1 = \int_{-\infty}^{\infty} \delta(x)\,dx.$$

The integral of the delta sequence members in Eq. A1.4 is as follows:

$$\mathcal{H} = \frac{1}{2} + \frac{1}{\pi}\,\tan^{-1} nx.$$

Members of this delta sequence and the corresponding Heaviside sequence are shown in Fig. A1.1.

As an example of the use of delta sequences, let us determine the derivative of the delta. If $\phi_n(x)f(x)$ tends to zero for $x \to \pm\infty$, we may integrate by parts to obtain

$$\lim_{n\to\infty} \int_{-\infty}^{\infty} \frac{d\phi_n(x)}{dx} f(x)\,dx = -\lim_{n\to\infty} \int_{-\infty}^{\infty} \phi_n(x) \frac{df(x)}{dx}\,dx$$

$$= -\frac{df(x)}{dx}\bigg|_{x=0} = -\frac{df(x)}{dx}\bigg|_{x=0} \equiv -f'(0).$$

We may express this, with the derivative of the delta called the *doublet*, ψ,

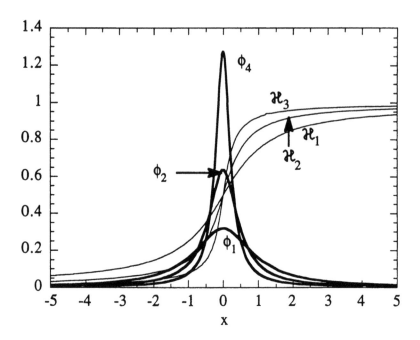

Figure A1.1 Members of delta sequence ϕ_n and the corresponding Heaviside sequence $\mathcal{H}_n$.

$$\psi = \frac{d}{dx}\delta(x),$$

as

$$\int_{-\infty}^{\infty} \psi(x)f(x)dx = -\frac{df(x)}{dx}\bigg|_{x=0}.$$

So the doublet sifts the negative of the derivative of a function.
    Members of the sequence

$$\psi_n(x) = \frac{d}{dx}\phi_n(x),$$

based on $\phi_n(x)$ from Eq. A1.4 are shown in Fig. A1.2.
    The sifting properties of the doublet may also be illustrated via the definition of the derivative and the sifting properties of the delta [A1.3].

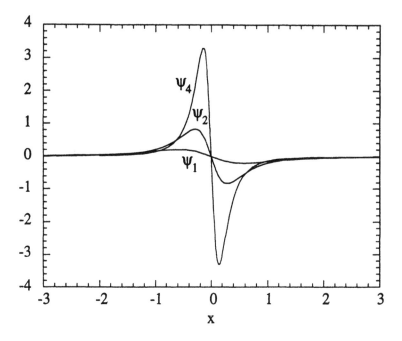

Figure A1.2   Members of doublet sequence ψ_n.

$$Z \equiv \int\limits_{-\infty}^{\infty} \frac{\delta(x+h)-\delta(x)}{h} f(x)dx$$

$$= \frac{1}{h}\left\{\int\limits_{-\infty}^{\infty} \delta(x+h)f(x)dx - \int\limits_{-\infty}^{\infty} \delta(x)f(x)dx\right\}$$

$$= \frac{1}{h}\left\{\int\limits_{-\infty}^{\infty} \delta(X)f(X-h)dX - f(0)\right\} = \frac{1}{h}\{f(-h)-f(0)\}.$$

$$\lim\nolimits_{h\to 0} Z = \int\limits_{-\infty}^{\infty} \psi(x)f(x)dx = -\frac{df(x)}{dx}\bigg|_{x=0}.$$

Expressions involving the nth derivative $\delta^{(n)}(x)$ may be developed using similar arguments.

$$\int_{-\infty}^{\infty} \delta^{(n)}(x)f(x)dx = (-1)^{(n)} \frac{d^n f(x)}{dx^n}\bigg|_{x=0}.$$

Physical interpretation of the step, delta, and doublet

In the spatial domain, the step function can be used to represent distributed loads upon part of the length of a beam subjected to bending. In that context, the delta represents a concentrated force, and the doublet represents a concentrated moment or torque.

In the time domain, the step function represents the stress history used in a creep test on a viscoelastic material, or the strain history used in a relaxation test. The delta can be used to represent an impact force.

Gamma function

The *gamma function* $\Gamma(x)$ is given for $x > 0$, by

$$\Gamma(x) = \int_0^\infty u^{x-1}e^{-u}du.$$

If x is an integer n, then $\Gamma(n + 1) = n!$. Some useful identities involving the gamma function are

$$\Gamma(x)\Gamma(1-x) = \pi/\sin \pi x, \quad \Gamma(x+1) = x\Gamma(x).$$

$$\int_0^\infty \frac{\sin x}{x^p}dx = \frac{\pi}{2\Gamma(p)\sin(p\pi/2)} \quad \text{for } 0 < p < 1.$$

$$\int_0^\infty \frac{\cos x}{x^p}dx = \frac{\pi}{2\Gamma(p)\cos(p\pi/2)}, \quad 0 < p < 1.$$

$$\mathscr{L}[t^n] = \frac{n!}{s^{n+1}}, \quad n = 0, 1, 2, \ldots; \quad \mathscr{L}[t^k] = \frac{\Gamma(k+1)}{s^{k+1}}.$$

Integrals which may be useful include

$$\int_0^\infty e^{-ax}\sin bx\, dx = \frac{b}{a^2+b^2}, \quad \int_0^\infty e^{-ax}\cos bx\, dx = \frac{a}{a^2+b^2}$$

Liebnitz' rule [A1.6]

$$\frac{d}{dx}\int_a^b F(x,t)dt = \int_a^b \frac{\partial F(x,t)}{\partial x}dt + F(x,b)\frac{db}{dx} - F(x,a)\frac{da}{dx}.$$

References

A1.1 Zemanian, A. H., *Distribution Theory and Transform Analysis*, Dover, NY, 1965.

A1.2 Schechter, M., *Principles of Functional Analysis*, Academic, NY, 1971.

A1.3 Hoskins, R. F., *Generalised Functions*, Ellis Norwood, Chichester, England, 1979.

A1.4 Sneddon, I., *The Use of Integral Transforms*, McGraw-Hill, NY, 1972.

A1.5 Tschoegl, N. W., *The Phenomenological Theory of Linear Viscoelastic Behavior*, Springer Verlag, Berlin, 1989.

A1.6 Hildebrand, F. B., *Advanced Calculus for Applications*, 2nd ed., Prentice Hall, Englewood Cliffs, NJ, 1962, p. 365.

appendix 2

Transforms

Transforms operate on a function to give another function. Transforms are useful in solving problems in various branches of engineering and science including viscoelasticity theory since they can be used to convert differential equations or integral equations into algebraic equations which are more easily solved. The most commonly used transforms are linear transforms [A2.1–A2.3]. Linear transforms $T[f(x)]$ obey the superposition rule

$$T\big[f(x)+g(x)\big] = T\big[f(x)\big]+T\big[g(x)\big].\qquad\text{(A2.1)}$$

Laplace transform

The Laplace transform of a function $f(t)$ is defined as

$$\mathscr{L}\big[f(t)\big] \equiv F(s) \equiv \int_0^\infty f(t)e^{-st}dt,\qquad\text{(A2.2)}$$

with s as the transform variable. Laplace transforms of selected functions are presented in Appendix 3.

The *convolution* of two functions f and g is defined as

$$\int_0^t f(t-\xi)g(\xi)d\xi.\qquad\text{(A2.3)}$$

The convolution theorem for Laplace transforms is

$$\mathscr{L}\left[\int_0^t f(t-\xi)g(\xi)d\xi\right] = \mathscr{L}\big[f(t)\big]\,\mathscr{L}\big[g(t)\big].\qquad\text{(A2.4)}$$

The derivative theorem for Laplace transforms is

$$\mathscr{L}\left[\frac{df(t)}{dt}\right] = s\,\mathscr{L}\left[f(t)\right] - f(0). \qquad (A2.5)$$

The proof of these theorems is by direct integration using the definition.

Fourier transform

The Fourier transform is defined as follows [A2.3], with the angular frequency $\omega = 2\pi\nu$, and ν as frequency.

$$\mathscr{F}\left[f(t)\right] \equiv F(\omega) = \int_{-\infty}^{\infty} f(t)e^{-i\omega t}dt. \qquad (A2.6)$$

The inverse transform is given by

$$f(t) = \frac{1}{2\pi}\int_{-\infty}^{\infty} F(\omega)e^{i\omega t}d\omega. \qquad (A2.7)$$

There are different conventions [A2.1] in the definition of Fourier transforms; some authors include 2π factors in the argument of the exponential to achieve symmetry; others define both the transform and the inverse transform with $1/\sqrt{2\pi}$ before the integral.

Consider several theorems in Fourier transforms.

Shift theorem
If

$$\mathscr{F}\left[f(t)\right] = F(\omega)$$

then

$$\mathscr{F}\left[f(t-a)\right] = e^{-i\omega a}F(\omega). \qquad (A2.8)$$

To demonstrate this, use the definition

$$\int_{-\infty}^{\infty} f(t-a)e^{-i\omega t}dt = \int_{-\infty}^{\infty} f(t-a)e^{-i\omega(t-a)}e^{-i\omega a}d(t-a)$$

$$\qquad (A2.9)$$

$$= e^{-i\omega a}F(\omega).$$

Convolution theorem

If

$$\mathscr{F}[f(t)] = F(\omega)$$

and

$$\mathscr{F}[g(t)] = G(\omega)$$

then

$$\mathscr{F}\left[\int_{-\infty}^{\infty} f(t')g(t-t')dt'\right] = F(\omega)G(\omega). \qquad (A2.10)$$

To demonstrate this, use the definition and exchange orders of integration.

$$\int_{-\infty}^{\infty}\int_{-\infty}^{\infty} f(t')g(t-t')dt'e^{-i\omega t}dt = \int_{-\infty}^{\infty} f(t')\int_{-\infty}^{\infty} g(t-t')e^{-i\omega t}dt\, dt'. \qquad (A2.11)$$

Use the shift theorem, and then

$$\int_{-\infty}^{\infty} f(t')e^{-i\omega t'}G(\omega)dt' = F(\omega)G(\omega), \qquad (A2.12)$$

as desired.

Derivative theorem

If

$$\mathscr{F}[f(t)] = F(\omega)$$

then

$$\mathscr{F}\left[\frac{df(t)}{dt}\right] = i\omega F(\omega). \qquad (A2.13)$$

The proof involves integration by parts, using the definition.

Hartley transform

The Hartley transform [A2.1] resembles the Fourier transform; it differs in that the kernel is cas($2\pi vt$) rather than exp($-2\pi ivt$).

$$H(s) = \int_{-\infty}^{\infty} f(x)cas(2\pi sx)dx.$$

The cas function is defined as cas $x = \cos x + \sin x$. It is a sinusoid of amplitude $\sqrt{2}$ shifted by one eighth of a period. The Hartley transform is a real transform and the transform is identical to the inverse transform.

Hilbert transform

The Hilbert transform is defined as:

$$F_{Hi}(x) = \frac{1}{\pi} \int_{-\infty}^{\infty} \frac{f(\xi)}{\xi - x} d\xi, \tag{A2.14}$$

and the inverse transform is

$$f(x) = -\frac{1}{\pi} \int_{-\infty}^{\infty} \frac{F_{Hi}(\xi)}{\xi - x} d\xi. \tag{A2.15}$$

The Hilbert transform is seen in the derivation of the Kramers–Kronig relations in §3.3. The Hilbert transform is also related to the concept of "instantaneous phase" via the following argument [A2.1]:

$$\text{if } f(x) = \exp(-\pi x^2)\cos 4\pi x,$$

then $F_{Hi}(x) = -\exp(-\pi x^2) \sin 4\pi x$, so that the oscillations in the transform share the same envelope as those in the original function. However, the phase of an oscillatory function is explicitly revealed only where it crosses zero. The phase ϕ at intermediate points can be expressed as

$$\tan \phi = F(x)/f(x).$$

References

A2.1 Bracewell, R. N., "Numerical transforms", *Science* 248, 697–704, 1990.

A2.2 Sneddon, I. N., *Fourier Transforms*, McGraw-Hill, NY, 1951.

A2.3 Zemanian, A. H., *Distribution Theory and Transform Analysis*, Dover, NY, 1965.

appendix 3

Laplace transforms

Some Laplace transform general properties are as follows [A3.1–A3.3]. Write the Laplace transform as

$$\mathcal{L}\{f(t)\} = F(s).$$

Linearity (superposition)

$$\mathcal{L}\{a_1 f_1(t) + a_2 f_2(t)\} = a_1 F_1(s) + a_2 F_2(s). \tag{A3.1}$$

Derivative property

$$\mathcal{L}\left\{\frac{df(t)}{dt}\right\} = sF(s) - f(0). \tag{A3.2}$$

Integral property

$$\mathcal{L}\left\{\int_0^t f(\tau)d\tau\right\} = \frac{F(s)}{s}. \tag{A3.3}$$

Multiplication by time

$$\mathcal{L}\{tf(t)\} = -\frac{d}{ds}F(s). \tag{A3.4}$$

Division by time

$$\mathcal{L}\left\{\frac{f(t)}{t}\right\} = \int_s^\infty F(u)du. \tag{A3.5}$$

Multiplication by an exponential

$$\mathscr{L}\{e^{-at}f(t)\} = F(s+a).$$ (A3.6)

Time shift

$$\mathscr{L}\{f(t-T)\,\mathscr{H}\,(t-T)\} = e^{-Ts}F(s).$$ (A3.7)

Scale change

$$\mathscr{L}\{f(at)\} = \frac{1}{a}F\left(\frac{s}{a}\right).$$ (A3.8)

Convolution theorem

$$\mathscr{L}\left\{\int_0^t f(\tau)g(t-\tau)d\tau\right\} = F(s) \cdot G(s).$$ (A3.9)

Some Laplace transform pairs are as follows:

f(t)	F(s)
$\delta(t)$	1
$\delta(t-a)$	e^{-as}
$\mathscr{H}(t)$	$1/s$
$\mathscr{H}(t-a)$	$\dfrac{1}{s}e^{-as}$
$t\mathscr{H}(t)$	$1/s^2$
e^{-at}	$\dfrac{1}{s+a}$
$\dfrac{t^{n-1}e^{-at}}{(n-1)!}$	$\dfrac{1}{(s+a)^n}$
$\dfrac{1}{a}\left(1-e^{-at}\right)$	$\dfrac{1}{s(s+a)}$
$\dfrac{\left[be^{bt}-ae^{at}\right]}{(b-a)}$	$\dfrac{1}{(s-a)(s-b)}$ for $a \neq b$

f(t)	F(s)
$\dfrac{\left[e^{bt} - e^{at}\right]}{(b-a)}$	$\dfrac{s}{(s-a)(s-b)}$ for a $\neq$ b
cosh(at)	$\dfrac{s}{s^2 - a^2}$
sinh(at)	$\dfrac{a}{s^2 - a^2}$
cos (at)	$\dfrac{s}{s^2 + a^2}$
sin (at)	$\dfrac{a}{s^2 + a^2}$
$\dfrac{\sin at}{t}$	$\tan^{-1}(a/s)$
t cos (at)	$\dfrac{s^2 - a^2}{\left(s^2 + a^2\right)^2}$
t sin (at)	$\dfrac{2as}{s^2 + a^2}$
$e^{-at} \sin \omega t$	$\dfrac{\omega}{(s+a)^2 + \omega^2}$
$e^{-at} \cos \omega t$	$\dfrac{s+a}{(s+a)^2 + \omega^2}$
t^n	$\Gamma(n+1)s^{-n-1}$
$t^n e^{-qt}$	$\Gamma(n+1)(s+q)^{-n-1}$

References

A3.1 Sneddon, I. N., *Fourier Transforms*, McGraw-Hill, NY, 1951.
A3.2 Swisher, G. M., *Introduction to Linear Systems Analysis*, Matrix, Cleveland, 1976.
A3.3 Spiegl, M. R., *Mathematical Handbook*, McGraw-Hill, NY, 1968.

appendix 4

Convolutions

The *convolution* of two functions f and g is defined as

$$C(t) = f(t) * g(t) = \int_0^t f(\xi) g(t - \xi) d\xi.$$

The range of integration for the causal variables of interest in viscoelasticity is from zero to t. Convolutions are also used to deal with spatially varying quantities for which causality is not an issue. In that case they are defined over a range of integration from $-\infty$ to ∞.

Convolutions have the *commutative property*, that is [A4.1],

$$f(t) * g(t) = g(t) * f(t).$$

The proof is by a change of variable, $x = t - \xi$. Other properties of convolutions are associativity,

$$f(t) * [g(t) * h(t)] = [f(t) * g(t)] * h(t),$$

and distributivity,

$$f(t) * [g(t) + h(t)] = f(t) * g(t) + f(t) * h(t).$$

There is also the shift property,

if $$C(t) = f(t) * g(t),$$

then $$f(t) * g(t - T) = f(t - T) * g(t) = C(t - T).$$

The integration can be carried out by graphical or by numerical means to aid in visualization or to perform calculations. Graphical convolution can be carried out as follows. Visualize the function $f(\xi)$, and keep it fixed. Obtain $g(-\xi)$ by flipping $g(\xi)$ about the vertical axis. Obtain $g(t_0 - \xi)$ for a particular time value by shifting $g(\xi)$ by an amount t_0 along the t axis. Determine the area under the product $f(\xi)g(t_0 - \xi)$ to obtain the convolution $C(t_0)$ for the particular time value t_0. Repeat the procedure for different times to obtain $C(t)$ for all values of t.

The convolution integral, as with other integrals, can be defined as the limit of a sum.

$$C(t) = f(t) * g(t) = \lim_{dt \to 0} \sum_{mdt=0}^{t} f(mdt)g(t - mdt)dt.$$

Here, dt is the time increment and m is an integer specifying how many time increments are taken. If $t = k\, dt$, this can be written

$$C(t) = \lim_{dt \to 0} (dt) \sum_{m=0}^{k} f(m)g(k - m).$$

The quantity

$$y(k) = \sum_{m=0}^{k} f(m)g(k - m)$$

is called the "convolution sum". The limits on the summation come from the causal nature of viscoelastic response. Since $f(k)$ is causal, $f(m) < 0$ for $m < 0$. Causality of $g(k)$ means that $g(k - m) = 0$ when $m > k$. So the product $f(m) g(k - m) = 0$ for $m < 0$ and for $m > k$, and it is nonzero only for $0 \le m \le k$.

Numerical convolution can be visualized by the "tape algorithm". Consider the numerical data written in sequence on tapes:

tape for f: f(0) f(1) f(2) f(3) ... .
tape for g: g(0) g(1) g(2) g(3) ... .

<u>f(m) sequence</u>
         f(0) f(1) f(2) f(3)
<u>g(3) g(2) g(1) g(0)</u>
g(−m) sequence

         ← f(0) f(1) f(2) f(3)
g(3) g(2) g(1) g(0) →
g(k − m), k = 1.

The values of the convolution are extracted as follows from sums of products of the data on the tapes. From the first diagram,

$$C(0) = dt\, f(0)g(0).$$

from the second diagram,

$$C(1) = dt\{f(0)g(1) + f(1)g(0)\}.$$

By sliding the tapes one more step,

$$C(2) = dt\{f(0)g(2) + f(1)g(1) + f(2)g(2)\}.$$

By continuing, one can obtain the numerical convolution of two functions over the full time domain. This procedure is identical to the graphical procedure described above.

The convolution sum can be written as a matrix equation

$$\begin{pmatrix} y(0) \\ y(1) \\ \cdots \\ y(k) \end{pmatrix} = \begin{pmatrix} g(0) & 0 & 0 & 0 \\ g(1) & g(0) & 0 & 0 \\ \cdots & \cdots & \cdots & \cdots \\ g(k) & g(k-1) & \cdots & g(0) \end{pmatrix} \begin{pmatrix} f(0) \\ f(1) \\ \cdots \\ f(k) \end{pmatrix}.$$

This form suggests a numerical method for inverting the convolution process. The reverse of convolution is *deconvolution*. One can solve for f as $f = g^{-1}y$, which requires an inversion of the matrix g. The inversion can be carried out numerically. Such a procedure can be used to obtain the creep function from the relaxation function (or vice versa) using the convolution relation between these functions developed in Chapter 2. Deconvolution is also used in digital optics to remove blur from images.

Reference

A4.1 Lathi, B. P., *Linear Systems and Signals*, Berkeley–Cambridge Press, Carmichael, CA, 1992.

appendix 5

Concepts for dynamical aspects

Sinusoids

Amplitude, frequency, phase

We use sinusoid functions to represent oscillatory stress and strain histories. Suppose we have $\sigma(\omega t) = \sigma_0 \sin \omega t$, in which t is time, σ_0 is the amplitude, and ω is the angular frequency. The sine function repeats every 2π radians. So $\sigma(\omega t + 2\pi) = \sigma(\omega t)$. The time T required for the sine function to complete one cycle is obtained from $\omega T = 2\pi$, or $T = 2\pi/\omega$. T is called the period, the number of seconds required for one cycle. The inverse of T, the number of cycles per second (also called Hertz [Hz]), is called the frequency, $\nu = 1/T$. The relationship between frequency and angular frequency is $\omega = 2\pi\nu$.

Consider two sinusoids, $A \sin \omega t$ and $A \sin (\omega t + \delta)$. The quantity δ is called the phase angle. In a plot of the two waveforms, the sinusoids are shifted with respect to each other on the time axis. Observe that the cosine function is $\pi/2$ radians out of phase with the sine function.

Complex exponential representation

Sinusoidal functions which represent oscillatory quantities in which phase is important are commonly written in complex exponential notation [A5.1].

A complex number may be written in rectangular representation as $Z = a + bi$ with $i = \sqrt{-1}$ (called j in the electrical engineering community). The magnitude is $|Z| = \sqrt{a^2 + b^2}$. The complex number can be written in polar form $Z = |Z| e^{i\theta}$. The connection is given by Euler's equation. Euler's equation is $e^{i\theta} = \cos \theta + i \sin \theta$, with $i = \sqrt{-1}$. θ is called the phase and represents an angle in the complex plane. Euler's equation may be rewritten

$$e^{-i\omega t} = \cos\omega t + i\sin\omega t.$$

The quantity $e^{i\omega t}$ is a complex number. The physical meaning is that the sine and cosine functions are $\pi/2$ radians out of phase, so that the imaginary term represents a quantity $\pi/2$ radians out of phase.

Reference

A5.1 Irwin, J. D., *Basic Engineering Circuit Analysis*, Macmillan, NY, 1989.

appendix 6

Phase determination by subtraction

Suppose that electrical signals are available, proportional to stress and to strain. The stress and strain are sinusoidal in time, but the material is visco-elastic, so there is a phase shift. Suppose further that the phase shift is small enough that it is difficult to determine via Lissajous figures (see Fig. 3.2). Consider a quantity Γ formed by subtracting a fraction f of one signal from the other. The signals are assumed to be normalized to unity.

$$\Gamma = \sin(\omega t + \phi) - f \sin \omega t. \qquad (A6.1)$$

$$\Gamma = \sin \omega t \cos \phi + \cos \omega t \sin \phi - f \sin \omega t$$

$$= \sin \omega t \cos \phi + \cos \omega t \sin \phi - \frac{f}{\cos \phi} \sin \omega t \cos \phi.$$

$$\Gamma = \left(1 - \frac{f}{\cos \phi}\right) \sin \omega t \cos \phi + \cos \omega t \sin \phi. \qquad (A6.2)$$

But

$$\sin(\omega t + \xi) = \sin \omega t \cos \xi + \cos \omega t \sin \xi.$$

Set

$$\Gamma = a \sin(\omega t + \xi), \qquad (A6.3)$$

and solve for a and ξ.

$$\left(1-\frac{f}{\cos\phi}\right)\sin\omega t\cos\phi + \cos\omega t\sin\phi = a\sin\omega t\cos\xi + a\cos\omega t\sin\xi.$$

$$(A6.4)$$

So

$$a\cos\xi = \left(1-\frac{f}{\cos\phi}\right)\cos\phi, \qquad (A6.5)$$

$$a\sin\xi = \sin\phi. \qquad (A6.6)$$

$$a^2\cos^2\xi + a^2\sin^2\xi = a^2\left(\cos^2\xi + \sin^2\xi\right) = a^2$$

$$(A6.7)$$

$$= \left(1-\frac{f}{\cos\phi}\right)^2\cos^2\phi + \sin^2\phi.$$

So

$$a = \sqrt{\left[\left(1-\frac{f}{\cos\phi}\right)^2\cos^2\phi + \sin^2\phi\right]}, \qquad \mathbf{(A6.8)}$$

$$\sin\xi = \frac{1}{a}\sin\phi. \qquad \mathbf{(A6.9)}$$

Here ξ is the phase angle in the new signal Γ which is obtained by subtraction of the normalized stress and strain signals. Since ξ is larger than ϕ, it is easier to measure.

We may consider several special cases. For example, if $\xi = 90°$, then sin $\xi = 1$ and a = sin ϕ, so

$$\left(1-\frac{f}{\cos\phi}\right)^2\cos^2\phi = 0,$$

$$\left(1-\frac{f}{\cos\phi}\right)\cos\phi = 0. \qquad (A6.10)$$

Since cos $\phi \neq 0$ (otherwise we would not magnify it),

$$f = \cos \phi, \tag{A6.11}$$

$$a = \sin \phi, \tag{A6.12}$$

if $\xi = 90°$, that is, the phase is substantially magnified.
Now consider the case $\phi \ll 1$.

$$a = \sqrt{\left\{ 1 - \frac{2f}{\cos \phi} + \frac{f^2}{\cos^2 \phi} \right\} \cos^2 \phi + \sin^2 \phi}, \tag{A6.13}$$

but

$$\left(\cos^2 \phi + \sin^2 \phi \right) = 1, \tag{A6.14}$$

so

$$a = \sqrt{1 - 2f \cos \phi + f^2}. \tag{A6.15}$$

This is still exact. Now for $\phi \ll 1$,

$$a \approx \sqrt{1 - 2f + f^2} = 1 - f. \tag{A6.16}$$

Then

$$\sin \xi = \frac{\sin \phi}{(1-f)} \approx \frac{\phi}{(1-f)}. \tag{A6.17}$$

If the phase angle ξ in the processed signal is also small, then the subtraction process magnifies the phase between the input signals as follows:

$$\xi \approx \frac{\phi}{(1-f)}. \tag{A6.18}$$

To summarize, one can magnify small phase differences by a subtraction process. This procedure has been used by geologists who are interested in the behavior of rocks and minerals as low-loss materials, at low frequencies of interest in seismology and geophysics [A6.1]. At low frequencies, resonance methods are impractical, so a subresonant method is used. In this subresonant method a standard sinusoid is subtracted from the desired signal to "fatten up" the hysteresis loop, improving phase resolution. Loss tangents of 0.001 can be measured this way provided the ratio of signal to noise is sufficient.

Reference

A6.1 Brennan, B. J., "Linear viscoelastic behaviour in rocks", in *Anelasticity in the Earth*, ed. F. D. Stacey, M. S. Paterson, and A. Nicholas, American Geophysical Union, Washington, DC, 1981.

appendix 7

Interrelations in elasticity theory

Formulae for isotropic elastic media

For an isotropic elastic material, the elastic constants are related [A7.1] as follows. Here E is Young's modulus, G is the shear modulus, B is the bulk modulus, v is Poisson's ratio, C_{1111} is an axial component of the elastic modulus tensor, and λ and μ are the Lamé constants.

$$G = \mu,$$

$$B = \lambda + \frac{2}{3}\mu, \quad B = \frac{2G(1+v)}{3(1-2v)},$$

$$B = \frac{E}{3(1-2v)}, \quad B = \frac{GE}{3(3G-E)},$$

$$v = \frac{E}{2G} - 1, \quad v = \frac{3B-2G}{6B+2G}, \quad v = \frac{1}{2} - \frac{E}{6B},$$

$$v = \frac{\lambda}{2(\lambda+G)}.$$

$$E = 2G(1+v).$$

$$E = 3B(1-2v) = \frac{9GB}{3B+G}.$$

$$C_{1111} = \lambda + 2G = 2G\left[\frac{v}{1-2v} + 1\right].$$

Reference

A7.1 Sokolnikoff, I. S., *Mathematical Theory of Elasticity*, Krieger, Malabar, FL, 1983.

Symbols

Latin letters

$a_T(T)$	time–temperature shift factor
a	constant
a	radius of contact region
$\mathbf{a}_r$	unit vector
$\mathbf{a}_\theta$	unit vector
$\mathbf{a}_z$	unit vector
A	constant
A	cross-section area
A	amplitude (Chapter 10)
$\mathbf{A}$	a vector
B	bulk modulus
B	a constant
b	cross-section dimension
C	a constant
C_{ijkl}	elastic modulus tensor
C_P	heat capacity, constant pressure
c	a constant
c	cross-section dimension
c	wave speed
D	Deborah number
D	electric displacement
d_{ijk}	piezoelectric modulus tensor
E	elastic Young's modulus
E	electric field
E^*	complex dynamic Young's modulus
$E'(\omega)$	dynamic Young's storage modulus
$E''(\omega)$	dynamic Young's loss modulus
$E(t)$	relaxation Young's modulus

É(t)	time-dependent part of relaxation modulus
E_e	equilibrium modulus after infinite time
Ei(x)	exponential integral function
f	porosity in Biot model
f	frequency
F	force
G	shear elastic modulus
G^*	complex dynamic shear modulus
G'	dynamic shear storage modulus
G''	dynamic shear loss modulus
G(t)	shear relaxation modulus
H	Biot fluid coupling parameter
$H(\tau)$	relaxation spectrum
H_0	drop height
H_1	height of rebound
h	axial length
i	$\sqrt{-1}$
I	moment of inertia
j(t)	creep function
J(t)	creep compliance
$J^*(\omega)$	complex dynamic compliance
$J'(\omega)$	dynamic storage compliance
$J''(\omega)$	dynamic loss compliance
J_e	equilibrium compliance
k	wave number
K	fluid flow permeability
K_{ij}	dielectric tensor
k_{ij}	dimensionless dielectric tensor
K	piezoelectric coupling coefficient
$L(\tau)$	retardation spectrum
L	rod length
m	a constant
m	mass
M	bending or twisting moment (torque)
M_{ext}	external torque
M_o	torque amplitude
n	a constant
p	a characteristic frequency
P	force on a ball
P(x)	fluid pressure
q	a strain rate
Q	quality factor
Q	Biot fluid ingress parameter
r	position vector
r	radial coordinate

R	rod radius
R	ball radius
R	Boltzmann constant
S_{ijkl}	elastic compliance tensor
SCF	stress concentration factor
s	Laplace transform variable
S	action integral
S	entropy
t	time
t_r	rise time
t_{imp}	impact time
T	a transform
T	kinetic energy of spin
T	temperature
T	transmissibility of vibration
T	tensile force in a guitar string
u	a variable
u	displacement
U	activation energy
U(x)	potential barrier
U(t)	linear displacement
V	shear force
V_1, V_2	phase volume fraction
v	wave speed
W	energy density
y	a spatial coordinate

Script letters

e(t)	normalized relaxation function
j(t)	normalized creep function
$\mathscr{F}[f(t)]$	Fourier transform of function f(t)
$\mathscr{H}(t)$	Heaviside step function
$\mathscr{H}_n(x)$	element of Heaviside sequence
$\mathscr{L}[f(t)]$	Laplace transform of function f(t)
ℓ	block length
$\wp$	principal part of an integral

Greek letters

α_{ij}	thermal expansion coefficient
α	wave attenuation
β	an internal variable
γ	rate of decay of stress
Γ	gamma function

Γ	normalized dynamic torsional compliance
Γ	normalized energy dissipation in a satellite
$\delta(\omega)$	loss angle: phase between stress and strain
δ	Biot unjacketed compressibility
$\delta(x)$	Dirac delta function
δ_{ij}	Kronecker delta
Δ	relaxation strength
Δt	increment of time
$\Delta \omega$	width of resonance amplitude vs. frequency curve
$\Delta \tau$	increment of time variable
ΔT	increment of temperature
ε	strain
η	a viscosity
η	imaginary part of a complex variable
$\phi_n(x)$	element of delta sequence
ϕ	phase angle
ϕ	an Euler angle for satellite motion
θ_0	complex angular displacement amplitude
$\theta(t)$	angular displacement
θ	phase
θ	an Euler angle for satellite motion
κ	Biot jacketed compressibility
κ	bulk compliance
λ	wavelength
λ	Lamé elastic constant
λ	coefficient of rolling friction
Λ	log decrement
Λ	geometry-dependent parameter in piezoelectric field–displacement relation
Λ	dislocation length
μ	Lamé shear elastic constant
μ	molecular weight of gas
ν	frequency
ν_0	a characteristic frequency
ν	Poisson's ratio
ω	angular frequency
ϖ	angular frequency variable of integration
ω_0, ω_1	natural angular frequency of vibrating object
ω_0, ω_1	angular velocity component of a satellite
ω_2, ω_3	angular velocity component of a satellite
ω_3	natural angular frequency of a higher mode
Ω^*	complex torsion resonance ratio
$\Omega(t)$	curvature in bending
Ω	resonance ratio
ψ	doublet functional

ψ	an Euler angle for satellite motion
Ψ	specific damping capacity
ρ	density
σ	stress
τ	time variable of integration
τ_r	relaxation time in single exponential model
τ_c	retardation time in single exponential creep model
ξ	a phase angle in signal obtained via subtraction
ξ	a variable of integration
ξ	real part of a complex variable
ξ	an amplitude
ζ	reduced time
ζ	a complex variable

Index

A

Activation energy, 36, 233–234, 321, 331–332
Active vibration control, 395–396
Adaptive materials, 42–43
Aging of materials, 41, 252
Aircraft, 392–393
Alfrey approximation, 118
Aluminum, 244, 253–254, 298
Andrade creep, 33
Anelastic, 3
Anisotropy, 39, 188, 342–344, 358
Arrhenius relation, 34–36
Ashby, 319, 329, 374
Asphalt, 264–266
Attenuation of waves
    analysis, 90–93
    measurement of ultrasonic attenuation,
        220–225

B

Baseball, 408
Bazant, 265
Bending,
    correspondence principle, linear beam,
        145–146
    direct construction, linear beam, 141–143
    nonlinear beam, 164–165
    plate, 175
    waves and vibration, 158–160
Berry, 290
BKZ relation, 45
Biot model, 299–306
Birnboim apparatus, 216
Boltzmann superposition principle, 15
Boltzmann superposition integral, 18, 344

Bone
    bone loss tangent, 244, 267–268
    bone structure, 355–356
Bordoni relaxation, 314
Bounds on stiffness, 343–345, 347–348
Bounds on viscoelasticity, 353–354
Box spectrum, see spectra
Brass, 244, 253, 298
Buildings, 388–389
Bulk modulus, 15, 40, 463
Bulk modulus from shear, tensile data,
        236–238
Bulk relaxation, 249–251
Bumpers, 405

C

Cadmium, 258
Capacitive displacement transducer, 192
Cavendish balance, 400
Cellular solids,
    deformation characteristics, 272–274
    structure, 343
    foam properties related to those of solid,
        360, 273
    viscoelasticity from fluid flow in pores,
        299–306, 364–366
Coefficient of restitution, 100, 406–408
Cole–Cole plot, 67
Complex compliance, 66
Complex modulus, 65
Compliance,
    creep compliance, 3
    elastic compliance, 1
    storage compliance, 66, 68, 70
    structural compliance in experiment,
        228
Composites, 274–275, 341–370, 394

Milton Keynes UK
Ingram Content Group UK Ltd.
UKHW031138141024
449569UK00024B/1252